Societies and Cities in the Age of Instant Access

The GeoJournal Library

Volume 88

The titles published in this series are listed at the end of this volume.

Societies and Cities in the Age of Instant Access

Edited by

Harvey J. Miller

University of Utah,
Salt Lake City, Utah, USA

Springer

A C.I.P. Catalogue record for this book is available from the Library of Congress.

ISBN-10 1-4020-5426-2 (HB)
ISBN-13 978-1-4020-5426-6 (HB)
ISBN-10 1-4020-5427-0 (e-book)
ISBN-13 978-1-4020-5427-3 (e-book)

Published by Springer,
P.O. Box 17, 3300 AA Dordrecht, The Netherlands.

www.springer.com

Printed on acid-free paper

Table of Contents

TRANSPORTATION

MOBILE INFORMATION SERVICES

Preface

In November 2005, the Department of Geography at the University of Utah, in collaboration with the Institute of Public and International Affairs and with the support of the Rocco C. and Marion S. Siciliano Forum: *Considerations on the Status of American Society*, hosted an international symposium on the topic of *Societies and Cities in the Age of Instant Access*. The symposium invited researchers and practitioners to convene and discuss the technologies of instant access and their potential impact on they way we conduct our lives and organize our cities, societies and economies. The symposium began with a keynote address by Howard Rheingold, author of *Smart Mobs: The Next Social Revolution*, and plenary addresses by internationally recognized scholars Michael Batty (University College London), Helen Couclelis (University of California, Santa Barbara), Anthony Townsend (Institute for the Future) and Susan Hanson (Clark University). Responding to an open call for papers were scholars and practitioners from a diverse range of disciplinary backgrounds, including geography, architecture, planning, communication, transportation and information science. This book is a result of that symposium: the chapters comprise a select group of the plenary and contributed papers.

I would like to thank Rocco and Marion Siciliano for their generous support of the forum and Steve Ott (Dean, College of Social and Behavioral Science) for his encouragement and enthusiasm. Lisa Clayton and Maryann Golightly (Department of Geography) worked tirelessly to make the symposium a success; I could not have done it without them. Thanks are also due to student volunteers from the Department of Geography: Scott Bridwell, Lina Cao, Tina Gillman, Matt Hansen, Tetsuo Kobayashi, Phoebe McNeally, Carrie Spruance and Chris Upchurch.

Harvey J. Miller
Salt Lake City, Utah, USA
July 2006

INTRODUCTION

1 Societies and cities in the age of instant access

Harvey J. Miller
Department of Geography, University of Utah, UT

> The Internet is like a twenty-foot tidal wave coming thousands of miles across the Pacific, and we are in kayaks.
> – Andrew Grove, CEO of Intel; quoted in Kaku (1997)

1.1 Introduction

The ability to access resources, people and information is central to most human phenomena. Much of human social and economic organization is predicated by accessibility and the artifacts that we have created to improve accessibility. Cities compress a multitude of lives into a compact space to reduce the time and energy required for people to obtain resources, seek opportunities and interact with each other. Transportation technologies facilitate physical accessibility, profoundly altering the relationships between places and creating new patterns of production, trade and consumption. Communication technologies facilitate information accessibility between people, similarly altering interpersonal relationships and activity organization. The combined effects of accessibility technologies such as cities, steamships, railroads, telephones, automobiles, airplanes and the Internet have profoundly changed human society, politics, culture and history.

At the dawn of the 21st century, accessibility is undergoing dramatic, epoch-making changes. We are on the verge of the "second information revolution," based on near-ubiquitous access to computing, information and people through portable and wireless *information and communication technologies* (ICTs). We have been liberated from the severe spatial and temporal restrictions imposed by wired communication technologies: rather than contacting a designated place from another place and hoping the desired person is there at same time, we can now access people directly at almost anyplace and anytime. Similarly, access to computing and information no longer requires being in a specific place and time: portable computing devices and wireless connectivity allow information to

H.J. Miller (ed.), Societies and Cities in the Age of Instant Access, 3–28.

be acquired when and where it is most useful. Feeding this is an explosion in the volume and scope of data collected about places and people through digital technologies.

We can already see changes created by the technologies of instant access. New possibilities for interaction created by mobile phones and the wireless Internet are making cities more dynamic, decentralized and complex (Townsend, 2000; Zook et al., 2004). New behaviors are emerging such as *flocking* (fluid social congregation based on communication in real-time) and *flash mobs* (self-organized public events for both irreverent and serious objectives) (Rheingold, 2002). ICTs, combined with the globalization of capital, production and trade, and an emphasis in many cultures on individual freedom and openness, are creating a new society and economy based on networks and flows (Castells, 2001). Yet, we are only at the beginning of the new era of instant access. The technologies of instant access are still in their infancy. When they mature, they will have the potential to transform dramatically our lives, cities, societies and economies much like the train and telephone changed our world in the 19th century and the automobile, airplane and Internet changed our world in the 20th century.

This book examines the nature of instant access technologies and their broader impacts on societies, cities, economies, community and daily lives. The authors of the chapters in this book come from diverse disciplinary backgrounds, including architecture, geography, transportation science, urban studies, communication studies and information science. They also range from academic scholars to practitioners. All were charged with the same question: will the technologies of instant access change your domain, and if so, how?

This chapter provides an introduction and overview of the book. The next section of this chapter briefly addresses a fundamental concept at the heart of this book, namely, what is accessibility? Section 3 follows with an overview of the technologies of instant access. Section 4 provides conceptual perspectives, including time geography, space-time convergence and telepresence, theories of competition and cooperation, and frameworks for examining interactions between real and virtual spaces. Section 5 provides an overview of the remainder of the book.

1.2 What is accessibility?

Accessibility is a complex and multi-faceted concept, but ultimately it concerns the ability of individuals to conduct activities within a given environment (Weibull, 1980). Accessibility is predicated on the ability to be present at locations where activities such as work, shopping, education, health care, recreation, socializing or entertainment occur. Traditionally, this involves the physical movement of people to activity locations. Consequently, in transportation and urban sciences accessibility refers to the ease with which distance can be overcome in physical movement (Couclelis, 2000; Miller, 2005c). Accessibility is traditionally a property associated with places and considered to be an enabling aspect that ultimately determines their value and functionality relative to other places (Couclelis, 2000). Standard methods of measuring place-based accessibility include travel distance or time between locations and network connectivity for roads, public transit and other transport systems. More elaborate measures the compare the cost of overcoming distance with the utility of interaction between locations (Handy and Niemeier, 1997, Miller, 2005c).

Increasing mobility and use of ICTs is calling into question a purely place-based conceptualization of accessibility. Place-based accessibility assumes that people must be physically present at a location where an activity occurs. ICTs allow people be *telepresent* and participate without a physical presence at the activity location (Miller, 2005c). Place-based accessibility also assumes that activities are tightly-coupled with places: there are privileged locations for work, recreation, shopping and so on. ICTs are creating a fragmentation of activities with respect to space (Couclelis, 2000). For example, with a laptop computer and an Internet connection substituting for resources traditionally associated with an office, a person can work at home, a coffee shop, a friend's home or even a public park. Similarly, a person can shop, socialize, recreate and learn in locations that are not traditionally associated with these activities.

Rather than using places as a surrogate for people, a people-based perspective focuses directly on the individual in space and time. Using time geography as a foundation (see below), people-based accessibility examines the individual, the activities that comprise her life, their distribution in space and time, the availability of resources to overcome spatial separation among activities, and the constraints imposed by required activities such as work. Using time as a common denominator, people-based accessibility measures can accommodate ICTs and activity fragmentation better than traditional measures based on physical movement between places. People-based accessibility measures are also sensitive to differences in activity schedules and the availability of transportation and ICTs resources among different social groups, age cohorts, cultures, gender roles and household organization (Couclelis, 2000; Kwan and Weber, 2003; Miller, 2005c).

1.3 The technologies of instant access

Three separate technologies are converging to dramatically alter peoples' accessibility to each other, information and things in the environment. These are *wireless communication, location-aware technologies* and *pervasive computing*. The convergence of these technologies will likely lead to a state of *ambient findability* in the near future.

1.3.1 Wireless communication

Major wireless technologies include (roughly in order of increasing spatial range) Bluetooth, WiFi, mobile phones and WiMAX. Bluetooth is a standard for *personal area networks* (PANs) or wireless communication over short distances (typically, 10 meters). PANs facilitate digital data exchange among portable computers, mobile phones and personal digital assistants (PDAs) and other ICTs, allowing the creation of ad-hoc networks for handheld devices and "body nets" that include wearable computers (Rheingold, 2002; Wikipeda, 2006b). WiFi is a standard for wireless *local area networks* (LANs) that allow data exchange at the architectural scale. WiFi typically connects mobile devices to base stations commonly known as wireless access points or "hotspots." WiFi can also support peer-to-peer connections between devices. Wireless LANs can be configured into mesh networks where each node can serve as a repeater to convey messages (similar to Internet packet relaying) and thereby cover large distances with wireless service (Wikipedia, 2006d). Mobile phones utilize a radio network comprised of cells or service regions anchored by base stations with ranges up to 30 km or more depending on local terrain, the built environment and atmospheric conditions; although in practice each region is much smaller (typically, 0.8 to 8 km depending on the network density) (Wikipedia, 2006a). Worldwide Interoperability for Microwave Access (WiMAX) allows high-speed data transfers with geographic coverage from base stations over ranges similar to mobile phone networks.

The growth of communication technologies such as the telegraph, telephone, the Internet and World Wide Web (WWW) and their transformative impacts on individuals, societies, economies and cultures is well-known and well-documented (see, for example Abbate, 1999; Castells, 2001; Rheingold, 2000; Standage, 1998; Winston, 1998). Although these changes have been dramatic, they have been limited by the need for wired connectivity that anchors people to specific, privileged locations for information access and communication. The growth of wireless communication technologies free the individual from the need to be in particular locations when accessing information or communicating with others. This removes a subtle but profound boundary between individuals' behaviors in real space and virtual space, allowing communication and information to be more fully integrated into daily lives, and vice-versa.

Another advantage of wireless technologies is their availability. Mobile phones and WiMAX can cover large distances at a very low cost relative to wired infrastructure. Both are enormous opportunities for regions of the world such as Africa or Latin America where wired infrastructure is poor. In addition, clients such as mobile phones are more affordable and user-friendly than desktop or laptop computers, allowing more people to take advantage of wireless connectivity (Rheingold, 2002; Walton, 2006).

Wireless technologies are emerging at a time when communication technologies are both expanding and converging. Although originally analog, communications technologies are now almost exclusively digital in many regions of the world. The transformation of communications technologies from analog to digital has coincided with a deregulatory climate that is privatizing government and public franchise telecommunications monopolies. This is leading to an open and international market characterized by an exploding heterogeneity of telecommunications services (Graham and Marvin, 1996). Digital technologies allow convergence since data and information, once digitalized, can be shared across multiple media. For example, third generation mobile phones can convey sound, text messages, graphics and video. They also allow access to the Internet and the WWW. Similarly, mobile phones have merged with PDAs and Internet clients to create so-called smart phones, and *voice over Internet protocol* (VoIP) converts analog sounds into digital packets that can be conveyed over the Internet. These trends are increasing the volume and spectrum of information available to people.

A remaining issue is the regulation of the electromagnetic spectrum. In general, this continues to be controlled as a form of property with designated frequencies reserved for analog broadcasting by licensees. In contrast, digital spread-spectrum technologies can allow communal sharing of the electromagnetic spectrum through digital packet-switching combined with smart receivers. These technologies can essentially eliminate capacity constraints on wireless communication (Lessig, 1999; Rheingold, 2002).

1.3.2 Location-aware technologies

Location-aware technologies (LATs) are devices that can report their geographic location in near-real time. Technologies for determining geographic location include radiolocation methods, the global positioning system and interpolation. *Radiolocation methods* exploit wireless communication systems and determine location using methods such as those based on the time, time difference or angle of the signals' arrivals at base stations from mobile clients. The *global positioning system* (GPS) exploits time differences of signals arriving from subset of a satellite constellation in Earth orbit. *Dead-reckoning methods* use distances and directions along a route from a known location to determine the current location. Other methods include acoustic, optical and magnetic tracking (see Grejner-Brzezinska, 2004).

The GPS is traditionally the most common LAT due to its high accuracy and low cost. GPS receivers are becoming small and light enough to embed in many other mobile technologies. However, the GPS requires line-of-sight with orbital satellites; this can be a problem in places with dense foliage or tall buildings, as well as inside built structures. Therefore, it requires other methods such as radiolocation and/or dead-reckoning for support.

An emerging LAT is *radiofrequency identification* (RFID) tags. Mobile RFID tags transmit data to fixed readers using either passive or active methods. Passive tags contain no power source and rely on the current generated by passing through the electromagnetic field from the reader, while active tags contain a power source to transmit the outgoing signal. Passive tags are cheaper, smaller and lighter, but have a very limited range and expensive readers that cannot track multiple tags simultaneously. Active tags are heavier and more expensive, but have a longer range and cheaper readers than can track multiple tags simultaneously.

An important distinction between GPS and RFID methods is where the location calculations occur. With GPS, the mobile client determines the geographic location and remains anonymous to the system. In contrast, RFID tags must self-identify to the reader since the reader conducts the location calculations. This means that RFID systems have a greater potential for surveillance (Morville, 2005).

LAT enable *location-based services* (LBS). LBS provide targeted information to individuals based on their geographic location though wireless communication networks and devices such as portable computers, PDAs, mobile phones and in-vehicle navigation systems (Benson, 2001). Information services include emergency response, navigation, friend-finding, traffic information, fleet management, local news and concierge services (see Spiekermann (2004) and Chapters 15 and 17 of this volume). LBS are widely expected to be the "killer application" for wireless Internet devices: some predict worldwide deployment levels reaching one billion devices by 2010 (Bennahum, 2001; Smyth, 2001).

LATs also enable *geosensor networks*. These are interconnected, communicating and georeferenced computing devices that monitor a geographic environment. The geographic scales monitored can range from a single room to an entire city or ecosystem. The devices are typically heterogeneous, ranging from temperature and humidity sensors to video cameras and other imagery capture devices. Geosensor networks are radically changing the nature of digital geographic databases, making them heterogeneous rather than homogeneous in terms of format, resolution, content and accuracy. They are also expanding digital geographic databases to include time since they can capture the evolution of the phenomenon or environment over time (Stefanidis, 2006; Stefanidis and Nittel, 2004; also see Chapter 12 of this volume).

1.3.3 Pervasive computing

Pervasive or *ubiquitous computing* refers to the ability to access information and computing power anywhere at anytime. This involves integrating computation into infrastructure and the built environment, rather than computers being distinct objects (Wikipedia, 2006c). Pervasive computing creates omnipresent, unobtrusive, background computing that free people from routine tasks in their daily lives (Bohn et al., 2005; Weiser, 1991, also see Chapter 13 of this volume).

Four main principles underlie pervasive computing. Pervasive computing *decentralizes* computational power among a variety of small devices rather than concentrate power in mainframes or servers. It also involves the *diversification* of devices to specialized appliances rather than generic computers. These include information access devices such as cell phones and PDAs, smart controls such as thermostats and home lighting controls, intelligent appliances such as point-of-sale terminals and home refrigerators, and entertainment systems such as digital music players, televisions and gaming systems. These devices are *simple*, intuitive and convenient with mature human-computer interfaces rather than complex and cumbersome like the personal computer, albeit at the expense of flexibility. These diverse devices require a high degree of *connectivity* and interoperability across shared networks and through common protocols (Hansmann et al., 2003).

Pervasive computing has the capability to deliver computing power to more things, places and people than traditional computing. Third generation cell phones have more computing power than mainframes from decades in the past. Equally as important, the simple, specialized devices are inexpensive relative to the complex, generic computers of previous generations, meaning that more people can utilize pervasive computing than traditional computing (Rheingold, 2002). Pervasive computing can enhance interpersonal and group decision-making by enhancing social interaction and information sharing among individuals (Dryer Eisbach and Ark, 1999) The real-time operations of large and complex enterprises such as hospitals and cities can be mirrored and viewed within simulated worlds, allowing greater understanding, participation and decision-making (see Gelernter, 1992). Pervasive computing can improve logistics through monitoring and controlling the entire life of the product cycle in real-time, increasing efficiency and facilitating innovation (Bohn et al., 2005). The ability to link information with places in the real world and access this information in real-time using mobile display technologies will dramatically change our conception of place and space (Spohrer, 1999).

Pervasive computing also raises serious social and ethical issues. Omnipresent computing means we will become dependent on their correct, reliable and adaptable functioning; even more dependent than current computing. There are also issues of control and responsibility. Smart objects will make decisions for us; will these decisions reflect loyalty to the user, to their manufacturer or the government? Who is ultimately responsible for these decisions? It will also be more difficult to "opt-out" of the constant accessibility created by the smart environment: pervasive computing can eliminate many of the borders constructed between public and private life (Bohn et al., 2005; Dryer Eisbach and Ark, 1999).

1.3.4 Ambient findability

The integration of wireless communication, LATs and pervasive computing are leading to a state of *ambient findability*. Ambient findability refers to a world where we can find anyone or anything from anywhere at anytime. This includes the ability to locate people and objects, as well as navigate through both real space and the virtual space implied by the network of the world's computers and communication devices (Morville, 2005).

As digital individuals, we are represented by parallel *data shadows* comprised of the transactions we create when sending email, surfing the web, using credit cards, talking on mobile phones, passing through the viewshed of video cameras, and so on. The information acquired and perpetuated through wireless communication, pervasive computing and LATs persists over time and cannot be controlled: seemingly ephemeral actions and events can be transformed from the "here and now" to "everywhere and forever" to the detriment of the person responsible (Grudin, 2002; Zook et al., 2004).

The ability to find anyone also raises the issue of *locational privacy*. Movements and activities in geographic space are a type of signature, revealing much about the individual and her life. *Geographic information systems* (GIS) can integrate locational tracking data with remotely sensed imagery, geo-referenced social, economic and cadastral data, point-of-sale data, credit card transactions, traffic monitoring and video surveillance imagery, and other geosensor network data, revealing an individual's space-time signature to others with harmful intent (see Monmonier, 2002). At the extreme, *geoslavery* can emerge where dominant individuals or entities monitor and exert explicit or implicit control over the physical location of other individuals (Dobson and Fisher, 2003).

1.4 Perspectives

Wireless communication, location-aware technologies and pervasive computing are revolutionary: they have the potential to transform our daily lives as well as broader socio-economic relations and organization. The exact nature of these changes is subject to speculation; indeed, this is the topic of this book. There are some powerful conceptual perspectives that can support these conjectures; these include time geography, space-time convergence and telepresence, theories of cooperation and reputation, and frameworks for understanding interactions between physical and virtual space.

1.4.1 Time geography

A powerful perspective for viewing the impacts of instant access on societies and cities is *time geography*, developed by Swedish geographer Torsten Hägerstrand. Time geography recognizes that human activities have both spatial and temporal dimensions: activities and people are available at specific locations for finite periods of time. Individuals must allocate scarce available time for activity participation, as well as travel to or communication with the location where the activity occurs. People have varying abilities to overcome these spatial and temporal constraints based on their available time budget as well as the resources available for trading time for space in physical movement or virtual interaction (Hägerstrand, 1970; also see Chapter 7 of this volume). Time geography is a constraints-oriented approach to understanding the relationships among human activities, transportation and ICTs: it highlights the necessary (but not sufficient) conditions for all human activities and interactions (Pred, 1977).

Time geography recognizes three major classes of constraints on human activities. *Capability constraints* limit the activities of individuals through their own physical capabilities and/or the resources they can command. Examples include the need to sleep for 6–8 hours per day at an appropriate location, the availability of a car and the ability to control your schedule. *Coupling constraints* define where, when and for how long an individual has to join with other individuals to produce, transact or consume. For example, having to attend a meeting or a lecture is a coupling constraint. *Authority constraints* impose conditions of access in particular space-time domains. For example, private shopping centers can restrict access to designated hours of the day and days of the week.

Capability, coupling and authority constraints interact with the spacing, timing and flexibility of activities to condition accessibility. Two major classes of activities are fixed and flexible. *Fixed* activities are those that cannot be easily rescheduled or relocated; examples include home and family obligations, scheduled work shifts and meetings. *Flexible* activities are those that are more easily rescheduled and/or can occur at more than one location. Examples include shopping, recreation, and socializing. There is a fuzzy boundary between these broad categories, but this conceptualization provides a powerful mechanism for understanding how the location and timing of activities such as home and work condition accessibility to other activities and opportunities. Fixed activities act as *space-time anchors* since other activities such as shopping and recreation must occur at the temporal gaps between fixed activities, and travel to and from flexible activity locations often occur using fixed activity locations as bases.

The role of space-time anchors and the ability to trade for space in movement is apparent in two central constructs in time geography, namely, the space-time *path* and *prism*. Figure 1.1 illustrates a space-time path among activity locations or *stations* in two-dimensional space and time, with time represented by the z-axis orthogonal to the plane. Note that the path is vertical when the individual is stationary at an activity location and becomes more horizontal when she is moving through space. The slope of the path is determined by the apparent movement velocity, i.e., the trade of time for space allowed by the individual's transportation resources within that environment. Coupling constraints lead to *bundling* of the space-time paths of individuals, typically at stations. Cylinders represent the activity stations in Figure 1.1, with the length of each cylinder with respect to the z-axis indicating its availability in time. Note that the two paths are bundling at a coffee shop.

Figure 1.2 illustrates a space-time prism: this is a direct measure of a person's accessibility to the environment and activities. Fixed activities and coupling constraints anchor a space-time prism since by definition these allow only one spatial possibility during their duration. For example, the first anchor in Figure 1.2 could be the person's home while the second anchor could be their workplace. At some time during the time interval between when the home activity ends and the work activity begins, the person wishes stop at some location to conduct an activity, say, meet a friend at a coffee shop. Given these anchoring activities and a maximum velocity of movement, we can construct the prism as the subset of locations in space and time that is available or accessible to the person. The region inside the

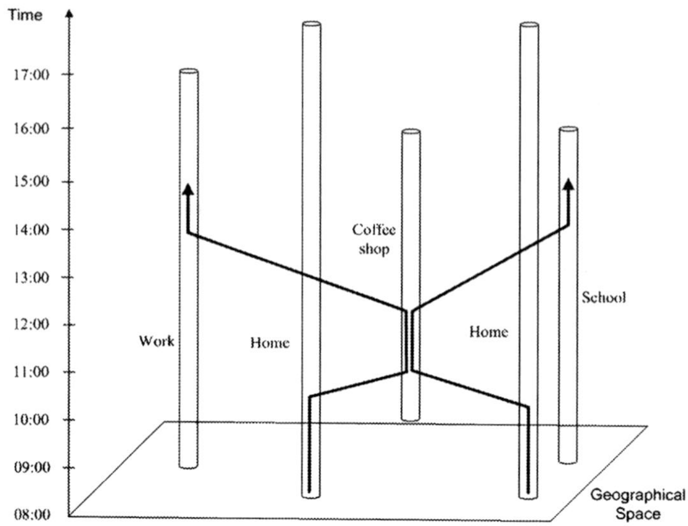

Fig. 1.1 Space-time paths and stations

prism comprises locations in space-time where an individual could be during that episode. An activity is not accessible to this person during this episode unless its location and duration intersects with the prism to a sufficient degree, with this determined by the minimum time required to conduct the activity. Similarly, two people cannot meet unless their prisms intersect to a sufficient degree. The projection of the prism to the two-dimensional plane is the *potential path area*: this comprises the region in geographic space where the person can be during that episode.

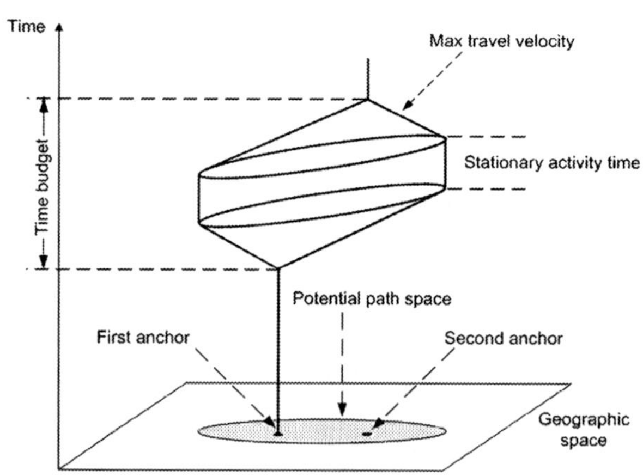

Fig. 1.2 A space-time prism

Classical time geography is a conceptual framework only: it does not have a rigorous, analytical foundation that can support the high-resolution measurements available through LATs. However, by temporally disaggregating the path, prism, and other time geographic relationships to a single instant in time (as opposed to intervals), it is possible to derive analytical statements that can support high-resolution measurement. This also allows a generalization of time geography to multi-dimensional space, including one-dimension (networks), two-dimensions (the plane) and three-dimensions (natural space) (Miller, 2005a).

Although temporal, classical time geography is a static theory in the sense that it assumes an unchanging transportation environment, in particular, a constant travel velocity. Also, while recognizing the possibility of interaction through technologies such as telephony, classical time geography focuses on the role of physical presence and movement in accessibility. The concepts of space-time convergence and telepresence extend time geography to include dynamics and virtual space.

1.4.2 Space-time convergence and telepresence

Changes in transportation technologies have led to a dramatic shrinking of the world during the past few centuries. For example, consider the time required for traveling from Portland, Maine to San Diego, CA, USA (Lowe and Moryadas, 1975):

- two years on foot in the 16th century
- 8 months on horseback in the 17th century
- 4 months by stagecoach in 1840
- 4 days by rail in 1910
- 4 hours by airplane in the late 20th and early 21st centuries

Donald Janelle invented the idea of *space-time convergence* to describe the dramatic impact of transportation technology on the organization of human activities in geographic space. We measure space-time convergence as a rate per unit time at which the travel time between two geographic locations decreases. For example, Janelle noted that during the period 1650 to 1950 C.E., the average space-time convergence rate for Edinbugh, Scotland and London, England was 29.3 minutes per year (Janelle, 1969).

It is important to note that space-time convergence is not equal everywhere; indeed, it is very uneven. For example, many parts of the world (e.g., much of sub-Saharan Africa) have not experienced space-time convergence. Some cities have excellent air connections; others are isolated. Some cities are so congested with traffic that they are experiencing space-time divergence. In his address to the 1999 Environmental Systems Research Institute (ESRI) User Conference in San Diego, CA, USA, Waldo Tobler noted that the world is *shrinking but shriveling*: while most geographic locations are becoming closer on an absolute scale and in a synoptic sense, the relative differences in transportation costs are increasing, particularly at local scales. This is creating a complex geography of accessibility.

The space-time convergence afforded by ICTs is more extreme than transportation-based convergence. Throughout much of history, being present meant that your body was in a specific location at a specific time. ICTs can remove these spatial and temporal constraints on interaction by allowing participation in an activity at a distant location through telepresence rather than physical presence. While these interactions may be less intense than physical interaction (at least until true virtual reality is developed) they can be achieved with greater convenience.

Janelle (1995, 2004) classifies communication modes based on their spatial and temporal constraints. Spatial constraints mandate either physical presence or telepresence, while temporal constraints mandate either synchronous or asynchronous communication. This leads to four communication modes as indicated in Table 1.1. *Synchronous presence* (SP) is corresponds to face-to-face (F2F) interaction: this requires coincidence in both time and space. *Asynchronous presence* (AP) requires coincidence in space but not time: examples include notes left on an office door. *Synchronous telepresence* (ST) requires only coincidence in time: telephones, radio and instant messaging services allow individuals to communicate among different places but only at the same time. *Asynchronous telepresence* (AT) does not require coincidence in space and time: this mode includes printed media, email, text messages and webpages (Miller, 2005b; also see Chapter 7 of this volume).

Many of our artifacts and social organization can be traced to an historical reliance on SP interaction. Cities were created to facilitate F2F interaction; so were our transportation systems, stadiums, theaters, classrooms and offices. More of our interactions are moving towards asynchronous and telepresent interaction. AT interaction is largely free of spatial and temporal constraints.

ST interaction is increasingly popular since it retains real-time interactivity without requiring expensive and time-consuming physical movement. Examples

Table 1.1 Communication modes based on spatial and temporal constraints (Janelle, 1995, 2004; Miller, 2005b)

Spatial & temporal constraints on communications		Spatial	
		Presence	Telepresence
Temporal	Synchronous	**SP** Face to face (F2F)	**ST** Telephone Instant messaging Television Radio Teleconferencing
	Asynchronous	**AP** Refrigerator notes Hospital charts	**AT** Mail Email Fax machines Printed media Webpages

include telephone conversations and teleconferencing. This is creating a new geography for ST activities that is based on advantageous times rather than geographic locations. Andrew Harvey and Paul MacNab have studied the constraints on ST interaction. The major coupling constraint on ST interaction is *interpersonal temporal accessibility*. This is a function of: i) the local time (i.e., time zone) of each participant; ii) their personal and social times; iii) the type of interaction. The type of interaction (business, social, recreation, shopping) determines which time periods of the participants' daily cycles are permissible for the interaction. Harvey and Macnab conclude that for ST interactions, the fundamental geographic concept of region needs reformulation into a spatio-temporal rather than a purely spatial construct. They suggest that while geographic proximity can still be important, accessibility is increasingly inseparable from time (Harvey and Macnab, 2000).

1.4.3 Cooperation and reputation

Accessibility is only part of the story. Humans may be able to access each other, but that does not necessarily mean they will do so in a benevolent manner. Access can foster competition as well as cooperation. Why do we cooperate? Why do human beings build cities, societies, corporations, governments and universities? Why are the technologies of instant access also the technologies of cooperation? In *Smart Mobs*, Howard Rheingold outlines a compelling argument about the emergence of cooperation based on reputation, and the role of ICTs in fostering cooperation and reputation (Rheingold, 2002).

A grand challenge in social science is determining the conditions under which cooperation can occur among competitive beings. A plausible explanation is *game theory*, invented by computer theorist John Von Neumann and his collaborator Oskar Morgenstern. Game theory concerns what a person should do in a decision problem where there are others with conflicting interests and whose decisions affect each other's outcome (Aumann, 1989; Flake, 1998).

There are two broad categories of games: zero-sum and non-zero sum. *Zero-sum* games are those where one player's reward exactly equals the other player's loss: one player's gain is the other loss. *Non-zero-sum* describes situations in which one person's benefit does not necessarily come at the expense of someone else. In many non-zero-sum situations, a person can benefit only when others benefit as well. An important property of non-zero sum games is that they allow for the possibility of *cooperation*: since the players' interests are not completely opposed, it may be beneficial for both to collaborate (Axelrod, 1997; Flake, 1998).

One of the most important and most studied games is the *prisoner's dilemma*. The two players in the game can choose between two moves, either *cooperate* or *defect*. The idea is that each player gains when both cooperate, but if only one of them cooperates, the other one will gain more. If both defect, both lose, but not as much as the player whose cooperation is not returned. The dilemma got its name from the following hypothetical situation: imagine two criminals arrested under the suspicion of having committed a crime together. However, the police only have enough proof to convict them on a lesser charge. The two prisoners are isolated from each other, and the police visit each of them and offer a deal: the one who "confesses" and offers evidence against the other one will be freed (Flake, 1998; Poundstone, 1992; Rapoport, 1989).

Table 1.2 is a typical statement of the prisoner's dilemma, with the prison sentences (in years) for each player in the parentheses listed in the format (Player A, Player B). Note these are penalties rather than rewards.

What should the players do? The dilemma is that the best thing for each is to defect. In fact, according to game theory you *should* defect. Therefore, both prisoners will full knowledge of these penalties and complete rationality will defect even though the outcome is worse than if both would have cooperated.

Prisoner's dilemmas are common in many real-world situations including price wars, arms races and honor systems such as public television and radio in the United States and some public transit systems (see Flake, 1998; Poundstone, 1992; Rapoport, 1989). Behavior in virtual space is also subject to similar dilemmas. For example, purchasing goods online can be a prisoner's dilemmas since the Internet can be used to shield identities and hence there is incentive to defect and not provide a valid payment or the purchased good (Rheingold, 2002).

The prisoner's dilemma indicates how the common good can be subverted by individual rationality (Flake, 1998). Collectively, prisoner's dilemmas can lead to the *tragedy of the commons*, a parable popularized by biologist Garrett Hardin in the 1960s (Hardin, 1968). This dilemma argues that people will despoil common resources through over-exploitation driven by self-interest since they believe others will do the same. Essentially, people will choose to defect and utilize a common resource to its fullest extent rather than cooperate by foregoing consumption to sustain the resource over the long-run (Rheingold, 2002). For example, Couclelis (2000) notes that despite the increasing proliferation and use of ICTs, travel demands are increasing at all geographic scales from local to global. She attributes this to the increasing number of activities across space and time, the increasing interpersonal contacts, and the relatively low cost and high comfort afforded by modern forms of transportation encouraging physical travel. This is essentially a prisoner's dilemma leading to a tragedy of the commons: individuals choose to defect (travel) rather than cooperate (substitute ICTs for travel) since they believe that others will do the same anyway. The result is air pollution, congestion, traffic accidents and consumption of non-renewable energy.

Table 1.2 A prisoner's dilemma

		Player B	
		Cooperate (don't confess)	Defect (confess)
Player A	Cooperate (don't confess)	(1, 1)	(5, 0)
	Defect (confess)	(0, 5)	(3, 3)

If someone should always defect when faced with a prisoner's dilemma, what does this imply for cooperation? People would be ruthless and rarely cooperate. It is possible that the tragedy of the commons can only be avoided by top-down, imposed rules to restrict self-interest, as Hardin suggests (Rheingold, 2002). However, there is an element missing from the prisoner's dilemma. What if players knew they were going to play the game repeatedly with the same players?

The *iterated prisoner's dilemma* (IPD) is the same as the original game, except that players play many rounds rather than a single round. Also, the players have complete knowledge of their own and their competitors' previous moves. Therefore, one can make decisions based on whether an opponent has been cooperative or not (Flake, 1998).

In 1980, University of Michigan political scientist Robert Axelrod solicited entries for a "round robin" tournament in which computer programs played the IPD for 200 rounds. The final score for a program is the sum of all scores in each confrontation. The intriguing result from Axelrod's IPD tournament is that one of the simplest strategies won: *tit for tat* (TFT), submitted by behavioral scientist Anatol Rapoport. TFT is based on a very simple principle: be nice, but punish defections. TFT offers cooperation in the first encounter with another player, but than does exactly what the opponent did in the previous round. TFT is a robust strategy since it is transparent and not greedy. It is nice by offering cooperation at first, but not so nice that it will not defect if confronted with a defection (Flake, 1998).

The results of Axelrod's tournament point to the importance of transparency and reputation in cooperation. The work of economist Mancur Olson in the 1950's and sociologist Elinor Ostrom in the 1990s suggests that cooperation can emerge when individuals can communicate easily, behavior can be monitored, there are multiple levels of sanctions, and these mechanisms are self-organized and well-matched to the context. The work of Axelrod, Olson and Ostrom suggest that tragedies of the commons can be avoided through bottom-up, self-organized and informal rules based on individuals' needs for positive reputation both for their own self-image as well as their successful participation in the given domain (Rheingold, 2002).

In *Nonzero: The Logic of Human Destiny*, Robert Wright argues that much of human history can be viewed as an evolving non-zero sum game with the quest for social status and power deriving from cooperative behavior driving human civilization to greater sophistication. Many of the institutions and technologies created by humans facilitate cooperation. Key among the technologies for cooperation is communication. Drawing on the work of game theorist Thomas Schelling, Wright notes that in a purely zero-sum world with no incentives for cooperation, there is no rational reason to communicate. Communication occurs since cooperation requires coordination, and the evolution of communication technologies is an attempt to unlock the rewards available through greater cooperation (Wright, 2000).

The technologies of instant access can facilitate cooperation by making available not only the person, but also their reputation. We can see this in online domains such as EBay where sellers and buyers have scores that influence their success or failure in that market. Trust emerges because buyers and sellers rate each other, essentially converting a single prisoner's dilemma to an iterated prisoner's dilemma. eBay is an example of an embryonic *reputation system*; these systems must become more robust and transferable across domains if the technologies of instant access are also to be the technologies of cooperation (Rheingold, 2002).

1.4.4 Interactions between physical and virtual space

The relationships between activities in physical and virtual space are complex and not easily explained by the simple "Death of Distance" argument that dominated the early literature on this topic (e.g. Caincross, 1997). Activities in physical and virtual space can be *substitutes* for each other: for example, online shopping or telecommuting can reduce the need to physically travel and conduct these activities at bookstores or offices. Virtual activities can also *complement* transportation: for example, reading a book review at Amazon.com can induce a person to travel to their local bookstore to purchase the book. Similarly, the ability to email colleagues can lead to more meetings. Physical and virtual activities can also *modify* activities in either realm by changing the location and timing of activities without a net increase or decrease in their frequency. An example is better coordination with family, friends and colleagues for gatherings and meetings through email and mobile phones (Krizek and Johnson, 2003; Miller, 2005b; Mokhtarian, 1990).

The ability to coordinate and manage activities at a distance, at any place, at any time and in real-time, is decentralizing control and coordination of human activities in real space. The increasing interactions among individuals in real-time allows the rapid reallocation and more intensive use of resources, speeding-up the metabolism of societies and cities. Activities are also becoming more fragmented across space and time as people can conduct activities such as working and shopping from any location with ICT access. These changes are creating new challenges for policy and planning. A hypercoordinated society or city creates the potential for positive feedback loops and non-linear dynamics, meaning that the society or city becomes more complex and less predictable. Decentralization of control and coordination threatens the foundation of policy and planning since this is based on centralized control and coordination. The lack of one-to-one correspondences between activities, places and times also confounds the planner's attempts to organize human activities by place and time. Finally, most planning tools are aggregate and intervene at a higher level than appropriate, namely, the individual (Couclelis, 2000; Townsend, 2000; Zook et al., 2004).

In his paper *Virtual Geography*, Mike Batty presents a conceptual framework for understanding the interactions between real and virtual spaces and how these can influence individual and collective behavior (Batty, 1997). Figure 1.3 illustrates the framework. *Place/space* refers to geographic space, i.e., the real world. Geography consists of places; these are often abstracted to spaces when representing, analyzing and modifying the real world through design, planning and policy.

Cspace refers to representations of geographic space in computer space, i.e., inside computers and their networks. This includes computer cartography, GIS and computer-based spatial analysis as well as imagined worlds such as those in gaming. Cspace has become central to both professional and casual manipulations and interactions with physical space: examples include urban planning and site selection using GIS and navigating through a city using MapQuest. Cspace amplifies these endeavors beyond simple calculation and number-crunching: the personal computer and its graphical orientation has fundamentally changed the way they are conducted.

Cyberspace is the new space that emerges from communication among remote computers. Through digital interaction, cyberspace supports a wide range of human activities, including production, consumption, recreation, learning, research and decision-making. Cyberspace does not have a one-to-one correspondence with real places, but it is not purely abstract: it is a network space implied by interactivity. However, cyberspace is also medium for sharing information about particular places.

Cyberplace refers to the substitution, complementation and modification of real world places by digital technology. Cyberplace consists of the wired and wireless digital infrastructure embedded in particular places, as well as the patterns and behaviors induced by this digital infrastructure.

Note the positive feedback loop represented by the cycle in the middle of the Figure 1.3. Real place and space are the driving forces, cspace is a prerequisite for the communications that dominates cyberspace and the act of using cyberspace involves the infrastructures of cyberplace. This feedback affects real world geographic space, by changing the pattern of our social, economic and cultural lives, our use of geographic space, eventually changing our

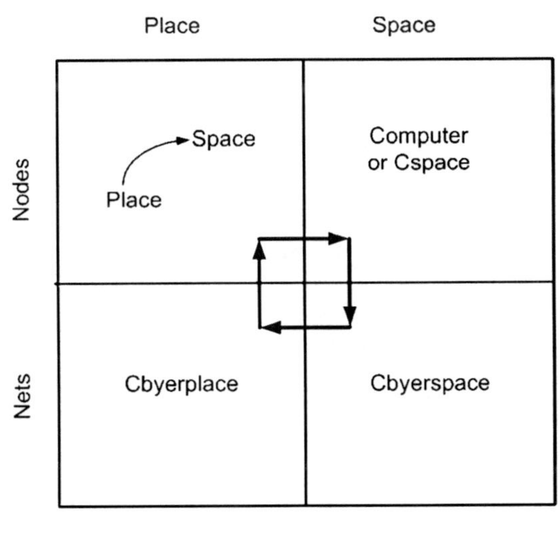

Fig. 1.3 The emergence of virtual geography

urban environments and leading to changes in our use of information technologies (cspace and cyberspace) and so on.

1.5 Overview of the book

Including the current introduction, seven major sections comprise this book, organized broadly around major domains potentially affected by the technologies of instant access and the use of these technologies in both professional and casual settings. The authors span a wide range of disciplinary backgrounds and professional orientations; this is appropriate for a multi-faceted topic such as instant access technologies and their impact on societies and cities.

1.5.1 Cities and the built environment

Section II leads with a chapter by Brian Cavanaugh on "The New Middle Landscape." Cavanaugh provides is a sweeping view of the impact of ICTs on landscapes, both in the real world and in our minds. The middle landscape is not a place on a map; rather it is an idyllic transformative landscape that balances the civilizing but oppressive city and the open but staid countryside. Thinkers such as Thomas Jefferson endorse this ideal, seeking to liberate technology from the dark, crowded and grimy European city and blend it harmoniously with the pastoral American landscape. This vision of the middle landscape has shaped the evolution of American urban design and planning. Cavanaugh argues that ICTs are creating a new middle landscape: an idyllic blend between the real and the virtual. While some may find this disembodied landscape disorienting, the spacelessness of the new middle landscape can be liberating.

Chapter 3 by Michael Batty and Andy Hudson-Smith introduces us to the strange, recursive worlds that can emerge through urban simulation. Cities are microcosms of societies that repeat themselves at different spatial scales and over different time horizons. This recursion can be taken to an entirely new level in the digital age where we can represent cities numerically, embed them within computers, and scale and distort their representations so that we can embed them within one another. Batty and Hudson-Smith illustrate these ideas with a large-scale digital representation of London, showing how we can generate different realizations of the city for different purposes. Embedding these representations within one another builds virtual worlds, allows movement from the material to the digital and back again, using the digital model to represent the material world in different ways, and finally fabricating the digital model in the real world. Digital representation opens a cornucopia of possibilities in representation and communication through a variety of devices which in turn can be embedded in the city, Escher-like, rapidly becoming the city itself.

In Chapter 4, David Scheer and Ryan Smith describe *building information modeling* (BIM), a technology poised to revolutionize planning, design and construction. BIM capitalizes on the technologies of instant access to allow sharing and collaboration in planning, design and construction. Rather than *representing* buildings in analog (blueprints) or digital (computer-aided design) formats, BIM *simulates* buildings: creating a virtual building before it is constructed in reality. This allows architects to change aspects of the building model and have them updated in real-time. It adds information legacy to the project, facilitates collaboration among players in all stages of the design and construction process, and allows more experimentation with design. BIM can also provide detailed input into planning models, simulate and test planning initiatives, facilitate performance-based planning and improve communication with stakeholders and the public. These new methods for planning, design and construction can dramatically change how we shape our cities.

Not so fast, argues Helen Couclelis in Chapter 5. The notion that ICTs annihilate distance and space in urban societies is a central part of the evolving mythology of the information age. Phrases such as "the death of distance," "anything, anywhere, any time" and "everything is now a click away" characterize a widespread belief that we are reaching an age where instant access to anything of importance is becoming a reality. If this is the case, the implications for the structure and function of cities are momentous. A number of respected scholars have even speculated that in the age of instant access, cities become redundant. Couclelis brings a healthy skepticism to this discussion. She begins with a critical review of a number of mistaken forecasts, unrealized hopes and fears, and unanticipated developments concerning the city of the early 21st century. She then proposes a simple thought experiment highlighting the fact that access is not synonymous with ubiquitous ICT. The chapter concludes with a sober discussion of why the city of the information age cannot also be the city of instant access.

1.5.2 Activities in space and time

The next major section of this book examines the relationships between ICTs and human activities. Chapter 6 by Susan Hanson and Danielle Fontaine provides a bridge between the discussions of the built environment in the previous section and the remaining chapters in this section. The impacts of ICTs on physical activity and health concern Hanson and Fontaine. They note that human beings require physical activity to maintain health, and until recently the demands of everyday life, including access, ensured that people were physically active. Technological innovations such as the telephone, the automobile, and the Internet have relentlessly eroded the amount of physical activity needed to gain access to information, social interaction, or urban activity sites. Their chapter examines recent trends in time use for insights on people's involvement in physical activity. They explore the relationship between ICTs and physical activity, including the relationship between physical activity and the built environment. Hanson and Fontaine outline a research agenda for understanding physical activity in an age of instant access.

The next two chapters in this section address the time geographic framework developed by Torsten Hägerstrand. In Chapter 7, Hongbo Yu and Shih-lung Shaw note the physical orientation of time geography and its focus on corporeal movement. Using Janelle's typology of communication modes as a foundation, they revisit and extend the basic concepts in time geography to include virtual activities afforded by ICTs and other technologies of instant access. Chapter 8 by Pip Forer, Otto Huisman and Chris McDowall also question basic time geographic concepts in light of instant access technologies. They argue that the simple and static concepts in classical time geography such as anchor points (or *markers* in their terminology), activity schedules and the static boundary between accessibility and non-accessibility implied by the prism need to be re-visited. Because of mobile information delivery, space-time constraints are no longer static and strict but dynamic and fluid. As motivating examples, they discuss the space-time constraints and the role of information in the decision-making and accessibility for a university student and a tourist.

In Chapter 9, André Skupin notes that paths through geographic space are also paths through an attribute space created by the characteristics of the local environment. Skupin illustrates the use of LATs and other geo-spatial technologies to transform a geo-spatial trajectory into a corresponding path through the socio-economic space implied by an urban environment. Fusing the geographic trajectory with digital geo-referenced data (such as census data) creates a corresponding path in multi-dimensional attribute space. This multidimensional path is spatialized through projection to low-dimensional space suitable for visualization. Visual exploration of the spatialized attribute path can provide novel perspectives on the nature of geographic space, including how it is traversed, transformed, and experienced by those that inhabit it. Other possible applications include creating a library of similar geographic trajectories for use in LBS to support trip planning, tourism, and field trips.

1.5.3 Transportation

As noted earlier in this introductory chapter, the interactions between physical space and virtual space are complex and multifaceted. The chapters in this section examine the interactions between ICTs and transportation, addressing issues regarding the use of instant access technologies to substitute, enhance and/or modify mobility in the physical world.

Chapters 10 and 11 examine the interactions between ICTs and travel behavior. Chapter 10 by Feng Zhang, Kelly Clifton and Qing Shen analyze the general relationships between ICTs and travel demand. They perform a statistical analysis of the 2001 National Household Travel Survey for the city of Baltimore, Maryland, USA and some ancillary land-use data. They examine both the individual effects of ICTs as well as the overall impacts of ICTs on travel. Their preliminary results suggest both complementary and substitution relationships between ICTs and travel depending on the specific technology and activity. However, as is typical with research on this complex topic, there are some subtle effects that require additional investigation. Verena Franken and Barbara Lenz examine a more specific question in Chapter 11, namely, will travelers use mobile information services regarding traffic conditions to make more optimal travel decisions? Their results suggest latent demand for traveler information services across different social classes. However, information accessibility is a critical factor.

Chapter 12 by Silvia Nittel, Stephan Winter, Arda Nurall and Trang Cao discuss an advanced travel information system that supports ad-hoc shared-ride trip planning in transportation networks. Shared-ride trips involve transportation clients (such as pedestrians) arranging on a short-term basis with transportation hosts (such as private automobiles or taxi cabs) for flexible travel provision. Noting the limitations of centralized services for performing the assignment between clients and hosts, they propose a novel approach considering the transportation network as an ad-hoc mobile geosensor network using a short-range, self-organizing strategy. Their results demonstrate that local communication strategies save communication costs and still deliver nearly optimal trips. Chapter 12 is a lucid illustration of the potential for instant access technologies to facilitate cooperation and coordination in activities that comprise our daily lives.

1.5.4 Mobile information services

The chapters in Section V focus on mobile ICTs and LBS. Chapter 13 by Karsten Weber, Ricarda Drüeke, Axel Schulz and Chapter 14 by Scott Bridwell address privacy issues surrounding these technologies and services. Weber, Drüeke and Schulz argue that the implementation of mobile ICTs will increase the tendency for state authorities and private companies to control public spaces. Their chapter explores the extensive debate about privacy in cultural and social studies and briefly describes the functions of privacy in public spaces. They also discuss a study regarding the behavior of students using mobile ICT in public spaces. They conclude that some behavioral changes already can be observed but that further research is needed to decide whether this will alter the notions and functions of public spaces significantly.

Chapter 14 explores the varying dimensions of locational privacy with the intent to derive standards that can be applied in mobile ICTs and LBS. Bridwell begins with a discussion of the general principles of privacy. Chapter 14 also reviews the developments of safeguards that implement these principles for both aspatial and spatial data and develops a general model of locational disclosure based on time geographic principles. The chapter concludes with suggestions for continued research on these topics.

Chapters 15, 16 and 17 address LBS more specifically. Chapter 15 focuses on technical aspects of LBS, while Chapters 16 and 17 examine their broader impacts. In Chapter 15, Seungmo Kang, Tschangho John Kim and Sung-Gheel Jang discuss the growing global market for LBS and the supporting technologies. They also discuss a concierge service model that provides a user with a minimum total cost route for both shopping and travel. The total cost includes the purchasing and the stopping costs at points-of-interest as well as the travel cost from origin to destination. Using efficient computer algorithms, the model can be implemented without technological barriers or additional costs. Similar to Chapter 12, this chapter is a clear illustration of the potential of instant access technologies to facilitate, and perhaps change, our daily lives.

Chapter 16 by Matthew Zook and Mark Graham note that LBS and related technologies creates a blending of digital and physical space as networked individuals navigate through contemporary urban life. They introduce the concept of *DigiPlace* as a way of thinking about the lived, hybrid space created as people negotiate through time, space and information. Zook and Graham use the example of mobile information services by the search engine Google in order illustrate ways in which DigiPlace can influence how people experience and move through space and place.

In Chapter 17, Daniel Sui analyzes LBS as a new type of media. LBS are creating multiple and paradoxical impacts at the individual, organizational/business and societal levels, raising challenging ethical questions. Using Marshall McLuhan's laws of media as a framework, Sui analyzes LBS in a synthetic manner that captures multiple perspectives, including positive and negative views. This robust framework ties together disparate strands of discussion in the literature and is a first step towards developing geospatial ethics for the applications of LBS.

1.5.5 Social and economic networks

Section VI of this book discusses the impact of instant access technologies on social and economic networks. The discussion earlier in this chapter suggested that ICTs can foster cooperation by facilitating coordination and reputation. Manifestations of these relationships are the social and economic networks that form between individuals and organizations.

Chapter 18 by Sucharita Gopal examines the social networks created by blogging. *Blogs* are a form of Internet publishing that can range from individual diaries to campaigns run by political groups, media and corporations. Blogs are characterized by frequent updates, the use of attribution and external links (*blogrolls*) integrated into each posting. Blogs evolve with every post since each is tied to a moment in time. Chapter 18 examines the growth and evolution of blogs using Hägerstrand's time geographic framework. Gopal analyzes two types of blogs: the first type is communities that coalesce around special interests (such as politics, law, technology, gadgets, religion) while the second is influenced by geography and oriented to local action as in political campaigns. This chapter illustrates how new communication channels are affecting the social geography of the Internet and how virtual communities are linked to local or regional geography.

Much has been written on the impact of instant access technologies on social behavior in relatively affluent societies and social groups. However, the use and social implications of these technologies can vary widely based on social class. Chapter 19 provides an alternative perspective. In this chapter, Adriana de Souza e Silva examines the social uses of the mobile phone in Brazil among different social classes. Brazil is an interesting case study since it is has one of the fastest cell phone growth rates in the world, as well as one of the greatest disparities in income and wealth. Chapter 19 focuses on how Brazil's social economic diversity entails different cell phone usage. Specifically, it shows how low-income populations appropriate technology in new and unexpected ways.

Chapter 20 by Yuko Aoyama and Samuel Ratick discusses the economic networks in the logistics industry and how instant access technologies are affecting these networks. *Logistics* refers to the planning and management of material, capital, service or information flows within an organization (Miller and Shaw, 2001). Logistics have a strong geographic orientation since it involves the procurement and distribution of material, capital, services or information from locations where it is produced (or stored) to locations where it is demanded, often under time pressures as exemplified by "just-in-time" systems. Therefore, the logistics industry is a excellent case for understanding the relations between ICTs and space. Aoyama and Ratick discuss the changes in the logistics industry and the increasing use of ICTs to meet competitive demands in this changing environment. They also propose an agent-based modeling framework for simulating and analyzing these dynamics.

1.5.6 Community

The final section of this book consists of two chapters that examine the role of instant access technologies in fostering political participation and education. Chapter 21 by Timothy Nyerges and Michael Patrick involves the use of ICTs for creating new types of meetings that improve participation in public decision situations. Using a case study in regional transportation improvement, they describe the design of Internet portals to facilitate involvement of large groups of stakeholders within asynchronous and distributed meeting arrangements called *virtual meetings*. In this context, they consider instant access to virtual meetings to include both access to a meeting and access to "voice" within a meeting. They describe instant access in terms of human-computer-human interaction at multiple levels. Chapter 21 concludes with research directions and next steps for supporting virtual meetings.

In Chapter 22, O'Neal Smitherman describes an ambitious pilot project to deliver rich media to the community in Muncie, Indiana USA. Muncie is the community now known to be the "Middletown" of the famed sociological studies initiated in the 1920's to examine the culture and conflicts of Middle America as it transitioned from the agricultural age to the industrial age. The *Digital Middletown Project* connects three community schools and a surrounding neighborhood to the local university's data network, allowing videoconferencing, electronic field trips, delivery of high definition entertainment and educational media, and educational games and simulations via wired and wireless networks. As in the past, Muncie, Indiana is serving as a laboratory for the transition of mainstream America; in this case, from the industrial age to the age of instant access.

1.6 Conclusion

As the quotation at the beginning of this chapter suggests, we know that the Age of Instant Access is coming and we know that the changes will be monumental. However, we do not know the precise nature of these changes, and similar to kayakers faced with a tidal wave, our ability to successfully deal with the coming age will depend on our and foresight and finesse. This book is an initial attempt to anticipate the changes that will occur when the technologies of instant access mature and profoundly alter our daily lives, economic institutions and social relationships and communities. The chapters in this book are diverse and speculative, but hopefully will provide the reader with a sense of the depth and breadth of the challenges and opportunities that will occur when the wave of instant access sweeps through our societies and cities.

References

Abbate, J. (1999) *Inventing the Internet*, Cambridge, MA: MIT Press.

Aumann, R. J. (1989) "Game theory," in J. Eatwell, M. Milgate and P. Newman (eds.) *Game Theory (The New Palgrave - A Dictionary of Economics)*, New York: W. W. Norton, 1–53.

Axelrod, R. (1997) *The Complexity of Cooperation: Agent-based Models of Competition and Collaboration*, Princeton, N.J.: Princeton University Press.

Batty, M. (1997) "Virtual geography," *Futures*, 29, 337352

Bennahum, D. S. (2001) "Be here now," *Wired*, 9.11, 159–163.

Benson, J. (2001) "LBS technology delivers information where and when its needed," *Business Geographics*, 9(2), 20–22.

Bohn, J., Coroamă, V., Langheinrich, M., Mattern, F. and Rohs, M. (2005) "Social, economic, and ethical implications of ambient intelligence and ubiquitous computing," in E. Aarts, W. Weber and J. Rabaey (eds.) *Ambient Intelligence*, Berlin: Springer-Verlag, 5–29.

Caincross, F. (1997) *The Death of Distance: How the Communications Revolution will Change our Lives*, Boston: Harvard Business School Press.

Castells, M. (2001) *The Internet Galaxy: Reflections on the Internet, Business and Society*, New York: Oxford University Press.

Couclelis, H. (2000) "From sustainable transportation to sustainable accessibility: Can we avoid a new *Tragedy of the Commons*?" in D. G. Janelle and D. C. Hodge (eds.) *Information, Place and Cyberspace: Issues in Accessibility*, Berlin: Springer, 341–356.

Dobson, J. E. and Fisher, P. F. (2003) "Geoslavery," *IEEE Technology and Society Magazine*, Spring, 47–52.

Dryer, D. C., Eisbach, C. and Ark, W. S. (1999) "At what cost pervasive? A social computing view of mobile computing systems," *Pervasive Computing*, 4, 652–676.

Flake, G. W. (1998) *The Computational Beauty of Nature: Computer Explorations of Fractals, Chaos, Complex Systems, and Adaptation*, Cambridge, MA: MIT Press.

Gelernter, D. (1992) *Mirror Worlds, or the Day Software puts the Universe in a Shoebox... How it will Happen and What it will Mean*, New York: Oxford University Press.

Graham, S. and Marvin, S. (1996) *Telecommunications and the City: Electronic Spaces, Urban Places*, London: Routledge.

Grejner-Brzezinska, D. (2004) "Positioning and tracking approaches and technologies," in H. A. Karimi and A. Hammad (eds.) *Telegeoinformatics: Location-based Computing and Services*, Boca-Raton, FL: CRC Press, 69–110.

Grudin, J. (2002) "Group dynamics and ubiquitous computing," *Communications of the ACM*, 45(12), 74–78.

Hägerstrand, T. (1970). "What about people in regional science?" *Papers of the Regional Science Association*, 24, 1–12.

Handy, S. L. and D. A. Niemeier (1997). "Measuring accessibility: An exploration of issues and alternatives," *Environment and Planning A*, 29, 1175–1194.

Hardin, G. (1968) "The tragedy of the commons," *Science* 162, 1243–1248.

Hansmann, U., Merk, L., Nicklous, M. S. and Stober, T. (2003) *Pervasive Computing*, second edition, Berlin: Springer.

Harvey, A. and Macnab, P. A. (2000) "Who's up? Global interpersonal temporal accessibility," in D. G. Janelle and D. C. Hodge (eds.) *Information, Place and Cyberspace: Issues in Accessibility*, Berlin: Springer, 147–170.

Janelle, D. G. (1969) "Spatial organization: A model and concept," *Annals of the Association of American Geographers*, 59, 348–364.

Janelle, D. G. (1995) "Metropolitan expansion, telecommuting and transportation," in S. Hanson (ed.) *The Geography of Urban Transportation*, 2ed., New York: Guilford, 407–434.

Janelle, D. G. (2004) "Impact of information technologies," in S. Hanson and G. Giuliano (eds.) *The Geography of Urban Transportation*, 3ed., New York: Guilford, 86–112.

Kaku, M. (1997) *Visions: How Science will Revolutionize the 21st Century*, New York: Anchor Books.

Krizek, K. J. and A. Johnson (2003). Mapping the terrain of information and communication technology (ICT) and household travel. *Proceedings, 82nd annual meeting of the Transportation Research Board*, CD-ROM.

Kwan, M.-P. and J. Weber (2003). "Individual accessibility revisited: Implications for geographical analysis in the twenty first century," *Geographical Analysis*, 35, 341–353.

Lessig, L. (1999) *Code and other Laws of Cyberspace*, New York: Basic Books.

Lowe, J. C. and Moryadas, S. (1975) *The Geography of Movement*, Boston: Houghton Mifflin.

Milller, H. J. (2005a) "A measurement theory for time geography," *Geographical Analysis*, 37, 17–45.

Miller, H. J. (2005b) "Necessary space-time conditions for human interaction," *Environment and Planning B: Planning and Design*, 32, 381–401.

Miller, H. J. (2005c) "Place-based versus people-based accessibility," in D. Levinson and K. J. Krizek (eds.) *Access to Destinations*, London: Elsevier, 63–89.

Miller, H. J. and Shaw, S.-L. (2001) *Geographic Information Systems for Transportation: Principles and Applications*, New York: Oxford University Press.

Mokhtarian, P. L. (1990). A typology of the relationships between telecommunication and transportation. *Transportation Research A*, 24, 231–242.

Monmonier, M. (2002) *Spying with Maps: Surveillance Technologies and the Future of Privacy*, Chicago: University of Chicago Press.

Morville, P. (2005) *Ambient Findability: What We Find Changes Who We Become*, Sebastopol, CA: O'Reilly Media.

Poundstone, W. (1992) *Prisoner's Dilemma*, New York: Doubleday.

Pred, A. (1977) "The choreography of existence: Comments on Hagerstrand's time-geography and its usefulness," *Economic Geography*, 53, 207–221.

Rheingold, H. (2000) *The Virtual Community: Homesteading on the Electronic Frontier*, revised edition, MIT Press.

Rheingold, H. (2002) *Smart Mobs: The Next Social Revolution*, Perseus Press.

Rapoport, A. (1989) "Prisoner's dilemma," in J. Eatwell, M. Milgate and P. Newman (eds.) *Game Theory (The New Palgrave - A Dictionary of Economics)*, New York: W. W. Norton, 199–204.

Smyth, C. S. (2001) "Mining mobile trajectories," H. J. Miller and J. Han (eds.) *Geographic Data Mining and Knowledge Discovery*, London: Taylor and Francis, 337–361.

Spiekerman, S. (2004) "General aspects of location-based services," in J. Schiller and A. Voisard (eds.) *Location-based Services*, San Francisco, CA: Morgan-Kaufmann, 9–26.

Spohrer, J. C. (1999) "Information in places," *Pervasive Computing*, 4, 602–628.

Standage, T. (1998) *The Victorian Internet*, New York: Berkley Books.

Stefanidis, A. (2006) "The emergence of geosensor networks," *Location Intelligence*, 27 February 2006; http://locationintelligence.net.

Stefanidis, A. and Nittel, S. (editors) (2004) *GeoSensor Networks*, Boca Raton, FL: CRC Press.

Townsend, A. (2000) ""Life in the real-time city: Mobile telephones and urban metabolism," *Journal of Urban Technology*, 7, 85–104.

Walton, M. (2006) "Is 'Wi-Fi on steroids" really the next big thing?" http://cnn.com; posted 31 March 2006.

Weibull, J. W. (1980) "On the numerical measurement of accessibility," *Environment and Planning A*, 12, 53–67.

Weiser, M. (1991) "The computer for the 21st century," *Scientific American*, 265(3), 66–75.

Wikipedia (2006a) "Cellular network," http://en.wikipedia.org/wiki/Cellular_network; last modified 11 June 2006.

Wikipedia (2006b) "Personal area network," http://en.wikipedia.org/wiki/Personal_area_network; last modified 14 June 2006.

Wikipedia (2006c) "Ubiquitous computing," http://en.wikipedia.org/wiki/Pervasive_computing; last modified 19 June 2006.

Wikipedia (2006d) "Wireless mesh network," http://en.wikipedia.org/wiki/Wireless_mesh_network; last modified 14 June 2006.

Winston, B. (1998) *Media Technology and Society*, New York: Routledge.

Wright, R. (2000) *Nonzero: The Logic of Human Destiny*, New York: Random House.

Zook, M., Dodge, M., Aoyama, Y. and Townsend, A. (2004) "New digital geographies: Information, communication and place," in S. D. Brunn, S. L. Cutter and J. W. Harrington (eds.) *Geography and Technology*, Dordrecht, The Netherlands: Kluwer Academic, 155–176.

CITIES AND THE BUILT ENVIRONMENT

2 The new middle landscape

Brian Cavanaugh
Victoria University of Wellington, Wellington, New Zealand

2.1 Introduction

> Late afternoon, Cambridge, England. I sit at the desk of a Xerox PARC researcher. Outside, through the grimy window to the street, I can glimpse the sun setting over stone spires. Simultaneously, through the electronic window before me, I see an empty office at Xerox PARC headquarters in Palo Alto, California. And, through the window of that distant office, that same sun is visible rising over the ochre Palo Alto hills. I am in the media space that has been constructed to weld two distant office buildings together by adding continuously open, two-way, electronic windows at both ends.
>
> – William Mitchell, *City of Bits*

Our sphere of influence is becoming significantly greater. The scenario described above by William Mitchell speaks to the growing intersection between the real and the virtual that characterizes the current condition in which we live, a condition that has broad reaching implications for our built environment. Each and everyday we are provided with news of technological advances in *telematics* and *telerobotics*[1]. And while the scientific and technological importance of these advances is obvious, little thought has been given to the impact they will have on the design of our cities. Through the minimization of material infrastructures in favor of the development of virtual infrastructures, it is clear that the means by which we construct and navigate our world is changing.

No longer is the perceptual boundary of the visual horizon that which defines the limits of our experience, and thus, our *place*[2] in the world. Developments in telematics and telerobotics are allowing us to reach beyond the horizon by collapsing the space of our world in on itself. It would seem that the transportation revolution has converged with the information revolution to create a condition without duration or distance. Spacelessness. This new 'geography' has been described as "transformational" and in a "constant state of becoming" (Sassen, 1991). The traditional relationships between the defined domains of the *city* (e.g. downtown and suburb), and between cities themselves, are all but

H.J. Miller (ed.), Societies and Cities in the Age of Instant Access, 31–37.

disappearing. In their place a ubiquitous new *middle landscape* is taking shape, a transitional zone at the intersection between the real and the virtual.

However, the increasing developments of these technologies call into question the very concept and nature of a *middle*. For, this new *middle* does not reside statically on a map, nor does it exist constantly in time. What then constitutes the middle landscape in a world in which physical boundaries no longer hold? What will constitute the form of the city of the *instant*, of the *now*, where proximities fluctuate from moment to moment? Has that which previously constituted the middle landscape superseded all other conditions, thus, creating a state of singularity, or critical expanse[3]?

2.2 Towards critical expanse

Since the nineteenth century, our position relative to the visual horizon has been changing to the point where today we find the relation having all but disappeared. It has been a gradual transition in which perceptual and physical boundaries have been slowly eroding. The nineteenth century witnessed the start of many technological revolutions that have lead to continued developments in optics (and by extension information exchange) and transportation (two areas critical to this discussion as both vision and duration play a significant role in the limit conditions of our experience). Architectural spectacles such as the Panorama were, in many respects, the virtual technologies of their time. The horizon was the boundary of the known world. Everything that was beyond its limit was new and unfamiliar, while everything contained by it was, by definition, familiar and known. Thus, it also came to be a metaphor for hope and the liberation for the burgeoning middle-class of the period. To travel beyond one's own visual limits (the horizon) was to escape the tyranny of their everyday existence[4].

What makes our relation to technology today different from that of the past, though, is the degree of connectivity and simultaneity. Over the last decade the rate of change and nature of these developments have intensified, if not all but converged. This ever-increasing mobility, both physical and virtual, has broken down the barrier of visual space. The limit condition of the horizon has been replaced with the temporal limit condition of technological speed. Distance has been replaced by duration. The usefulness of the horizon, and thus, the perspectival orientation that aids us in our recognition of the *here* and *now*, has all but diminished.

Most illustrative of this point are the recent developments in the field of telerobotics, specifically two milestone events in planetary exploration and medical surgery. The first was NASA's Mars Pathfinder mission in 1997, the second was the first major telesurgery performed in 2001.

The first event occurred on July 4, 1997. A small remotely controlled robotic lander began to transmit images and data from a distant world 119 million miles from earth. Pathfinder was made up of a landing module that served as a base of transmission, and a micro-rover named Sojourner. The rover was remotely as well as autonomously controlled. It received primary navigation commands from the Jet Propulsion Laboratory (JPL) and then had the ability to make autonomous navigation decisions based on immediate environmental conditions. While there was a perceptible lag-time due to transmission speed over such a large distance, human agency was projected to a physically remote location, extending via telepresence the boundaries of our *habitable* world. The NASA scientists, as well as the viewing public, were experiencing, through mediated sensory data, a distant world beyond the material limits of physical space. As artist Eduardo Kac wrote after the first still images were published:

> It is clear that the aesthetic dimension of this historic event introduces telepresence to the population at large, pointing to a future when personal telepresence will be an integral part of our daily lives. As our presence on the Red Planet increases via telerobotics, and eventually with humans, one can easily foresee webcams enabling us to look at the Martian surface on the internet with the same ease and regularity as today we see the skyline of several North-American cities (Kac, 1997).

The second event occurred on September 19, 2001, in which the first major trans-oceanic telesurgery was performed. Although during the later part of the nineties there were a number of minor telesurgical procedures performed, all were routine in nature and over relatively short distances. The September 19th operation was groundbreaking in that telerobotic technology allowed two surgeons at New York's Mount Sinai Medical Center to successfully remove a diseased gall-bladder from a 68-year old women in Strasbourg, France, over 4,000 miles away. The control console in New York was linked to a robot at Louis Pasteur University, via high-speed fiber-optic cables that run under the Atlantic. Unlike in the previous example, the issue of lag time was critical to this procedure, as it meant life or death for a human patient. Even with only a 160-millisecond gap (80 milliseconds each way for a transmitted signal), the doctors had to operate more slowly in order to insure safety. However, because of the high-resolution nature of the transmitted images they could be confident in what they were seeing (and doing), thus, were able to compensate for the delay. Dr. Michael Gagner, one of the surgeons who performed the operation said:

> The detail was so great, our vision was as good as if we were in my own operating room. . . For them (the Strasbourg team), they told it was like magic – nobody else was in the room, yet here this surgery went on (Choi, 2001).

In and of themselves, these are isolated achievements. However, when considered together we can see that the experiential boundaries of proximity, physical distance, and time being made ever more transparent. The *here* can now be *here* and *there* simultaneously with no apparent distance or duration between, creating a state of telepresence. It will not be long before these technologies call into question the primacy of physical space as a main qualifier of our world making, and thus introduce a state of *critical expanse*, where any and all previous limit conditions on expansion have been eliminated. It is in this new condition that the new middle landscape will take shape.

2.3 The (new) middle landscape

The middle landscape has come to embody both reality and ideal in our thinking about the city. It is a concept intricately wrapped up in the history of the United States and it's impulse towards (westward) expansion; representing the search for a pastoral ideal that Leo Marx described as the "green hollow", situated between the chaos of both the city and the countryside (Marx, 1964). We have labored to create an environment that exists between civilization and wilderness. It is, in many ways, what has characterized the history of urban form in the U.S. more than any other ideal. However, the middle landscape, while the product of pastoral idealism, is also the residual condition of a prioritizing of other destinations. It is the 'stretch of urbanity' between the downtown and the suburb 'foregrounding urban process over form' (Lerup, 2000). It is a physically indeterminate, in-between landscape ripe for speculation. Found in most, if not all, American cities, it is characterized by the background typologies and topographies of light industry, manufacturing, and transportation. Simultaneously subordinate and liberated, it is always allowing for changing interpretations, ever in a state of transformation.

Common to both the ideal and the reality of the middle landscape is its inherent relation to technological development and infrastructure. The middle landscape of the pastoral ideal is represented by a hybridized context of 'raw' open land and technology. In Marx's account of Thomas Jefferson's vision for a pre-industrialized United States, he writes:

> From Jefferson's perspective, the machine is a token of that liberation of the human spirit to be realized by the young American Republic; the factory system, on the other hand, is but feudal oppression in a slightly modified form. Once the machine is removed from the dark, crowded, grimy cities of Europe, he assumes that it will blend harmoniously into the open countryside of his native land. He envisages it turning millwheels, moving ships up rivers, and, all in all, helping to transform a wilderness into a society of the middle landscape (Marx, 1964).

The middle landscape becomes the resulting transformational condition between what were once distinct and separate domains, the city and countryside now superimposed by technological advancement. While in its reality, the middle landscape of the (contemporary) city is the residual territory given over to its necessary infrastructure. It must, by definition, be indeterminate and able to form where needed. Not a territory transformed, but rather a transformational territory - a third domain.

In both cases the middle landscape is seen as a mediating condition, the transitional zone between two distinct and different domains. It is in this consensus of the two notions that we can conceptualize the intersection between the real and the virtual that has come to characterize our current condition. A *new* middle landscape where the materiality of place intersects with "the technologies and organizational forms that neutralize place and materiality" (Sassen, 1998). Jefferson's vision of a "society of the middle landscape" realized. However, in this new conception, the wilderness is the trans-territorial spacelessness created by virtual technologies.

2.4 Proximity and world making

The story of technology is largely one of a renunciation of distance and time. For a large part, the formation of cities has been to facilitate communication. The early industrial city was dependent upon physical movement of people, goods, and services. Proximity was a critical condition. Travel times had to be minimized in order to facilitate the development of economic and social interaction. Over time, as the means of communication (and by extension transportation) have increased so has the expansion of the city. Today, cities within the global economy act as command points in the organization of the world economy and market places. It should then be argued that the key element, historically speaking, to the restructuring of urban form has been the development of technologies that have increased the means, quality, and availability of social and economic connectivity.

Distance, proximity, and connectivity have been epistemological cornerstones of western thinking. For, only that which has been immediately 'within-our-world' helped constitute 'being-in-the-world' (Heidegger, 1962). Soon we will need to relinquish ourselves from our association of distance with *fiction*. To some degree all knowledge of the world comes with some mediation, or *distance*. However, in the case of telematics and telerobotics, we not only acquire a representation of reality, but we are able to have agency within that representation. The state of telepresence these technologies afford is a reciprocal condition that involves both the observer and the observed. At this point it does not matter what is mediated or actual because all becomes the same. They become equal partners in the process of our world making. Whether or not the actual environment remains unchanged is insignificant. Versions of the world rather than the world itself are what are important. As Nelson Goodman writes, "World making begins with one version and ends with another" (Goodman, 1978).

2.5 Conclusion

The technologies of telematics and telerobotics operate more or less independently of physical place, terrain, geography, and the built landscape. It is because of the primacy of vision in our culture that the developments of these new technologies have such a profound effect. Our ocularcentrism coincides with our philosophical conflation of *Being* with objects[5]. Generally speaking, our visual input has, until recently, corresponded directly to our physical surroundings. Now a *stereoscopic* spatial condition has developed which calls into question the mitigating ability of our traditional visual based thinking. In this condition, mediated reality can actually supersede *actual* reality. What then is the condition that is replacing the traditional paradigm of the city, what does this landscape look like? Will it be characterized by local centralities or global nodality?

What this discussion reveals is the critical transformation that is occurring with respect to the city. The city, traditionally constituted of distinct domains, is ceasing to be developed as a thing and beginning to be developed as a condition (Michitaka et al., 1998). This may seem to be a crisis to some. Artist and theorist Francis Dyson has written that space implies certain possibilities, "immersion, habitation, being-there, phenomenal plentitude, unmediated presence." She goes as far as to say that, "Without space there can be no concept of presence within an environment, nor, more importantly, can there be the possibility for authenticity that 'being-in-the-world' allows" (Dyson, 1998). However, within the spacelessness of the new geography, we may find liberation.

Notes

[1] *Telematics* refers to the technologies and services using both information technologies and telecommunication technologies for collecting, storing, processing, and communicating information. *Telerobotics* refers the technology that allows a robot to accept instructions from a distance, generally a trained human operator. The human operator can thus, perform live actions in a distant environment and through sensors gauge the effects. (See Goldberg, 2000).

[2] By *place*, I mean to suggest the condition described by J. Nicholas Entrinkin. He writes:

> It is difficult to imagine the existence of an active subject in a world that contains no "here." In order to create room for such a subject we require two irreducible parts to the concept of place: place as the relative location of objects in the world, and place as the meaningful context of human action (Entrinkin, 1991).

[3] This conception of a new, or expanded, middle landscape is greatly informed by Ray Kurzweil's notion of technological singularity and Paul Virilio's notion of critical expanse, they write:

> Progress will keep accelerating to a point in which the rate of change becomes infinite, a singularity. (Boutin, 2001)
>
> And yet critical space, and critical expanse, are now everywhere, due to the acceleration of communications tools that *obliterate the Atlantic* (Concorde), *reduce France to a square one and a half hours across* (Airbus) or *gain time over time* with the TGV, the various advertising slogans signaling perfectly the shrinking of geophysical space... (Virilio, 1997)

[4] This conception of the world lead to the invention of the Panorama, along with its offshoots such as the Diorama and the Stereopticon. By all accounts the Panorama in particular provided a spectacular, and in some cases destabilizing, experience. It was a building whose planning drew in visitors off the street through a darkened corridor and then onto a raised viewing platform. From this platform the visitor was presented with a 360 degree pictorial representation of some far away locale, historical event, or some other such view. Given the sensibilities of the period the effect was dramatic. These spectacles in many ways foreshadowed the condition of instantaneous telepresence that exists today. (See Oettermann, 1997).

Alexander von Humbolt's account of the Panorama described it as a 'substitute for traveling through distant regions.' He writes:

> Panoramas are more productive of effect than scenic decorations, since the spectator, enclosed, as it were, within a magical circle, and wholly removed from all disturbing influences from reality, may the more easily fancy that he is actually surrounded by a foreign scene (Humbolt, 1850).

[5] Frances Dyson writes, "... of all the attributes of objects, visibility and extension are primary, thus vision and occupation of spaces are deeply implicated in the constitution of existence." (Dyson, 1998).

References

Boutin, Paul (2001) "Kurzweil's Law," *Wired*, April, p. 204–205.

Choi, Charles (2001) "First trans-oceanic, robot-aided surgery," UPI Science News, September 9, 2001, http://www.geocities.com/choi9998/UPI/upi-sept19.html.

Dyson, Frances (1998) "'Space', 'Being', and other Fictions in the Domain of the Virtual," *The Virtual Dimension*, New York: Princeton Architectural Press, p. 33.

Dyson, Frances (1998) *The Virtual Dimension*, p. 28.

Entrinkin, J. Nicholas (1991) *The Betweenness of Place: Towards a Geography of Modernity*, Baltimore: John Hopkins University Press, p. 10.

Goldberg, Ken, ed. (2000). *The Robot in the Garden: Telerobotics and Telepistemology in the Age of the Internet*, Cambridge: MIT Press.

Goodman, Nelson (1978), *Ways of Worldmaking*, Indianapolis: Hackett Publishing, p. 97.

Heidegger, Martin (1962) *Being and Time*, Oxford: Black Publishers, trans. John Macquarrie & Edward Robinson, pp. 138–148.

Humbolt, Alexander von (1850) *Cosmos: A Sketch of the Physical Description of the Universe Volume 2*, New York: Harpers & Brothers, trans. E. C. Otté p. 98.

Kac, Eduardo (1997) "Live from Mars," http://www.ekac.org/MARS.html.

Lerup, Lars (2000) *After the City*, Cambridge: MIT Press, p. 157.

Marx, Leo (1964) *The Machine in the Garden: Technology and the Pastoral Ideal in America*, New York: Oxford University Press, pp. 21–22.

Marx, Leo (1964) *The Machine in the Garden*, p. 150.

Michitaka, Hirose and Chihiro, Minato (1998) "The Future of Virtual Reality and Telepresence," *InterCommunication*, no. 25, Summer, pp. 18–29.

Mitchell, William (1995) *City of Bits: Space, Place, and the Infobahn*, Cambridge: MIT Press, p. 32.

Oettermann, Stephan (1997) *The Panorama: History of a Mass Medium*, New York: Zone Books.

Sassen, Saskia (1991) "Juxtaposed Temporalities: Producing a New Zone," *Anytime*, New York, Anyone Corporation, p. 117.

Sassen, Saskia (1998) *Globalization and Its Discontents: Essays on the New Mobility of People and Money*, New York: The New Press, p. 182.

Virilio, Paul (1997) *Open Sky*, London: Verso, p. 9.

3 Imagining the recursive city: explorations in urban simulacra

Michael Batty, Andy Hudson-Smith
Centre for Advanced Spatial Analysis, University College London, London, UK

> "In a riddle whose answer is chess, what is the only prohibited word?"
> I thought a moment and replied, "The word *chess*"
> "Precisely," said Albert. "*The Garden of Forking Paths* is an enormous riddle, or parable, whose theme is time; this recondite cause prohibits its mention."
>
> Jorge Luis Borges (1964, written circa 1941)
> *Labyrinths*, New Directions Publishing Co.,
> New York, p.27

3.1 Introduction

And so it is with the riddle of recursion and the recursive city. In a strict definition, recursion is the process where the same set of rules or function is applied to some phenomena over and over again, successively, in such a manner that the phenomena is repeatedly transformed as these rules are sequentially applied. Such a process implies that time passes even if the rules appear to operate near instantaneously. Recursion is normally observed not in the operation of its process but in its output, in its patterns whose signatures show the function of the recursive process in time or space or both. Indeed, this is usually the way we identify that recursion is at work, by detecting similarities in successive outputs. In cities for example, evidence of recursion might be seen in the physical patterns of its development where certain modules repeat themselves in different places, at different times, evidence enough of similar processes at work. But recursion may not show itself physically. The way populations in cities behave in space and time, the way individuals, groups, institutions organize their activities, the way economic and social functions determine modes of living and work can all show evidence of such patterning. The social world is full of clichés that imply such recursion: "History repeats itself, first as tragedy, second as farce", the famous maxim of Karl Marx; "Great fleas have little fleas upon their backs to bite 'em, And little fleas have

H.J. Miller (ed.), Societies and Cities in the Age of Instant Access, 39–55.

lesser fleas, and so ad infinitum", from Augustus de Morgan, parodying Jonathan Swift's early 18th century quip: "So, naturalists observe, a flea Has smaller fleas that on him prey, And these have smaller still to bite 'em, And so proceed ad infinitum";[1] Zeno's Paradox of the Tortoise and Achilles, and so on. . .

In this essay, we will explore the recursive city in terms of its physical patterning and the use of these as icons in the digital age. Our key thesis will be that for the first time, due to our newfound abilities to represent the city digitally, we now have tools at hand which enable us to explore it in a kaleidoscope of different ways, at different scales, in different times, in different contexts. This symposium is about the age of instant access which has been fashioned during the last 200 years as first mechanical analog technologies, and more recently digital technologies have given us opportunities to communicate ever faster (Gleick, 2000). In terms of replicating our own responses to this emerging spatial world, we are now able to intersect the digital with the material, in Negroponte's (1995) terms, to intersect 'bits' with 'atoms', and this is giving us new ways of copying what traditionally has always been material. Moreover, the abstraction that the digital world provides enables us to replicate it at different scales, to distort it, and to embed in ways that we can manufacture quickly, almost instantaneously. In fact, although recursion can occur in time as history repeating itself, there is a real sense that as the world speeds up, as we get faster, in the age of instant access, space and time collapse to a point, a singularity where recursion converges with simultaneity.

This essay is an experiment and a speculation. As we have already implied, there are many types of recursion for what is recursive can be perceived individually as well as collectively. Here we will focus simply on spatial recursion and although we will begin with some abstractions about patterns, our central focus will be upon how the material city as we see it in three dimensions can be represented in digital form. Once we have such a representation, we can scale it, we can transform it, we can repeat it in countless ways and we can begin to embed each replica within any other, thus providing windows on the world that enable us to switch between diverse representations. Why one would want to do so is simply predicated on the very plausible assumption that if we can represent the world in different ways, we gain insights that add to our knowledge and experience (Hofstadter, 1979). Moreover we are able to construct model worlds for different purposes, for different audiences; in short, we are able to use our new found digitality and virtuality to communicate more effectively.

We will focus our discussion around the idea of 'models' which we define as simplifications through abstraction. This approach is as old as the hills and as we all know, there are many models but they are usually distinct. Individuals usually create one model, not many, at least at any one time. The power of recursion for digital representations however enables us to design models that are subtly different, that merge into one another; in short to define models of models, 'simulacra' as Baudrillard (1994) calls them. Here we will build our recursive city as models within models, emphasizing not the abstraction of two-dimensional urban patterns but the iconic world of three-dimensions. In what follows, we will begin with simple recursion, some definitions. We will then illustrate our simulacra from a large scale digital model of central London, Virtual London we call it (Batty and Hudson-Smith, 2005). We explore how to embed various media into the model illustrating a kind of recursion. Then we launch into embedding the model into a virtual world, Escher-like, where we have models within models within models. But these transformations help us to stay with the material world, distributing the model at different scales on different devices from CAVEs (**C**ave **A**utomatic **V**irtual **E**nvironments), to hand-helds, printing the digital model as hard copy, augmenting traditional models with these new virtual realities. Our quest is to impress on our world the idea of recursion, the idea that instant access means access to plural realities, virtual and otherwise. We are but at the beginning.

3.2 Simple recursion: recursion defined

In formal terms, a common definition of recursion is the definition of an object or objects in terms of previously defined objects of the same class, while in mathematics, a recursive function is a function that is defined in terms of itself, usually so that there is some transformation of the output implied by the function. A good example is $f(n) = n^* f(n-1)$ for any natural number $n > 1$ which, as you can see, leads to the factorial expression $n! = n(n-1)!$. Repeatability but also transformation marks recursion of this kind and in cities, this transformation is often one of scale. For example, households organize their living at the smallest scale around different functions which are spatially diversified in the dwelling, dwellings form neighborhoods which require some kind of central place for exchange, a local market, and thence districts are composed of neighborhoods. So on up the hierarchy to the largest cities which have a structure that mirrors the tiniest dwellings in terms of the way space is organized. It is little surprise that such hierarchical organization mirrors a tree-like form in which smaller spaces are embedded in larger but with the same sorts of functions, suitably transformed for smaller populations and markets and for more routine purpose as one travels down the hierarchy: this is the cornerstone of human geography, central place theory.

Hofstadter (1979) has attempted a much more general analysis of recursion suggesting that this kind of logic is the one that underpins the most elaborate music, art, and logic. He builds on the idea of the 'strange loop', elements that repeat themselves by transforming and repositioning, reflected in and through 'tangled hierarchies'. The idea that recursion is asymmetric dominates in that its transformation through hierarchy over and over again with as many repeatable modules as one cares to define, soon destroys the logic of symmetry. Hierarchies blur into lattices (Alexander, 1966) as the modules that form the system spread epidemic-like through the medium. As an aside, however, effective recursion where the experiences are meaningful imply some strictness of control over these strange loops, tangled hierarchies they may be but hierarchies that are structured, not random. We illustrate this below in the animation of Escher's *Print Gallery*.

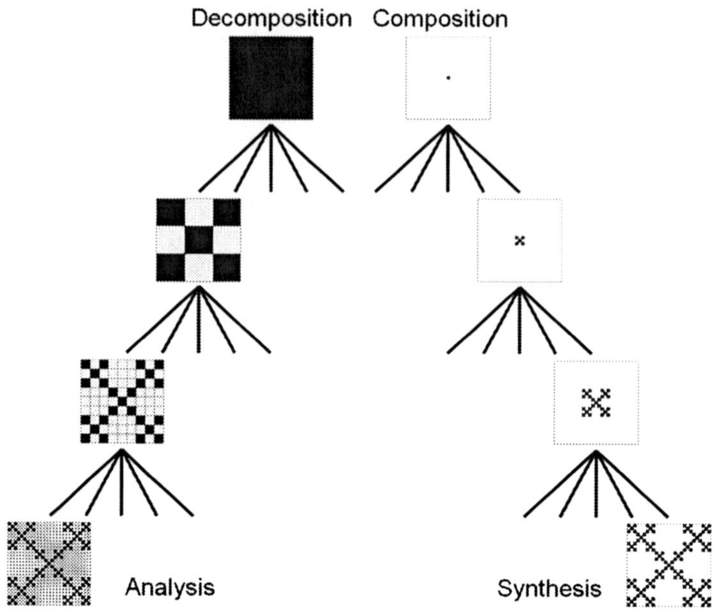

Fig. 3.1 Hierarchical recursion generating fractal self-similarity: top-down (left) to bottom-up (right)

We can travel the strict hierarchy from top to bottom, successively subdividing space in the manner shown in Figure 3.1 where we generate a fractal structure revealing different levels of detail but following the same pattern as we move to ever finer scales. But we can also grow the same structure from the bottom up which we show in parallel. There is a strict mathematical form to such recursion from the top down where a unit is successively subdivided according to a standard pattern – in this case where an array of 3×3 cells, 5 of them (at the centre and NW-NE-SE-SW) are filled as the scale successively decreases. Seen from the bottom up, we can grow the city in the same way, repeating the basic 5 cells from the module of 9 and positioning it at successively higher levels as the city grows from a single seed. That such models appear to replicate how actual cities grow has not escaped our attention but our purpose here is simply to impress the power of recursion.

The Sierpinski carpet which the patterns in Figure 3.1 have been called (Mandelbrot, 1983) are only transformed through a simple scaling but more elaborate designs can be formed if the patterns are rotated. In fact, simple rotation does not change the Sierpinski carpet very much but were the basic module somewhat more irregular – jigsaw-like for example – then one could devise many kinds of tessellation which show not only a spatial recursion but a translation towards infinity. In Figure 3.2, we begin to show how we can scale and distort in this way, with the final figure based not on a 3×3 module but on the simpler 2×2 cellular

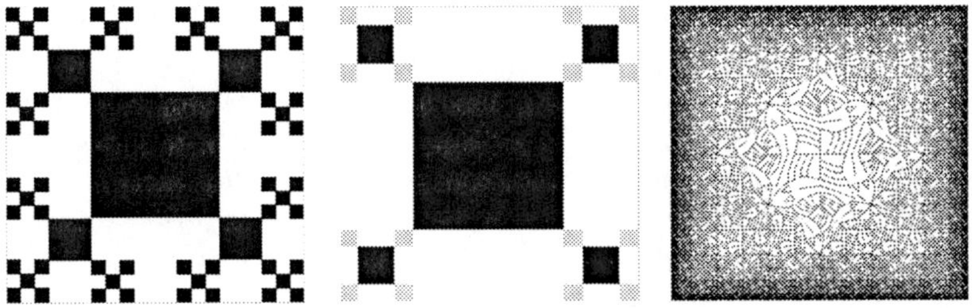

Fig. 3.2 Recursive Pictures: Transformation and scaling generating escher-like fractal self-similarity
(from http://w3imagis.imag.fr/Membres/Fredo.Durand/Quad/escher.gif)

array – a quad-tree used here to generate an Escher-like pattern. What we are not able to show through scaling and transformation is how worlds sit within worlds which are only accessible by peeling away the current image for this involves recursion through time. But in Figure 3.1 were we to embed the recursion in one specific location, starting with one world and subdividing this successively into our 5 from 9 cell pattern without translation from the starting point, we would generate a sequence of truly embedded worlds. The best example that we are able to give here is also taken from Escher's 1956 lithograph *Print Gallery* set up as a movie within a movie so that the viewer can access a world within a world as a picture within a picture. The temporal recursion in Figure 3.3 is generated by ***CTRL + click*** on. . .

Fig. 3.3 Escher and the droste effect
(from http://escherdroste.math.leidenuniv.nl/)

Hofstadter (1979) describes the *Print Gallery* as the tightest of strange loops, "an Escher vortex where all levels cross" – a tangled hierarchy in which ". . . we see. . . a picture gallery where a young man is standing, looking at a picture of a ship in the harbor of a small town, perhaps a Maltese town to guess from the architecture, with its little turrets, occasional cupolas, and flat stone roofs, upon one of which sits a little boy, relaxing in the heat, while two floors below him a woman – perhaps his mother – gazes out of the window from her apartment which sits directly above a picture gallery where a young man is standing, looking at the picture of a ship in the harbor of a small town, perhaps a Maltese town – What? We are back on the same level as we began though all logic dictates that we cannot be." (page 715).

Our purpose in this short section has not been to develop techniques that we will use in the more realistic virtual realities introduced in the rest of this paper but to show how we can embed worlds within worlds and how both space and time are woven into one another when we enable such embeddings. But there is another important dimension to our focus and this involves the idea of a model. In a sense, the basic module in Figure 3.1 is a model of how space is organized, and when we begin to populate a larger space with various transformations of the way parts of the space can be modeled, we are in a sense developing a model of a model: this is our urban simulacra. At the top of the tree, the model covers the whole space and as we successively disaggregate, we place a model of the model at finer and finer scales. If the model is not translated in space but simply placed behind the previous one, then we require a temporal recursion to access it in the same way as our Escher-animation works. And if we loop this forever, then we generate an infinity of recursions. Of course, the digital world lets us use the trick that we only require one model behind another to scale back to the first to give the impression of infinite regress as the animation of *Print Gallery* above suggests. But this is for the future. Our examples which follow will not take the argument this far for to begin, it is sufficient to indicate a couple of levels of recursion to show the power of such simulacra.

3.3
An urban simulacra: virtual London

The first digital models of cities were thought about as soon as the digital computer was invented half a century or more ago. After all, pioneers such as von Neumann and Turing impressed upon the world the philosophy of the universal machine in which the computer could do anything that in principle might be reduced to a string of binary numbers. Computer graphics was an early quest although it has taken 50 years for us to accept the idea that pictures are computable. The first realizations of cities inside a computer in iconic rather than in more abstracted mathematical form were mooted in the 1960s with the Skidmore, Owings and Merrill wire frame model of Chicago an early exhibit of these possibilities (Jankel and Morton, 1984). In fact, the development of computer graphics and computer games has always been tied to examples involving cities with some of the first experiments in recursion being developed almost unwittingly in gaming environments: adventure games, MUDs and MOOs ('**M**ulti-**U**ser **D**imensions' and '**M**ulti-user dimension **O**bject **O**riented' which are both text-based virtual environments) invite a degree of recursion.

Our digital model of central London is in the first instance a spatial skeleton, literally a shell in which the 3D geometry of the city is used as a container for as many attributes of its spatial tags as we are able to collect. These tags or identifiers are elements of the geometry, the buildings and streets but also the terrain and the utilities, and the watercourses, indeed anything that has an x-y-z coordinate that can act as a marker for some significant quality of the city. The model is thus a geographic information system viewable as a database in a variety of ways. Its geometry is not the only lens through which it can be viewed but it is of course the most suggestive for the widest audience. And like all geometry, like our Escher pictures before, this represents the most obvious way in which a digital representation can be transformed and scaled to a different physicality and reality. There are of course many different digital realities which depend on capturing attributes in different ways but the commonality of digital representation enables different renditions to be linked together. In principle, one rendition can be placed within another and recursion generated through looping.

To illustrate the model, we show first how the skeleton or shell is built in layers from different geometries. We start with the digital terrain, then layer an aerial image onto this to give a feel for the actual ground cover, adding a digital street map and then extruding building parcels to heights which are given by remotely sensed LiDAR (**Li**ght **D**etecting **A**nd **R**anging) data. Into this shell, we can insert more detailed renderings produced by others and we show this construction process in Figure 3.4. A rather different realization of the city can be produced using much more superficial realisms, digital photographic media, which is captured in panorama form. These are wrap-around images which can thence be embedded into the shell to provide a different experience of the real city. Navigating the model is usually accomplished as fly-throughs but at any point, panoramas can be embedded within the scene, thus providing a degree of realism hard to manufacture in the shell itself. In this way, the user can move from one world to another and back and this form of recursion is simply one of switching between different media. Yet it does provide a plurality of experience and with two different media, each capable of being scaled geometrically, the possibilities are endless. In Figure 3.5, we show for example how we can embed panoramas within panoramas: as the user moves around the panorama, other panoramas at different scales which enable one to hop between different spaces, reveal themselves as globes floating in the space. We can embed this into the shell of the full model, thus using this media as a way of navigation as well as a pointer to how our experience of the city can be enriched in alternate ways. We are not able to show this here but we will demo it within the lecture. It represents our first Escher-like world in which we can interpenetrate

Fig. 3.4 Digital representation: building the shell

one digital media with another both in terms of moving from one to the other in time as well as moving from one to the same at a different scale in space.

There are several other ways in which we can begin to enrich the experience in this recursive fashion. We can link digital to traditional media building augmented realities: material models can be endowed with digital media. We can also deliver models in different forms of media – on desktops, on hand-held devices, in VR (**V**irtual **R**eality) theatres, CAVEs, as tabletop holographs and so on, and in this way, we are able to build plural experiences of what the city is like. Some of these we will exploit below as we continue to embed ourselves and our realities ever deeper into the digital domain.

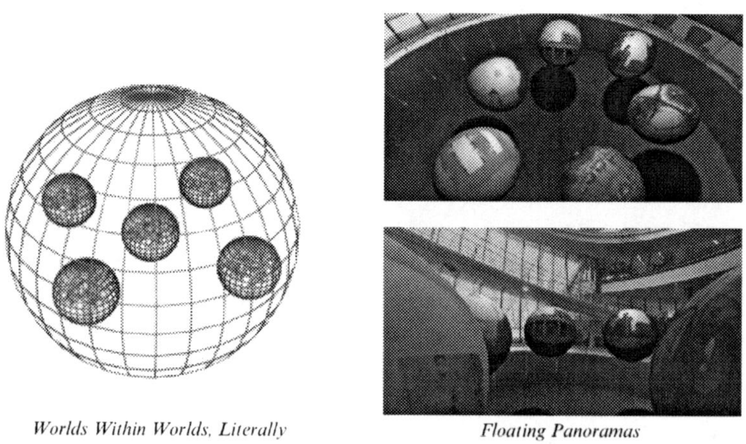

Fig. 3.5 Scaling one form of digital media to enable navigation through space

3.4 Generating virtual worlds

The simplest recursion with our GIS (Geographic Information System) shell[2] is to embed a scaled down version of the model inside a component of the model and so on, ad infinitum, but for the experience to be worthwhile, it must be meaningful. One obvious way would be to place the model inside one its buildings which may in reality be an exhibition space. The large-scale material model of London used by the City – the Piper Model – is in fact located in the Guildhall and we could easily scale our model in the model Guildhall. But this of course is pure serendipity. The digital world allows us to create different kinds of digital environment which we can populate with the same or similarly transformed modules and it is the nature of these different environments that enriches our experience. The simplest alternative to the digital model which we can represent and navigate through on the screen itself is an environment once removed which is called a 'virtual world'. Such a world is in and of itself a rather different 3D environment in which the user can appear as an avatar and behave in different ways that may or may not mirror real behaviors. It is into this kind of world that we can import our more traditional visual media – our GIS shell, our panoramas, still pictures of the city and so on, and we can thus arrange them to impart various experiences which are different from navigating through the shell and its panoramas directly.

In a sense, this throws the kind of media for visualization into the spotlight with the earliest versions of such environment entirely immersive. The world we will use here is in fact a networked world in which the environment – the virtual world – exists somewhere on the net – and those interacting with the world – from anywhere on the net – appear as avatars within the shared environment. The world can be configured as an exhibition space – a virtual room or a maze of rooms through which you might navigate – with the spaces sequenced to develop an experience of the phenomena. In the digital world, one can hop from place to place without any friction of distance and some of the earliest virtual worlds – *AlphaWorld*, for example, juxtaposed real and virtual space in such away that unusual effects were generated (Dodge, 2001; Schroeder, Huxor, and Hudson-Smith, 2001). In our example here, we have constructed such a world as an exhibition space focused on bringing together various pictures of the City of London and arranging these so the avatar or avatars can move from one specialized space or room to another through a portal or gateway – a kind of wormhole or visual hyperlink. We show the mechanics of this virtual world in Figure 3.6 which is developed using *Adobe Atmosphere* and in which users can appear as the avatar of their choice, beginning to interact with each other, and discussing the environment which they are sharing (Hudson-Smith, 2002).

If we port our GIS Shell and our panoramas into such a world, then the number of possible recursions explodes. In dealing with such models, there are many who are uncomfortable with navigating the raw digital model and thus we can place the model in the space as though it is a traditional model. Avatars can cluster

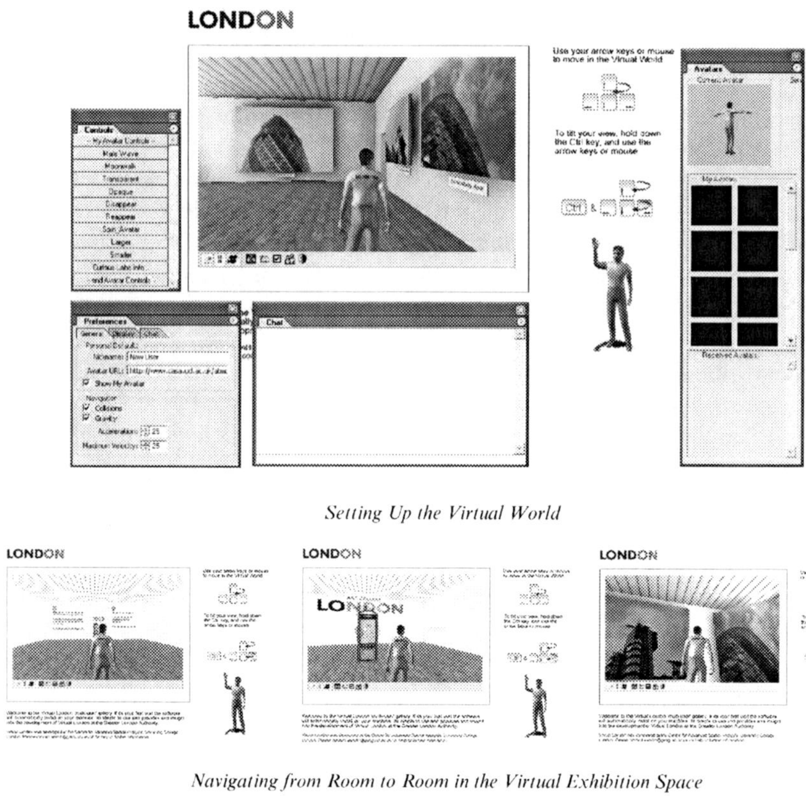

Fig. 3.6 Avatars enter the virtual city in a closed exhibition space

around it just as people view a traditional model and generate discussion about key issues. The model is still digital but with the great advantage that avatars can now manipulate it, move buildings virtually and reconfigure the scene in a way that cannot be done with a physical model. We might add our panoramas to the scene, intersecting and hopping from room to room within our panoramas, entering a building in the tabletop model and moving through the space this way. The possible manipulations are endless in that we can further embed the virtual world within another and so on, but once the loop becomes more than two or three steps deep, it loses its meaning as the variety of experiences overwhelm. In Figure 3.7, we show how we port our GIS shell into the virtual world and how we can move buildings, engaging in virtual urban design. In a sense, our virtual world becomes a virtual design studio but networked in such a way that in principle at least, a very large number of participants who are part of the city itself are able to engage in design and discussion.

We will make one final foray into this kind of virtual world which threatens to tangle the web still further. We can set up the environment of the virtual world

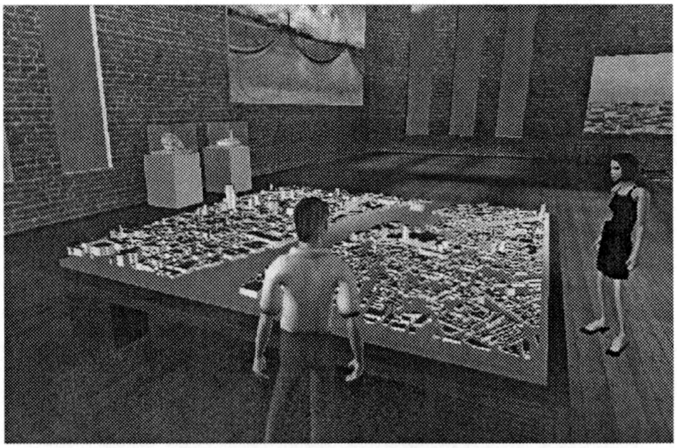

The GIS Shell in a Virtual World

Manipulating the Shell Virtually

Traditional Design and Negotiation

Fig. 3.7 Virtual and real exhibition spaces

An Avatar Entering a Digital Panorama

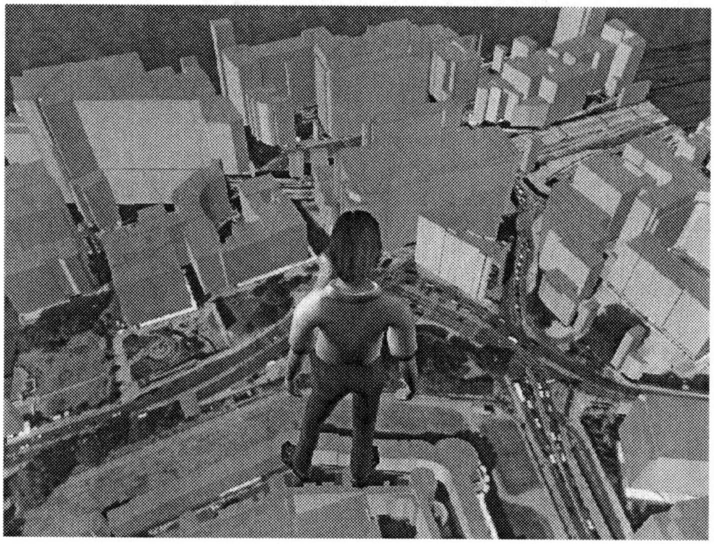

An Avatar Entering the GIS Shell of Virtual London

Fig. 3.8 Navigating out from the exhibition space

itself as the GIS shell or as a panoramic backcloth, being no longer restricted to the exhibition-like space. The world is thus the model and as we enter the model itself, we appear as an avatar. In this way, we can scale ourselves to fit the scene or dominate King-Kong like as in Figure 3.8. Or we can scale it so we populate our rooms with panoramas which enable us to move seamlessly through the city. Think of all these possibilities as being embedded within one another and you are soon drawn back to Hofstadter's (1979) argument that although the best experiences are based on strange loops, these loops must be tightly configured.

It is hard to generalize the possible recursions that we have hinted at here for we have begun from a very specific position, that of working with traditional physicality in its more superficial digital representation. We could have begun with

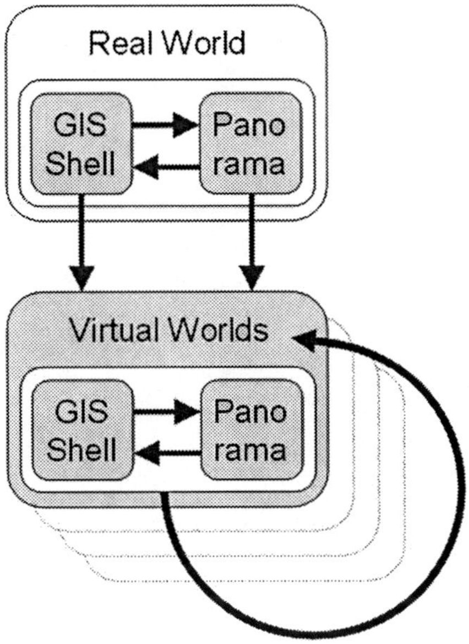

Fig. 3.9 Recursively embedding the virtual city

much more abstract spaces changing our level of abstraction accordingly but one way of visualizing the loops we have developed is in the logic chart in Figure 3.9. Here we pose two types of direct digital representation – the GIS shell and the panorama(s) – which can be woven into one another as we did in the previous section. This we call the 'Real World' although it digital in its presentation. The 'Virtual World' is where we can exercise true recursion and here we show how we can embed the real into the virtual from level to level but with the real world components being used to form the virtual world itself. In a sense, our starting point in the real world is important although the logic suggests that we could drop the top level of the flow and simply refer to all worlds as virtual, the real being just one case of a multiplicity of worlds, all being in some sense virtual in the true spirit of Baudrillard (1994).

3.5 Completing the loop: back to the material world

The last stage in our exploration of urban simulacra involves returning to the material world. Just as we can move from the real to the virtual, we can move back again in the sense that just as every model has inputs from the material world, it has outputs which manifest themselves materially, in the first instance on the screen from which we can print some visual copy. But printing is just one realization of the virtual environment – the model – in material terms and we are so accustomed

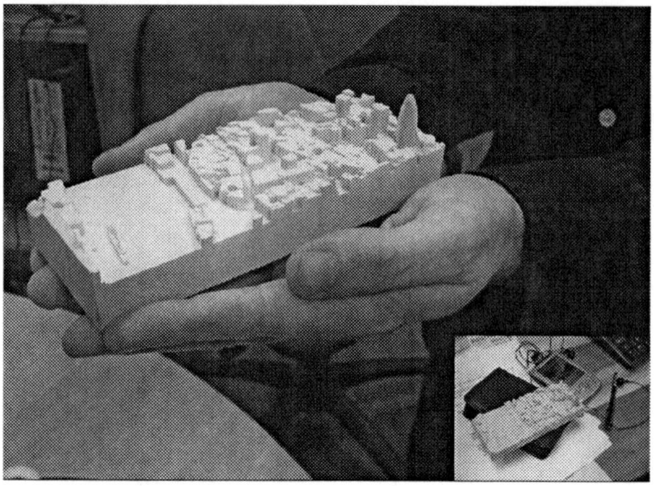

Fig. 3.10 Printing the fabricated reality

to printing in two dimensions on paper that we are often amazed by the fact that we can print in three. But our GIS shell, for example, has all the information which is necessary to print a 3D model by either attaching a milling machine to our computer or a machine for printing layer upon layer – for subtracting material from a 3D material or by adding slices as layers (Gershenfeld, 2005). This process of fabrication is the simplest possible and in Figure 3.10, we show how this can be done for a segment of our Virtual London model.

The fabrication we produce is a somewhat simpler form than the 3D geometry that we constructed the initial model from. It does not contain any detail or interior although this is simply a matter of configuring our printer into a form that would enable us to print such detail and engage in an automated process of constructing the printed output. But what is missing is all the attribute data that pertains to the city itself for the geometry is only a set of tags or coordinates on which hangs the real flesh of the city's skeleton. Our fabricated reality is thus somewhat different from our observed reality and a major question is 'how close are we ever able to get to our starting point?'. We will address this question in a somewhat speculative way in the conclusion but here it is worth indicating the nature of the loop we are proposing diagrammatically as we show in Figure 3.11.

Even at the level of the geometry, the observed and fabricated realities are somewhat different but in principle they could be the same if we had a good enough printer which we able to scale up to the city level. This sounds somewhat bizarre but in principle, even with fabrication at a much smaller, more toy-like scale, it is possible to merge the two in such a way that the 'sensing' of both real and fabricated leads to the same. In this way, then the model becomes a true rendition of the real, much truer than simply assuming that the digital and non-digital coincide in the computer. Of course cities are not geometric skeletons but are full of processes and

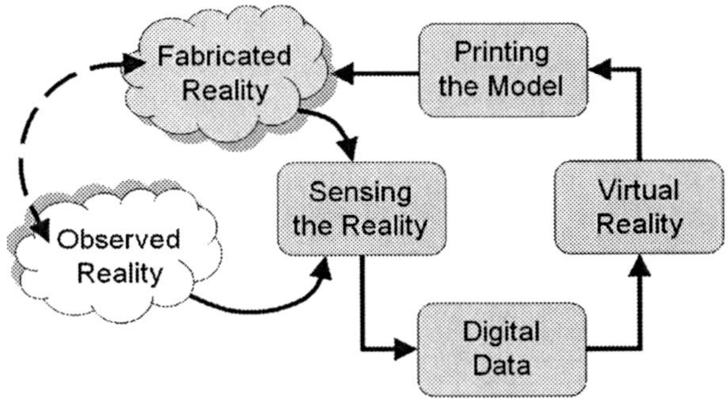

Fig. 3.11 Closing the loop: the ultimate recursion

behaviors that work within these structures and occasionally import and expend energy which changes these same structures. Once again, if our observed reality extends to what goes on within the geometry, we would be building models of how people and activities behave as our virtual reality and then we might fashion some analogs of how the printed version of the virtual reality might encompass some working parts. However this tests the limits of what we are able to do and although the implied recursion and implicit symmetry of Figure 3.11 might seem some sort of quest to aim for, in the last analysis the real and the virtual are intrinsically different, despite what Baudrillard (1994) has argued. Nevertheless the recursion set in motion by completing the loop in Figure 3.11 brings us full circle to the very idea of 'computer' which we will extemporize in our conclusion.

3.6 The future: Turing's dream

The notion of connecting the loop in such a way that what *is* being computed is what *has* been computed is central to much science and science fiction which depicts how we ourselves are beginning to merge with machines. From William Gibson's (1984) ***Neuromancer*** on, there has been a steady stream of science fiction that shows a future in which we cannot live without computers and of late, various movies such as ***The Matrix*** have taken this to the logical conclusion of a visually recursive world. In fact, all we have been demonstrating here is yet another perspective on the idea that computers are universal machines, a fact first deduced mathematically by Turing whose notion of a machine that was able to compute itself would hold the prospect for artificial intelligence. In this sense, computers are universal machines, Turing machines, and the notions of successive embedding and looping that we have posed in this essay, impresses on us the fact that we are now able to intersect different perspectives of the world in ways that we can control and manipulate much more effectively than hitherto.

If our real cities were actual working models, then it is clear that once translated into virtual realities, then this is just one way of illustrating their template from which we would be able to produce an identical working model. The reality and the model would thus be one and the same. In fact, our models, both real and virtual, would be the computer, with the computer producing or fabricating itself as a kind of artificial life. The analogy might seem far-fetched but we yet to attempt to show how the city itself is becoming computable (Batty, 1997). This will be another essay but linked to this one. There are many other levels of recursion in which we can embed computable elements into the city as hardware and software (which in the Turing context are the same) although in the material world, the virtual will always remains at arms length.

Our speculations might seem inconclusive and contestable for in this essay, we are acutely aware that we do not have a good model or template for getting to grips with the extent to which the new digitality and virtuality enables us to develop layers and layers of urban simulacra. Hoftstadter's (1979) analysis is as near the mark as we have found but this needs extending into contexts like cities where the medium is extensive and where multiple opportunities exist for different types and styles of embedding. This is for the future. In the meantime, what we have shown is how we can twist virtual realities for diverse purposes and in doing so, change our perspective, thus providing a medium for generating new insights into the way cities are structured and can be changed through the power of recursion.

Acknowledgements

Stephen Evans worked on the model and we thank him for his insights. The work was funded in part by the Greater London Authority, London Connects, and ESRC under their e-Social Science Program. Ordnance Survey and ESRI Inc. generously donated data and software.

Notes

[1] This quote comes in various guises. Lewis Fry Richardson, the climatologist, following De Morgan and Swift, summarized his own work in a 1920 paper entitled 'The Supply of Energy From and To Atmospheric Eddies' with the wonderful poem: "Big whorls have little whorls That feed on their velocity, And little whorls have lesser whorls, And so on to viscosity" (Richardson, 1920, 1993).

[2] The term shell is being used here to refer to the wider software context in which the GIS is used to produce the 3D model which is then scaled down and embedded in the virtual world in a form that is no longer a GIS but a CAD model.

References

Alexander, C. (1965), A City is not a Tree, Architectural Forum, 122 (1), 58–61 and 122 (2), 58–62.
Batty, M. (1997) The Computable City, International Planning Studies, 2, 155–173.
Batty, M. and Hudson-Smith, A. (2005) Urban Simulacra, Architectural Design, 75, 42–47.
Baudrillard, J. (1994) Simulacra and Simulation, University of Michigan Press, Ann Arbor, MI.

Borges, J. L. (circa 1941) The Garden of Forking Paths, in Labyrinths: Selected Stories and Other Writings (Edited D. A. Yates and J. E. Erby, 1964), New Directions Publishing Co., New York, 19–29.

Dodge, M. (2001) Explorations in AlphaWorld: The Geography of 3D Virtual Worlds on the Internet, in D. Unwin and P. Fisher (Editors) Virtual Reality in Geography, Taylor and Francis, London, 305–330.

Gershenfeld, N. (2005) FAB: The Coming Revolution on your Desktop – From Personal Computers to Personal Fabrication, Basic Books, New York.

Gibson, W. (1984) Neuromancer, Ace Science Fiction, New York.

Gleick, J. (2000) Faster: The Acceleration of Just About Everything, Vintage, New York.

Hofstadter, D. (1979) Gödel, Escher, Bach: An Eternal Golden Braid, Basic Books, New York.

Hudson-Smith A. (2002) 30 Days in ActiveWorlds: Community, Design and Terrorism in a Virtual World, in R. Schroeder (Editor) The Social Life of Avatars, Presence and Interaction in Shared Virtual Environments, Springer-Verlag, Berlin, 77–89.

Jankel, A., and Morton, R. (1984) Creative Computer Graphics, Cambridge University Press, Cambridge, MA.

Mandelbrot, B. B. (1983) The Fractal Geometry of Nature, W. H. Freeman, San Francisco, CA.

Negroponte, N. (1995) Being Digital, Vintage, New York.

Richardson, L. F. (1993) The Collected Papers of Lewis Fry Richardson: Volumes 1 and 2, edited by O. M. Ashford, H. Charnock, P. G. Drazin, J. C. R. Hunt, and P. Smoker, Cambridge University Press, Cambridge, UK.

Schroeder, R., Huxor, A., Hudson-Smith, A. (2001) Activeworlds: Geography and Social Interaction in Virtual Reality, Futures, 33, 569–587.

4 Download my building: How building information modeling will transform our cities

David R. Scheer, Ryan E. Smith

Center for Integrated Design and Construction, College of Architecture + Planning, University of Utah, UT

4.1 Introduction

> "From preliminary site-selection criteria to interior finishes, and everything in between... the design and construction process of tomorrow is finally here today..."
> (AIA Architect 2005)

The design and construction industry is in the midst of a historic shift in its methods, precipitated by the advent of an information technology known as Building Information Modeling (BIM). With this technology, all of the information needed for a project's design, construction and operation is contained in digital models.

At the present time, BIM technology is not fully developed. It is fragmented, with different and often incompatible products designed for architects, engineers, estimators, project managers, fabricators, constructors, facilities managers and other industry sectors. However, the advantages offered by BIM to building owners provide a strong incentive to adopt BIM across the Architecture, Engineering and Construction (AEC) industry. This development will demand deeply collaborative project teams and enable direct, automated connections between design information and fabrication and construction processes. As a result the industry is likely to see new entities emerge that blur the traditional separation of design, construction and facilities management.

BIM will also impact urban and regional planning. BIM provides many types of data that are useful for determining a project's impacts on larger scales. The ready availability of this information in digital formats will enable new methods of planning modeling and simulation based on building-specific data for simulations to test planning initiatives. To understand how this may come about, a brief introduction to the concepts of BIM is a useful starting point.

H.J. Miller (ed.), Societies and Cities in the Age of Instant Access, 57–70.

4.2 The "I" in BIM

4.2.1 Information rich models

For over 500 years architecture has been practiced by visualizing and designing three-dimensional buildings through orthographically projected two-dimensional representations. The representations are typical; meaning one drawing or note has implications across the project as ubiquitous for a typical condition in the design. This method of representations is fraught with errors because is not specific to the highly customized nature of an individual building. The advent of Computer Aided Drafting (CAD) nearly 30 years ago automated the process of two-dimensional drawing, but did not offer any fundamental changes in the way architecture was produced.

By stark contrast Building Information Modeling is a paradigm shift in the way in which buildings are designed and built because it demands real information be placed into the database of the model in order to construct a virtual building before it is constructed in reality. Buildings are therefore not represented but simulated. This necessitates that a large amount of information to be put into the model by a host of various key players in a design and building project such as the client, architect, engineers, fabricators, contractors, and builders. Currently, BIM is being used primarily as a design tool within the architect's office. Its potential for facilitating collaboration between key stakeholders in a project remains largely unexplored. However in an ideal BIM, the model is not a centrally located object, but a series of interconnected databases that have been produced by members of the design team through purpose built modelers that communicate pieces of a project and together potentially make a simulation of the building.

> "[BIM] is significantly more than transferring electronic versions of paper documents. It's more than pretty 3D renderings with construction documents as a separate function. It's about information use, reuse, and exchange. . . " (Davis, 2003)

4.2.2 Information parametric

The ability to change aspects of the models for simulation and have them updated in real time relies on the parametric nature of BIM. Parameters in a BIM may be changed and the BIM automatically reconfigures the entire project to reflect the changed parameter(s). This for example allows a single object to be manipulated for a base set of information. The information is then updated once the object is edited for geometry or finish, etc. Parametrics make for a powerful model that can take on a high degree of customization, without inventing a new system of objects. This trend, termed mass-customization (as opposed to mass production), has been observed by many in other industries: ". . . manufacturing is shifting from standardization and economies of scale to flexibility and economies of scope,

Fig. 4.1 Shoe manufacturers have initiated mass customization programs where buyers may customize their shoe and the pair is made to order at small increase in cost. mass-customization.blogs.com

using limited equipment to produce a range of products." (Buntrock, 2001) Mass-customization represents a strong direction of BIM in translating the information from the model directly into digital computer numerical control machinery for manufacturing of building components. Although mass-customization is done on a very small scale today in the AEC industry, it is the conceivable future method of production of because buildings are individual, unique and customized units of construction. (Gibb, 1999), (Fig. 4.1)

4.2.3 Information interoperability

In order to allow all of the purposed built modelers from key players to work together, interoperability of various platforms is necessary. Open source Building Information Modeling software is then the desired format in order to allow not only building modelers, but also planning and geographic modelers such as Geographic Information Ssystems (GIS) to potentially interface the building modelers. BIM in principle is very similar to GIS in that information is linked to objects. In a GIS the information is paired with two dimensional points, lines and polygons, while in BIM the objects are linked to three dimensional solids and surfaces. Like GIS, the BIM is fundamentally a database. Various other kinds of information it contains can be input and retrieved such as graphics, text, spreadsheets, and schedules. Figure 4.2 is a screen shot of a commercially available BIM platform that show both object and associated data being manipulated to update a model. Two-dimensional information can be taken from the model, but are only one representation of the information. Information rich models for the development of architecture, means the format of the information needs to be transparent, ubiquitous, immediate, consistent, and common. (Scheer et al., 2005) Ideally, with information as such an increased level of collaboration is possible reaching for what the architecture, engineering and construction (AEC) industry has longed for, integrated practice. (Figure 4.3)

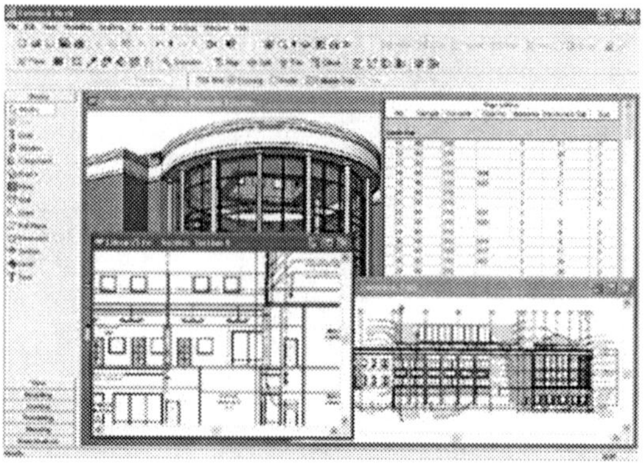

Fig. 4.2 Screen Shot of Revit BIM Modeler illustrating both object and data associated to build the model

> "The key prerequisite to achieving these innovations, however, is not more digital technology. It is creating new partnerships between owners, designers, and builders; developing organizational cultures and educational programs that support them; and inventing new delivery processes to leverage them." (Sanders, 2004)

4.2.4 Information legacy

The benefits to architects of utilizing this technology are great. BIM as a concept presents the opportunity to continuously add information to a project, from the beginning programming stages through design, bidding, construction, and management of a building during it operational life without a loss of information. In a traditional process design drawings are separate from the preparation of production

Fig. 4.3 BIM model of the Disney Concert Hall before construction used as a tool for collaboration, coordination, and integration. Courtesy of Gehry Technologies

drawings that will be used to ultimately build the building. BIM allows the information from schematic design and design development working phases in a typical architectural process to continue on into the construction drawing production phase so the objects and associated information have a legacy. This further increases efficiencies and ensures clients' needs are being met throughout the project.

4.2.5 Information collaborative

One of the greatest benefits is the potential to facilitate collaboration between key members of a building project. The AEC industry in the U.S. has been likened to the Biblical story of the Tower of Babel, speaking a tangle of languages and unable to communicate with each other. Design and construction is extremely complicated and mistakes and errors in construction documents costs the industry over $15B annually. "The reality of making architecture is far more complex than any diagram can convey."[1] (Kieran et al., 2004) BIM offers the prospect members of a project team having the same information the ability to communicate at an unprecedented level in design and construction. (Fig. 4.4)

> "Aspects of design and construction cause nearly a third of U.S. construction dollars to be spent on mistakes, inefficiencies and delays. BIM presents an unprecedented measure of design precision that benefits both the environment and the building industry itself."[2] (Bernstein, 2006)

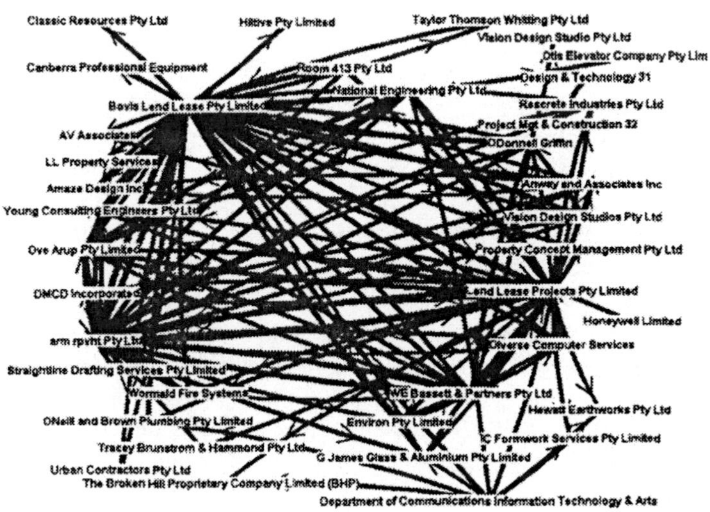

Fig. 4.4 Diagram of communication during a month of the construction of a building

4.2.6
Information simulation

The potential of BIM to offer this savings (beyond productivity gains) in design and construction inefficiencies lies in BIM as a simulation tool. Simulation is the ability to virtually model buildings and their components in design while also testing scenarios for cost, scheduling, construction sequence, energy performance, structural behavior, code compliance, egress, etc. Decisions concerning cost can be estimated more accurately at the 2design stages of a project when changes are economical. Purpose built modelers can test design scenarios for energy consumption, massing, scale, and daylight. The potential is also present to have the BIM model test urban and regional planning scenarios in conjunction with GIS on urban impacts such as energy, transportation, etc.

4.3
BIM and planning

4.3.1
Simulation types

Although most of the potentials of GIS and BIM integrated modeling have are yet been realized, their implementation is clearly within the capabilities of the technology. As described above, BIM can be used for a variety of simulations. The most widely discussed simulations have to do with properties of the building that most directly concern their designers, builders and owners. These include structural and thermal performance, cost control, code compliance and construction sequencing. There are many other types of simulation that could be performed using BIMs that would be of interest to urban and regional planners. These are broadly of two types.

The first type would provide more accurate data as input for planning models. The latter presently rely on generalizations and assumptions about the contribution of an aspect of the built environment to a particular urban or regional issue, such as traffic flows, air quality or utility demand. BIM-based simulations can provide precise data on the contribution of a specific project to these and other systems. This type of BIM-based simulation could be used to improve environmental and transportation studies as well as facilitate disaster planning.

The second type involves using BIM to simulate the impacts of planning initiatives on the built environment before they are implemented. The idea here is exploit the parametric nature of a BIM to generate building forms directly from parameters contained in a proposed regulation. Within the BIM model, these parameters would be used to generate building masses and other types of site development and permit their visualization. Carried out over an area of a city, a virtual model of development conforming to the proposed regulation could be created. This raises the possibility of extending the use of performance-based regulation and making it more effective in achieving its goals.

As mentioned above, BIMs are often purpose-built. It is neither realistic nor necessary to envision a universal BIM from which any and all information about a project can be derived. If one knows the end purpose of a BIM, one can design it to contain the specific types of information needed. This makes the creation of BIMs for the purposes discussed here less time-consuming and expensive than it might appear. Once a project has been modeled in a BIM, information can be added or removed as needed. Some models may be provided by applicants for building permits or other authorizations. In others, planners may add information needed for a specific purpose to the designers' BIM. In still other cases, it may be necessary to construct a BIM containing the information needed for a specific planning purpose. Disaster planning and other efforts involving existing structures would fall into this category.

4.3.2 BIM-derived data as input for planning models

The concepts underlying BIM are very similar to those of GIS. In both cases, a spatially situated graphic symbol provides a link to an arbitrary amount of data associated with an entity at the location indicated by the graphic symbol. In GIS, these objects are typically on the urban scale: streets, parcels, buildings, etc. In a BIM model, these objects are building components: walls, columns, fixtures, etc. It is somewhat deceptive to look at a BIM "model" on a computer screen. One seems to be looking at a conventional "3D" image of a building, representing objects in a simulated space. While it does serve this purpose, it is much more. Actually, it is more useful to think of the objects as keys to data and the virtual space as a means of arranging this data. The similarity of the concepts behind BIM and GIS provides the key to joining the two systems to provide seamless information across the spectrum of scales.

The word "building" is used here as shorthand for a site and the structures placed on it. BIMs include site information as well as "building" information. When an architect designs a project in BIM, she necessarily includes information about the site: parking, landscaping, utilities, etc. It is important to bear in mind that it is not only the building(s) but also the entire development of a site that are included in a BIM.

The following are some of the types of information that are typically contained in a BIM that are useful as input for planning models:

- Building height
- Building footprint area
- Building footprint configuration
- Area of each floor
- Number of residential units or square feet of commercial space

- Number of parking spaces
- Areas of pervious and impervious surfaces
- Volume of various materials such as concrete or asphalt paving that affect heat absorption and re-emission
- Areas covered by shadows cast by structures and planting on the site at different times of the day and year.

The following table shows some examples of typical BIM parameters and the planning parameters they can be used to calculate:

Table 4.1 BIM parameters versus planning parameters

	Planning Parameters					
	Setbacks	Max. height	Lot coverage	FAR	Pervious surface	Req'd parking
BIM parameters						
Element position	•					
Element area			•	•	•	•
Element volume				•		
Element elevation		•		•		
Element material					•	
Element type	•		•		•	
Element height		•		•		

Other types of useful data could be incorporated into a BIM:

- Material color and reflectivity/albedo
- Energy consumption by building systems
- Amounts of combustion waste products from furnaces, incinerators and other sources
- Volume of water used by plumbing fixtures and building systems
- Volume of waste water generated by plumbing fixtures and building systems
- Amount of embodied energy in a project's materials
- Volume of material to be disposed of when the project is demolished. This could be refined by classifying this material according to types of disposal, e.g. recycling, landfill, hazardous waste, etc.

Most BIM data is in a form that allows it to be used in computations. BIM data can be output in the form of a spreadsheet, for example, and manipulated mathematically. Furthermore, the information associated with a particular BIM object is customizable, just as GIS data is. New fields can be added according to the user's needs. Any information that is available for an object can be associated with it in the BIM model. Thus, the aggregate impact of individual building components on the building's surroundings can be calculated.

Such data can be used either as direct input for planning models, or to generate values needed as input for these models. As an example of the former, the number of parking spaces can be used by transportation planners to determine the amount of traffic the project will add to the surrounding streets. An example of a generated value would be using the BIM's surface area of a parking lot to calculate runoff volume. Another would be combining heat capacity information with the BIM's volume of concrete to model the project's contribution to the urban heat island. Similarly, BIM data can be used in modeling the impact of a project on water and sewer systems, electrical power demand, air quality and other urban and regional systems.

Furthermore, since this data can be extracted from a BIM from the earliest stages of a project, alternatives can be tested while it is still feasible to make design changes. This will allow architects and planners to modify a project to achieve certain goals without adding significant cost or delay. Developers often resist project modifications because they have already invested a considerable amount of money before the modifications are requested. To the extent that this resistance impedes the implementation of planning initiatives, BIM can make planning more effective by facilitating the review of projects earlier in their design.

4.3.3 Performance-based planning

The prescriptive nature of many planning regulations places well-known burdens on developers and architects. Because planners cannot envision all the possible situations that may arise when a regulation is written, unintended and undesirable consequences often occur. Although every project is unique, traditional regulations must impose a degree of uniformity or become unwieldy and difficult to understand and administer. Regulations are enacted to achieve certain results for public benefit. If the desired results could be stated directly and designers allowed to use their ingenuity to achieve them, everyone would benefit. Planners' goals would be directly realized and project designers would have greater freedom in meeting these goals. This is the idea behind performance zoning, which has been applied chiefly to new development planning. Applying performance criteria directly to infill and other types of localized development has been hampered by the difficulty both of determining the criteria an individual project must meet and administering a system in which every project is treated as a special case. BIM may help solve both of these problems through simulation.

An accurate simulation allows the performance of a project to be predicted. BIM can generate project-specific data that will increase the accuracy of planning models that predict the behavior of many urban and regional systems. The following table shows some BIM-derived data and the urban or regional systems they affect.

Table 4.2 BIM-derived data versus Urban/Regional Systems

	Urban/Regional Systems								
	Air quality	Heat island	Sunlight pene-tration	Traffic	Sanitary sewer	Storm sewer	Solid waste	Viewshed	Power distri-bution
BIM-derived data									
Building systems data	•				•	•	•		•
Impervious area		•				•			
Heat retention of materials		•							
Massing		•	•					•	
Shadow area		•	•						
Parking capacity	•			•					
Occupancy type	•			•					•
Plumbing fixture data					•	•			
Building operations data	•			•	•	•	•		•
Material color and reflectivity		•	•					•	

In a performance-based planning scheme, the planning authority would use models (simulations) that relate project parameters (given by BIM) to large-scale effects to evaluate the performance of each project they review. Performance criteria would be established and the simulations would determine if a project meets the criteria. This approach allows the simultaneous consideration of several parameters that affect a given urban system as well as dealing effectively with project parameters that affect several systems. For example, the number of parking spaces in a project also affects traffic and impermeable area and thus storm water and heat island effects. A full understanding of the traffic impacts of a project requires knowing not only how many cars are parked there, but also the number and frequency of trips generated. Information on building occupancy and operation contributes to this. Assessing a project's contribution to the urban heat island effect involves the complex interaction of several variables: the amount of heat stored in the site, how rapidly it reradiates the albedo of the building and site, the contribution of shading to reducing the amount of heat absorbed by the site, etc. The data available from the BIM allows all of these factors to be taken into account in determining the project's overall contribution to the urban heat island.

A BIM rarely contains information about all of the items on the left side of the table. Using BIM for planning simulations will require additional information to be included on one or more of these. While this is readily accomplished given the ease of adding data to a BIM, it would be desirable to have some uniformity among the various BIM platforms in the type and format of such data. Precisely what information a BIM contains (apart from basic parameters of building components) currently varies widely depending on the software and the user. BIM standards are only beginning to be developed at this time. One important data standard is called Industry Foundation Class (IFC). An IFC is a standardized software description of a building component or system that can be exchanged across many computer applications. IFC's do not currently exist for most of the data types listed in the table above. Planners should be involved in the development of these standards so that the data needed for planning simulations are included in the models in an exchangeable format.

4.3.4 BIM simulations to test planning initiatives

As Hugh Ferriss's famous drawings showed the effects of proposed building envelope regulations in New York City, BIM can generate models of the physical environment that would result from adopting a given set of planning initiatives. Unlike Ferriss's drawings, these models could take into account several parameters at once, simulate changes in the study area over time and quickly compare the effects of different alternatives.

The process of deriving data from a BIM can in a sense be reversed: parameters can be input to BIM software to produce virtual buildings that conform to these parameters. This possibility arises from the fact that the elements of a BIM are parametric. A column, for example, is characterized by such parameters as height, width, depth, wall thickness and material. The column is generated from these parameters by a script that interprets them as a column (as opposed, say, to a beam). In normal use, these parameters are determined directly by the designer. It is also possible, however, to determine these parameters by calculation or by looking them up in a table. By this procedure, building elements can be generated from alphanumeric inputs and algebraic and logical expressions. It is via such inputs and expressions that planning parameters would generate "buildings".

4.3.5 BIM in disaster planning

Many of the most devastating effects of natural and man-made disasters are the result of building failures. During an earthquake, for example, virtually all fatalities are the result of collapsing structures. The economic loss caused by the recent hurricanes on the Gulf coast is largely due to the destruction of and damage to building stock. Planning for such disasters could be greatly improved by knowing beforehand the extent of the damage each structure is likely to sustain. The information needed to predict a building's fate in a disaster can be contained in a BIM.

Earthquakes. The response of building structures in an earthquake of given magnitude has been the subject of much study in recent years. The performance of a given structure based on its construction, height, age, underlying soils and ground movement is increasingly well understood. Although it is possible to simulate the actual movement of a building during an earthquake, this level of detail is not necessary to obtain information about a building that is useful in planning. Buildings could be classified according to their construction, height and age, for example, and then placed on a soils map. A simple algorithm could then be applied to arrive at a probability of damage or collapse. Construction type and building height could be obtained from a purpose-built BIM. Predicting the structure's response beyond specifying its overall construction would be possible with further refinement of the BIM.

Hurricanes and floods. The violent winds associated with hurricanes damage buildings in ways that can be predicted using the same building data mentioned above in connection with earthquakes. The likelihood of a wind of given speed tearing off a roof or blowing out a window can be predicted fairly well based on overall design parameters that can be contained in a BIM. An estimate of the extent of damage a building would sustain due to high winds can be obtained on this basis using BIM.

The worst damage caused by the recent hurricanes was due to flooding, however, not high winds. BIM could help predict the damage caused by flooding in several ways. The elevations of a building's floors are obviously a major factor in determining the amount of damage a building sustains and this information can be contained in a BIM. For a given height of the flood waters, a BIM can calculate the quantity of each type of damaged material. Many materials such as plaster, gypsum wall board, woodwork, cabinetry and flooring cannot survive even brief immersion and must be replaced after a flood. BIM can be used to calculate the quantities of these materials in a building and combined with the costs of removing and replacing them to arrive at an estimate of the value of this type of damage. By giving the volume of the damaged material, BIM can help plan for its disposal. The extent of structural damage caused by flooding has several contributing factors: the depth of the foundations, the building materials, the duration of their submersion, the nature of the soil beneath the structure and others. At this time, there is not a great deal of data on the effects of submersion on different structural materials. Were such data to become available, they could be incorporated into a BIM and used to predict the damage to these materials.

4.3.6
BIM and the planning process

Public participation in the planning process is both crucial and difficult to achieve in a meaningful way. One major obstacle to effective public participation is the difficulty many people have understanding architectural drawings and planning documents. While traditional illustrations and renderings help people to visualize a project, the information they contain is limited and often highly selected to show a project to its best advantage. Using a BIM as a visualization tool can overcome this difficulty by offering much more information about it as well as the opportunity for even ordinary citizens to navigate through the model and view it from any angle they choose. BIM systems offer various kinds of user-friendly ways to observe and comment on a model with web-based interfaces and simple mark-up tools. A BIM can be posted on a web site for public review and anyone with a web browser can view and move through the model and leave comments on it.

Another set of obstacles to effective public input stem from the fact that, using traditional methods, three-dimensional representations needed by non-expert audiences to understand a project are expensive detours from the main design process. Renderings, whether done by hand or by computer, are separate from the process of designing and developing a project. The "real" project exists in two-dimensional architectural drawings and planning documents. Renderings are usually produced separately, by hand or using an entirely different type of software from that used for design. Renderings thus constitute a separate service from basic design and an extra expense for a developer, who therefore prefers to make as few as possible during a project. Continually updating these images, or showing alternatives, is very expensive indeed and is very rarely done outside of high profile public projects. Using BIM, however, three-dimensional images of a project can be generated directly from the model as often as necessary without extra cost from the earliest stages of a project. Design alternatives can be explored and shown easily. The public can thus follow the development of a project, look at options and have input based on representations of the actual project, not separate (and sometimes misleading) pictures. This can be done very early in the design process when little money has been invested and changes are relatively easy to make. BIM may thus facilitate effective public planning processes while helping developers make timely and cost-effective decisions.

4.4 Conclusion: implementation challenges

The possible uses of BIM suggested above are all clearly within the capabilities of the technology. However, major obstacles must be overcome in order for the technology to be implemented in the suggested manner. The challenges to be met originate in the current state of BIM, and other in the realm of planning.

As mentioned above, BIM data formats are not presently standardized. Data exchange among BIM, GIS, planning models and other software is therefore a major obstacle to many of the ideas presented above. This is a technical problem that has several components to its solution. Proprietary and incompatible data formats that are used by some BIM and GIS software are one problem. But before uniform data formats can be developed, there must be agreement among many parties as to what data will be contained in a "standard" BIM and how additional data will be handled. This work has hardly begun as there is little incentive to pursue it. Today's building industry is fundamentally based on two-dimensional drawings and, at the end of the day, BIM software must create and exchange these drawings to be marketable. Key members of a building project (architects, engineers, building owners, contractors) have differing priorities for BIM functionality and software vendors must try to balance these. The demands of the current market generally outweigh possible future applications, however useful. Although BIM will eventually become the basis of a new paradigm for the building industry, it will take time. In the meantime, solutions must be developed on a case-by-case basis.

Planners face a similar dilemma. They need solutions now for problems they face every day. The time and money needed to try out new methods are hard to find. As more architects adopt BIM, opportunities to try some of these ideas will arise. Academic research clearly has a role to play here as well. Interdisciplinary research among planners, geographers, architects, scientists and programmers is needed to explore the uses of BIM in planning and to develop useful tools to address pressing problems in the built environment.

References

AIArchitect (2005). *"Interoperability Panel Warns 'Get It or Quit It' "*. American Institute of Architects National Convention and Design Exposition. AIArchitect May 2005. www.aia.org/aiarchitect.

Bernstein, Phillip G. (2006). *"Green Building Information Modeling"*. Green Building Solutions. American Plastics Council 2006. www.greenbuildingsolutions.org.

Buntrock, Dana (2001). *Japanese Architecture as a Collaborative Process.* Spon Press 2001, (pg.105–106)

Davis, Dianne (2003). *"BIM (Building Information Modeling) Update"*. American Institute of Architects Knowledge Community. September 2003. www.aia.org/tap.

Gibb, Alastair (1999). *Off-site Fabrication.* Wiley 1999.

Kieran, Stephen and Timberlake, James (2004). *Refabricating Architecture.* MIT Press 2004, (pp. 53)

Sanders, Ken (2004). *"Why Building Information Modeling Isn't Working... Yet."* Architectural Record. Digital Practice Feature. McGraw Hill Construction. September 2004.

Scheer, David and Smith, Ryan (2005). *"The Sky is Not Falling: Building Information Modeling."* American Institute of Architects Utah Chapter 2005 Conference. Park City, Utah.

5 Misses, near-misses and surprises in forecasting the informational city

Helen Couclelis
Department of Geography, University of California, CA

5.1 Introduction

By choosing 'Societies and cities in the age of instant access' as the theme of this volume, the editor is inviting us to reflect on this and other related information-age slogans, to sort out the facts, to confront the hype, and to concede the truths that lie behind the articulation of every myth. This paper focuses on the *city* of the information age, for the simple reason that instant access for societies first and foremost means instant access in cities.

The city of instant access? It doesn't take much thought to appreciate the irony of the phrase. Ask the rush-hour commuter approaching a ladder dropped on a freeway lane, or the soccer mom on her daily circuit from school to ballpark to gym to library to shopping. In their wisdom, the editors chose a title that is sure to provoke and make us reflect on both facts and meanings. What developments prompted this highly popular notion of instant access? What does 'access' mean in the age of ICT? Access where, when, to what, for whom, by what means, for what purpose? This paper offers some thoughts on these and other issues, risking, in so doing, to fall prey to its own myths. Let us then begin with two quick myth-busting stories.

Exhibit One: I am writing this as the images of the disaster that Hurricane Katrina wrought in September of 2005 are still fresh in the national memory. For a few days in the late summer of that year, the legendary city of New Orleans, the thirty-third largest city of mighty America, the Mardi Gras city, the international tourist attraction, the Big Easy on the Gulf Coast, was one of the most inaccessible places on earth. Adding insult to injury were the information-age offers for help: Call this number to register for disaster relief – or register on-line. The irony was surely not lost on those still waiting for essential services to be restored several weeks after the disaster. Certainly, to those glued to their television screens, whether elsewhere in the USA or out there in the depths of Indonesia, access to the CNN

H.J. Miller (ed.), Societies and Cities in the Age of Instant Access, 71–83.

correspondents and their cameras appeared instant all right. But nothing seemed 'instant' to those stuck in the drowning city: not the feeding of the hungry, not the care of the sick and dying, not the reunification of the families torn apart, not the apprehension of the opportunistic criminals, not the help that seemed too little, too late – too inaccessible.

Exhibit Two: This one has nothing to do with cities, but it has everything to do with the 'death of distance', the popular slogan underlying most arguments about instant access. It is part of the account of the ill-fated 1996 Everest expedition, as reported by Krakauer (1997). Rob Hall was an experienced mountaineer leading one of two competing groups of climbers that were foolishly pushing for the summit under rapidly deteriorating weather conditions. It was on the scramble back down from the top that eleven people perished, Hall among them. But here is the poignant detail: From a few hundred feet below the Everest summit, Hall was talking to his wife in New Zealand on his satellite phone, just as he was dying for being too weak to walk a few hundred more feet to the relative safety of a ledge where he could have spent the night. For Hall, space and time had collapsed for the activity of communicating with another human being, while at the same time the friction of distance had become practically infinite (Couclelis, 1996).

What age of instant access?. . . Paradoxes such as these give us pause and make us wonder what we may mean by 'instant access' in today's societies and cities. Clearly, the rapidly diffusing, increasingly diverse and sophisticated information and communication technologies (ICT) afford practically instant access to all manner of information, including information carried by the voices of distant collaborators and loved ones. But the *city* of instant access? The city of asphalt, steel, and concrete, populated by millions of very material human bodies?.. Does the phrase characterize what is already happening around us, is it an anticipation of things soon to come, does it express the utopian vision of a few incorrigible technocrats, or is it an opportunity for us to reassess the meaning of access in the ICT-age city? More specifically: can we expect physical accessibility to become less of an issue as the increasingly ubiquitous, nimble bits of information flood the ponderous atoms that our cities and bodies are made of?

The thesis at the core of this essay is as follows: We now live in two parallel worlds, one made of atoms, the other made of bits. The two are intimately intertwined, and becoming ever more so. Sometimes they work together, sometimes they do not. Each claims us as its own, but ultimately, we are hostage to the world of atoms. The paper begins with a critical review of a number of mistaken forecasts, unrealized hopes and fears, and unanticipated developments concerning the information-age city that make us smile with the wisdom of hindsight. These predictions were made by thoughtful people and often expressed beliefs that were widely held at the time. It then proposes a simple thought experiment underlining the fact that access is not synonymous with ubiquitous ICT.

The paper concludes with a sober discussion of why the city of the information age cannot also be the city of instant access.

5.2 The cracked crystal ball: Seven foundation myths of the informational city[1]

> "In many ways, if cities did not exist, it now would not be necessary to invent them"
> Naisbitt and Aburdene, 1991

> since "In urban terms, once time becomes instantaneous, space becomes unnecessary"
> Martin Pawley, 1997

This section borrows heavily on Graham (2004, p. 4–5), whose Introduction provides a very perceptive historical overview of the 'anything-anywhere-anytime' dream going back to the early days of the information age. The scholars cited are no fools. The quotes reproduced below are taken out of their original context, and the hyperbole is often part of an ironic commentary rather than being an expression of the author's own thoughts. Some of them may exaggerate, for rhetorical purposes, speculations that the author considers to be at least partially well founded. Either way, these quotes reflect beliefs that are somehow 'in the air' in the urban societies of our age, amplified by the media and accepted as fact (or soon-to-be fact) by substantial segments of the public. All of them are based on assumptions of 'instant access' – to places, to jobs and other activities, to services, or to other people – and paint a picture of a near future where, in one way or another, cyberspace liberates us from the burdensome materiality of congested cities, the obligatory face-to-face contacts, and even from our own ponderous bodies.

Myth 1: The dissolution of cities

> "The coming of the 'global village' will inevitably mean that "the city as a form of major dimensions must inevitably dissolve like the fading shot in a movie"
> Marshall McLuhan, 1964

This early prediction by one of the prophets of the information age envisions a network of mostly modest settlements encircling the globe, closely interacting with one another through ICT flows of Amazonian volume. It is too bad then that all the currently available data and projections speak of explosive urbanization rates in virtually all of the world's developing regions.

Myth 2: The extensible individual

> "... it might be possible to locate on a mountain top and to maintain intimate, real-time and realistic contact ... All persons tapped into the global communications network would have ties approximating those used in a given metropolitan region"
> Melvin Webber, 1964

A similar scenario underlies this speculation – albeit at a smaller spatial scale – though the emphasis here is on the amenity-seeking individual rather that the settlement. Again, the implication is that the city becomes redundant. The word 'approximating' reflects a caution on the part of this author that is seldom found in similar writings. This is just as well, considering that a major use of telecommunications networks is to arrange physical gatherings of people, including massive political demonstrations, typically within urban contexts. (Rheingold, 2002).

Myth 3: A much-reduced need for mobility

> "Why would you want to drive for an hour, get stuck in traffic, and be scolded by your boss when work is a few keystrokes away from the comfort of your home office?"
>
> Shafraaz Kaba, 1996

The hope that telecommuting will be the answer to gridlock, noise, pollution, traffic accidents, wasted time and money, and the variety of physical and mental ailments plaguing our commuting society is as old as ICTs themselves. What we have instead are continuing steady increases in traffic and congestion. There are many reasons why most people don't choose to telecommute even where the opportunity is available. Most of these have to do with the fact that work is a multifaceted activity, involving practical, psychological, socioeconomic, and cultural considerations that go far beyond the information-processing tasks that telecommuting might efficiently support (Mokhtarian, 1994).

Myth 4: Physical networking will be substituted for by virtual

> The diffusion of ICT into cities will mean that "the city of the past becomes a paradoxical agglomeration in which relations of immediate proximity give way to interrelationships over distance"
>
> Paul Virilio, 1993

The best commentary on this speculation is provided by the proliferating international conferences of the ICT industry, most of which tout the 'unrivalled networking opportunities' they afford attendees. As a quick search on the Internet will reveal, those whose careers are devoted to facilitating networking over distance are spending considerable amounts of money, time and effort traveling to distant cities in order to establish 'relations of immediate proximity' with their peers. Wellman et al. (2001) have found that most Internet contact is with people who live within an hour's drive, which is exactly what is the case with old-fashioned face-to-face relations. While there is much we still don't know about the evolving effects of ICT on social relations, it is quite clear already that no direct substitution of virtual for physical interactions is taking place.

Myth 5: Spatial homogenization

> "... the era of the computer and the communication satellite is inhospitable to the high density city. With the passage of time will come spatial regularity; the urban system converges on, even if never quite attains, complete areal uniformity"
>
> Anthony Pascal, 1987

This is a more subtle version of the 'dissolution of cities' argument (Myth 1). High urban densities, the argument goes, are a relic from the days when people needed easy physical access to one another in order to conduct business. With ICT permitting equally easy, instant access to everyone anywhere, the need for close physical proximity (and thus high densities) goes away. And yet, tomorrow's world metropolises, emerging global cities such as Kuala Lumpur, Sao Paolo and Shanghai, are developing some of the densest urban centers ever seen. Even in our own suburb-loving nation, high-density downtown living has recently emerged as an increasingly attractive option.

The following quote, by a businessman asked why he chose to remain in Manhattan even though many of his clients have moved away, summarizes what is wrong with the 'death of high densities' argument: "I can get 24-hour service for anything in New York... It's terribly important. You want a consultant or technician, you can get them at a moment's notice. I have access to services and people I could never have had if I had moved." (Kotkin and Siegel, 2000, p. 10).

Myth 6: The transmission of place

> "... digital living will include less and less dependence upon being in a specific place at a specific time...." " Electronic windows" will allow one from Boston to "see the Alps, hear the cowbells, and smell the (digital) manure in the summer." While in Boston, one would, in a way, be "very much in Switzerland".
>
> Nicholas Negroponte, 1995

Technically speaking, this possibility is not too far-fetched. It sounds a bit like the IMAX experience, except that it is in real-time and is enhanced by olfactory sensations. The technology for putting something like this together already exists. But is this really "the transmission of place"? Can you climb that distant peak in the Alps and return to your chalet sunburned, dead tired, hungry and elated?

Myth 7: Casting away our material bodies

> Omnipotent subjects would be telepresent on the other side of the planet – engaging in virtual trading, virtual games, virtual tourism, virtual consumption, even virtual sex.
>
> Robins, 1995

This is the ultimate fantasy: to liberate ourselves from the burden of the 'meat' that is our body in order to expand the range and reach of the pleasures that this very body allows us to enjoy. The de-materialization and tele-transmission of bodies is a staple of modern science fiction ("Beam me up, Scottie!"). Sure, we can engage in e-commerce and virtual tourism or 'visit' pornographic web sites. But are we any closer to casting away our material bodies than our cave-dwelling ancestors were? Probably not – especially considering the growing obesity epidemic that makes it so hard for so many of us to drag ourselves as far as the DVD rental store around the corner.

5.3 Cities in the age of instant access?... A thought experiment

An Alien Power has descended upon Earth and turned all wheels, past, present and future, from round to square. Cars, trains, bicycles, and so on, cannot move, planes cannot land, and engineers will take decades to develop and implement new practical modes of transportation. Still, we are ahead of the pre – invention-of-the-wheel folks because we have ubiquitous ICT. What happens to *cities* under the circumstances?

Imagine a sadistic Professor giving the above question as part of an examination. The suggested responses are:

a) No cities (back to the land)
b) Very spread out settlements
c) Same like today
d) Very small, compact cities
e) Massive, compact cities

The thought experiment probes the twin assumptions (let's call them the null hypotheses here) that the drastic reduction in ordinary transportation options suggested by this scenario will i) result in the massive substitution of physical movement with ICT, and will ii) thereby encourage the development of new forms of settlement based on virtual rather than physical accessibility. And underlying this all is the notion that, because of the instant access afforded by ICT, physical distance no longer really matters. Thinking through the question potentially involves all seven of the myths (and their refutations) briefly discussed in the previous section. On the one hand, we may consider a world of extensible, even dematerialized individuals, relishing in their instant access to virtually transmitted places–individuals who also have gladly given up physical networking for the ease and reach of the virtual variety. In such a world there would obviously be not much need for physical mobility, leading to the dissolution of cities and to spatial homogenization as people are free to scatter over the distant countryside in search of the ideal homestead location. Responses (a) and (b) would be appropriate in this case. Both describe visions of frontier spirit and individual freedom that have been entertained before, and both celebrate the anti-urban bias that has characterized American society from its inception to this day. Response (b) in particular brings to mind Frank Lloyd Wright's 'Broadacre City', a vision of households living on one acre of land each and serviced by fleets of hovercraft (appropriately non-wheel-bearing). While this fantasy may not seem too far from the familiar exurban sprawl already filling up our metropolitan countryside, it glosses over the obligatory long-distance car trips and continuous truck deliveries that make these lifestyles possible.

On the other hand, there is the specter of masses of material human bodies, unable to access the places, activities and people for which or whom ICT can offer no real substitute. This is the other side of the story, whereby the loss of physical transportation as we know it results in a need for physical proximity that can only be met – if at all – by the densest conceivable urban structures. Responses (d) and (e) are the ones that may make sense here. Small, compact cities? We've seen these in the distant past, in the days of much simpler economies, societies, technologies, and lifestyles. Can an ICT-based society function in settlements barely one mile across? How will our digital devices be produced and distributed? How will our insatiable demands for consumer goods be met? What will happen to the global production chains and international divisions of labor? How will the infrastructure needs of a sophisticated, ICT-dependent society be served without the minimum number of customers needed to support them? Response (e), the massive, compact cities, might make more sense: perhaps something like Soleri's design of the 'Hyperbuilding', the kilometer-high structure concentrating over one square kilometer a complete city of one million people (Soleri, n.d.). This idea of putting millions of people in one spot is actually an older dream, whether motivated by extreme environmentalism, militant functionalism, or some dark desire for surveillance and control. Thus Le Corbusier's 1923 design for a 'Contemporary City of Three Million' would stack that number of people in a handful of tightly clustered blocks of monstrous scale – machines for living and working meant to clear the earth of the ever-expanding footprint of the less affluent classes. There was also the Reverend Louis Tucker's (1929) story whereby New York City was destroyed and replaced by 'The Cubic City', a single gigantic building two miles high and two miles wide (Tucker, 1929). Eight hundred stories high, it comfortably held a population of eighty million and boasted a giant rooftop airfield and zeppelin moorings (again, no wheels!) And let's not forget King Camp Gillette's late nineteenth-century socialist utopia of a highly efficient, gigantic city in the vicinity of Niagara Falls, which would hold the entire population of North America within a cluster of forty thousand gigantic structures (Gillette, 1894).

Regardless of the proponents' motivations, all these super-compact futuristic designs should have the effect of drastically reducing physical distances within the 'city'. Whether they would also increase accessibility is another matter: a moment's reflection will show that congestion from the flows of millions of pedestrians and millions of tons of goods within these structures might render any kind of movement all but impossible. Perhaps then we should go back to the conceptually more plausible answers favoring the dissolution of cities hypothesis – but which are directly contradicted by the empirical reality of accelerating urbanization and increasing central densities in most of the world today. So that won't work either. . . There thus seems to be no good answer to the dilemma of our hypothetical wheel-less society, no matter how many ICT resources they may deploy.

5.4 Trying to make sense of it all: Why so many smart people get it wrong

While it would take much more than a fanciful thought experiment to prove the point, there is every good reason to believe that ICT cannot substitute for the access provided by high urban densities and good physical transportation. Arguments to the contrary are based on a number of different fallacies, all sharing an overemphasis on the exciting new possibilities opened up by ICT and a failure to consider their limitations. I will briefly outline some of the most common of these fallacies under the rubrics of substitution, equalization, isolation, human nature, and materiality.

Substitution

In the early days of modern ICT, considerable excitement was generated in the planning and transportation literature, not to mention the media, by the expectation that easy-to-use, widely available, and versatile modes of telecommunication would provide valid substitutes for millions of hitherto necessary trips. Traffic congestion, accidents, parking shortages, and the host of related economic losses, health issues, and environmental problems would be greatly reduced (see Myth 3 above). If these earlier hopes appear naïve today, this is not necessarily because expected substitutions did not take place, but because these were accompanied by increased demand for travel at other points of the global system of interactions. Research on the effects of new technologies on older ones has established that such effects can be complex, subtle, and often counter-intuitive. Concerning the relationships between ICT and transportation, Mokhtarian and Meenakshisundaram (1999) have shown that at least three different kinds of interactions are taking place. These are: (1) substitution, whereby ICT-enabled remote access to people and the remote performance of tasks directly replace physical trips; an example would be telecommuting, the great albeit largely frustrated hope of those anticipating significant reductions in commuting traffic; (2) amplification, whereby ICT helps generate more trips, as when email, voice mail, or text messaging are used to facilitate the arrangement of face-to-face meetings; and (3) synergy, whereby transportation and ICT work together to produce novel ways of doing things, or new forms of activity. There are many examples of this phenomenon in the area of intelligent transportation systems (among several others), as when GPS-enabled cars signal their positions to police or rescuers. However, intelligent transportation systems may also present several instances of the phenomenon of amplification, since increased efficiency on roadways very often results in more discretionary trips being generated.

Reaching well beyond transportation, another form of the substitution fallacy is also reflected in Myth 6 ('The digital transmission of place') and Myth 7 ('Casting away our material bodies'). Because their most prominent feature is the wholesale substitution of physical with virtual sensations, these myths will be discussed below under the rubric of 'materiality'.

Equalization

Graham (2004) calls 'cyberlibertarianism' the tendency to view ICT, and the Internet in particular, as a major force for the equalization of opportunity and the spread of democracy and mutual understanding across the globe and across social, political, and ethnic divisions. What Mitchell (1995) has called the 'electronic agora', the notion of an open marketplace of ideas where every Jane and Joe can have a voice, has seduced throngs of scholars as well as politicians. Thus: "*I foresee a new Athenian Age of democracy, made possible by the global information highway... It will allow us to exchange ideas within a community and among nations.*" (Former U.S. Vice-President Gore, 1994; cited in Robins 1999, p. 45). Similar sentiments have sometimes been expressed by academics. Dertouzos (1997), a senior scientist at MIT's Media Lab, expressed the hope that "*A common bond reached through electronic proximity may help stave off future flare-ups of ethnic hatred and national breakups.*" Ironically, these very same hopes had been voiced some one hundred and fifty years ago on the occasion of the advent of the telegraph. Standage's (1998) fascinating little book on the social parallels between the age of the Telegraph and the age of the Internet provides quotations from the mid-1850s onward that bear an uncanny resemblance to the ones cited above and others along similar lines. Thus: "*The highway girdling the earth is found in the telegraph wires.*" (Tribute to Samuel Morse, the Father of the Telegraph, 1871; cited in Standage, 1998, p. 181). And: "*It is impossible that old prejudices and hostilities should longer exist, while such an instrument has been created for the exchange of thought between all nations of the earth.*" (Briggs and Maverick, 1863). It is unlikely that history will be kinder to the information-age prophets of equality, democracy and international harmony than it has been to their Victorian-era predecessors.

Myth 5 ('Spatial homogenization'), which predicts the disappearance of high-density urban areas and the progression towards 'complete areal uniformity', is the geographical version of the equalization fallacy. It fails to take into account the complex relationships between urban and regional organization, socioeconomic structure, and ICT access. Digital divides continue to exist not only on the basis of income and education, and between rich countries and poor, but also between geographic centers and peripheries at all scales. Today's telecoms infrastructure closely reflects the (current or forecasted) spatial distribution of paying customers, who happen to be concentrated in CBDs and other densely populated urban areas.

Areal uniformity in the structure of human settlements would imply not only corresponding areal uniformity in ICT availability, but also uniformity in the spatial distribution of activities as well as in the ability of populations to use the technologies for their diverse work- and non-work purposes. From a technical perspective, worldwide coverage through wireless networks is a distant but not impossible dream (Aichelle et al., 2006), though closing the diverse digital divides caused by socioeconomic and cultural differences will be a lot harder. As for the spatial distribution of activities, it is hard to imagine urban areas converging towards areal uniformity in – say – the location of retail stores and warehouses, or of transit stops and airports. ICT or not, the physical, functional and socioeconomic reasons why urban structure cannot be uniform are quite explicitly spelled out in geographic theory (Isard, 1975).

Isolation

Utopias, including ICT-based ones, nearly always pertain to isolated worlds: lost valleys, realms deep inside the earth, undiscovered islands in the middle of the ocean (Manguel and Guadalupi, 2000). In geography, von Thünen (1966) seminal treatise on the hypothetical three-product region lying on the isotropic plain – endowed with a perfect market and perfectly informed, rational farmers – was entitled *The Isolated State*. Implicitly, many of the speculations on life in the ICT age assume isolated individuals, groups, or cities – forgetting the tight inter-linkages at all scales that characterize our global society. Of course, ". . . it might be possible to locate on a mountain top and to maintain intimate, real-time and realistic contact" (Myth 2) if the rest of the world's economy and society continues as usual. And sure, even a (very small) wheel-less city might be possible if the rest of the world is allowed to keep its wheels. Unfortunately for such speculations, the 21st century is a bad time for Utopias. Never before has the world been so interconnected and interdependent.

Human nature

The early humans who created these stunning paintings in the depths of the caves of Altamira and Lascaux left us a second legacy along with their art –a sense that human nature has not changed much in forty thousand years. Then as now humans longed for community, physical closeness, and the sharing of emotional and spiritual experiences. It is unlikely that ICT will change that within a generation or two. No matter how many hours a day our children may be spending on MySpace.com, their best times still involve the physical presence of others. Indeed, early fears about the isolating effects of cyberspace proved unfounded: there is little evidence that Internet use is growing at the expense of traditional kinds of social relations. On the contrary, cell phones, e-mail and lately also Internet telephony (VoIP), by greatly facilitating spontaneous and frequent contacts among family and friends, are positively influencing these bonds (Wellman et al., 2001). As for the web, its role in bringing strangers physically together for all kinds of personal and social purposes – from marriage, travel, and community service to massive demonstrations and terrorist attacks – has been amply documented. In business, issues of trust and efficiency as well as personal gratification place a very high value on face-to-face interactions, and are largely the reason why high-density CBDs are thriving. Even where possible and encouraged, telecommuting is expanding at a much slower rate than many had hoped, largely because the workplace is also a locus of social relations that workers are very reluctant to give up (Mokhtarian, 1994). It is indeed hard to think of an urban activity, be it shopping, learning, or entertainment, which does not involve valued face-to-face interactions that the convenience of on-line access cannot replace.

Materiality

This is the ultimate hurdle, and also the ultimate source of ICT-fueled fantasies. To soar free in cyberspace, to be instantly anywhere and everywhere, to play and work and commune with others at will without the laws of nature getting on the way – what a powerful and seductive dream this is. Already today there are dozens of multi-player, Internet-based games that, for some people, blur the lines between the person and his or her cyberspace avatar. But haven't we all, as children, slain dragons, explored lost worlds, been fearless warriors and fair princesses? Then as now, these worlds were real until the dinner bell rang. There is no way around the fact that we have material bodies with an ever-expanding range of material needs – from basic food, clothing and shelter to the drugs that keep these bodies functioning longer than ever before, and to the astounding range of industrial products (including, of course, our very material cell phones and computers) that we have come to depend on. There is little prospect that our disembodied cyberspace doubles will ever be able to liberate us from these irreducibly embodied needs.

5.5 Conclusion

Ultimately, virtually all the myths of the 'city of instant access' falter because of two fundamental errors: they underestimate both the resilience of human nature, and the unyielding constraints of materiality. Most other fallacies may be reduced to these two.

This essay, like most others along similar lines, reflects the views of a middle-class, computer-literate member of a technological advanced society. But for the majority of people in the world today, discussions about instant access are totally irrelevant, if not outright incomprehensible. Access to clean water may be hours away, and access to loved ones may be years away for those who cannot afford high-speed modern transportation. There is reason to hope that some day, ubiquitous ICT may help redress many of the current inequities. Unfortunately, the whole of human history, including some three thousand years of utopian thinking, gives us strong reason to believe that new inequities will keep taking the place of older ones, as they always have. Human nature is not just about the need for face-to-face contacts and the enjoyment of the world around us; it also has to do with violence and fraud and callousness and ruthless competition, and with an obsession with categorizing humanity into 'us' and 'them'. It is naïve to think that ICT can change that, when the dark aspects of human nature can appropriate these technologies as effectively as the forces for good. A cell phone can be used to save a life, or to explode a bomb. Focusing too much on the 'promise' of ICT may cause us to lose sight of what is at stake out there.

The fallacy of underestimating materiality is even more glaring. There is an inescapable duality between the world of bits and the world of atoms. E-shopping and truck deliveries? E-mails and meetings? Electronic files downloaded and printed out? Text messaging and teenagers flocking? (Rheingold, 2002). Someday someone may come up with the First Law of ICT, stating the mathematical relationship between bits exchanged and atoms moved, transformed or recombined elsewhere. Much further down the road, someone may invent a machine that can instantly transmutate atoms into bits, and vice-versa. Until then – as the saying goes – you can order pizza over the Internet, but it takes a physical trip by some-body to have it delivered (Couclelis, 2004).

Note

[1] All quotes in this section are borrowed from Graham (2004), p. 5

References

Aichelle, C., Fickenger, R., Fonda, C., Forster, J., Howard, I., Krag, T. and Zennaro, M. 2006. Wireless networking in the developing world: A practical guide to planning and building low-cost telecommunications infrastructure. Electronic book, available from *www.lulu.com*

Briggs, C., and Maverick, A. 1863. *The story of the telegraph.* Rudd & Carleton: New York.

Couclelis, H. 1996. The death of distance. *Environment and Planning B: Planning and Design* 23: 387–89.

Couclelis, H. 2004. Pizza over the Internet: e-commerce, the fragmentation of activity, and the tyranny of the region. *Entrepreneurship & Regional Development* 16 (1): 41–54.

Gillette, K.C. 1894. *The Human Drift.* New Era Publishing Co.: Boston. (Reprint: Delmar, N.Y.: Scholars' Facsimiles & Reprints, Inc., 1976, pp. 88–112).

Graham, S. (Ed.) 2004. *The Cybercities reader.* Routledge: New York.

Isard, W. 1975. *Introduction to regional science.* Prentice Hall: New York.

Kaba, S. 1996. 'Building the future: an architectural manifesto for the new millennium'. *Web Architecture Magazine*, April.

Kotkin, J. and Siegel, F. 2000. *Digital geography: the remaking of city and countryside in the new economy.* Indianapolis: Hudson Institute

Krakauer, J. 1997. *Into Thin Air.* Random House: New York.

Manguel, A. and Guadalupi, G. 2000. *The Dictionary of Imaginary Places.* Harcourt Brace: New York.

Mitchell, W.J. 1995. *City of bits: space, place, and the Infobahn.* MIT Press: Cambridge, MA.

Mokhtarian, P.L. 1994. Modeling the choice of telecommuting: setting the context. *Environment and Planning A* 26: 749–66.

Mokhtarian, PL, and R. Meenakshisundaram. 1999. Beyond tele-substitution: disaggregate longitudinal structural equation modeling of communications impacts. *Transportation Research Part C – Emerging technologies* 7(1): 330–352.

Negroponte, N. 1995. *Being Digital*. Knopf: New York.

McLuhan, M. 1964. *Understanding media: the extension of man*. Sphere: London

Pascal, A. 1987. The vanishing city. *Urban Studies*, 24, pp. 597–603.

Rheingold, H. 2002. *Smart mobs*. Perseus: Cambridge, MA.

Robins, K. 1995. Cyberspace and the world we live in. In M. Featherstone and R. Burrows (Eds), *Cyberspace/Cyberbodies/Cyberpunk*, Sage: London, pp. 135–56.

Soleri, P. n.d. *Arcosanti, an urban laboratory*: the Hyperbuilding. *http://www.arcosanti.org/media/publication/hyperBuilding.html*

Standage, T. 1998. *The Victorian Internet*. Berkley Books: New York.

Tucker, A. W. 1929. The Cubic City, *Science Wonder Stories*, (nv) September.

Virilio, P. 1993. 'The third interval: a critical transition'. In V. Andermatt-Conley (ed.) *Rethinking technologies*. University of Minnesota Press: Minneapolis, pp. 3–10.

Thünen, J.H. von, 1966. The solated state; an English edition of Der isolierte Staat (1910). Translated by Carla M. Wartenberg. Edited with an introduction by Peter Hall, New York, Pergamon Press: New York.

Webber, M. 1964. The urban place and the non-place urban realm. In M. Webber, J. Dyckman, D. Foley, A. Guttenberg, W. Wheaton and C,. Wurster (Eds.), *Explorations into urban structure*. University of Pennsylvania Press: Philadelphia, pp. 79–153.

Wellman, B. 2001. Computer networks as social networks. *Science* 293: 14 September, pp. 2031–2034.

Wellman, B., Quan Haase, A., Witte, J. and Hampton, K. 2001. Does the Internet increase, decrease, or supplement social capital? Social networks, participation, and community commitment. *American Behavioral Scientist*, 45:3, pp. 436–455.

Activities in Space and Time

6 Does instant access promote sedentary behavior? Putting physical activity on the instant-access-in-cities agenda

Susan Hanson, Danielle Fontaine
School of Geography, Clark University, Worcester, MA

6.1 Introduction

People need physical activity (PA), preferably daily physical activity[1] to maintain and improve health (Astrand, 1986; Powell and Blair, 1994; Sparling et al., 2000). Until fairly recently, the demands of everyday life ensured that most people were physically active. Carrying out daily activities entailed burning calories (from preparing food and washing clothes to opening the garage door); similarly, moving between different activity sites required people to expend energy (walking to the bus stop or to the grocery store, or climbing stairs). In short, access involved PA. With the advent of "labor saving devices" and information technology (IT), PA is no longer an automatic part of an individual's daily routine; instead, most people must make a conscious decision to be physically active.

In this paper we focus on physical activity in an age of instant access. In doing so, we wish to initiate a conversation between two large but heretofore non-conversing literatures. One has to do with IT and instant access and their impacts on urban form and societal change; the other has to do with PA and the built environment (BE) and asks how changes in the BE might be affecting PA. Specifically, we seek to explore how IT—a defining variable of instant access—influences people's participation in daily PA and consequently their overall health and well-being. The research base on which we can draw to address this question directly is extremely thin; our goal is therefore to pull together the few bits of research that do exist in order to develop a conceptual framework to address the relationship between IT and PA and to outline a research agenda for this increasingly important area. Prevailing folklore suggests that the growing use of IT, including the substitution of IT for physical access, is somehow causally related to declining levels of PA and to the

H.J. Miller (ed.), Societies and Cities in the Age of Instant Access, 87–101.

increasing proportions of Americans who are overweight or obese (Utter et al., 2003; Okie, 2005). Our analysis suggests that the relationships between IT and PA are far more complex.

In section 2 we review patterns of time use as they shed light on trends in levels of PA, including PA and access to activity sites. We next examine the rather thin literature on IT/instant access and time use and consider the implications for PA. In section 4, we discuss the rapidly growing body of research on the relationship between PA and the built environment, and in section 5, we consider how IT/instant access, together with the BE, might affect PA. We conclude by sketching out a research agenda for understanding physical activity in the age of instant access.

6.2 The changing use of time

Labor-saving devices and information technologies have significantly altered the ways people allocate their time; in particular, these technologies are often blamed for changes in time allocation from physically active pursuits to sedentary activities (Okie, 2005), a shift in time use that has negative consequences for human health.

Several decades of time use studies enable the assessment of trends in the amount of time people spend in various activities. Such an assessment is subject to a number of caveats, however. Particularly salient in the context of our focus on PA in an age of instant access are the following: i) Results are sensitive to the definitions of activities and to the specificity of activity codes, which can vary from study to study; ii) For the most part, time use studies have not captured data on IT use or screen time as separate, identifiable activities, and longitudinal data on IT use/screen time are especially scarce; iii) Studies that collect data on PA have not linked PA to location via geocodes, making it difficult to study the relationship between the built environment and PA;[2] iv) Data collection procedures in time use studies have not accommodated multi-tasking associated with IT use, which can pose a problem as IT use is often combined with another activity.[3] With these caveats in mind, we describe a few time use trends that have implications for PA in an age of instant access.

As part of its study summarized in the report, *Does the Built Environment Influence Physical Activity* (2005), the National Research Council (NRC) reviewed trends in time use for the past 50 years. Figure 6.1 shows changes in time use among women in the U.S. (1965–1995), and Figure 6.2 shows changes for men over the same period. Among women, the most significant change has been the decrease in the amount of time devoted to housework, perhaps explained in part by the modest rise in the time that men are spending in household work. For both women and men, the time allocated for eating has been declining while the time spent watching TV and in recreation (leisure) has been increasing. A different data source - Neilson Media Research[4] - suggests that average daily TV viewing

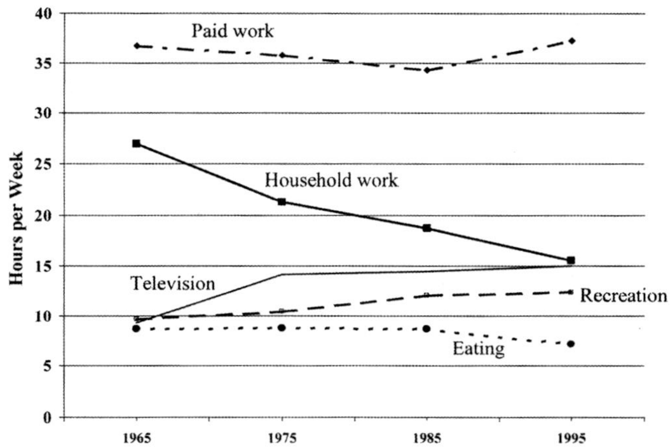

Fig. 6.1 Changes in time use among women in the U.S., 1965–1995 (Brownson and Boehmer, 2004)

has increased even more substantially than the amount indicated by the time use studies, growing from 4.8 hours per day in 1950 to 7.8 hours per day in 2000.

Although time spent in work has declined modestly for men and has changed relatively little for women, the proportion of the labor force in occupations requiring very little PA has increased dramatically, from 12% in 1950 to 28% in 2000 while the proportion of the labor force in high-activity occupations (e.g., construction worker, restaurant server) has declined somewhat since 1990.[5] While the decline in women's time spent in housework indicates a decrease for them in PA, the

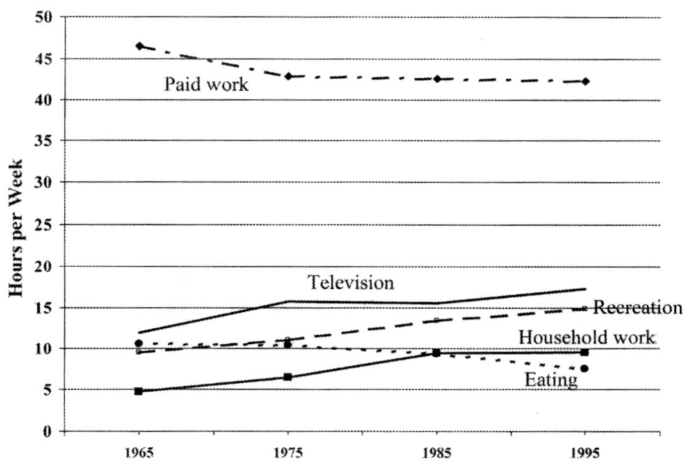

Fig. 6.2 Changes in time use among men in the U.S., 1965–1995 (Brownson and Boehmer, 2004)

increase in leisure time for both women and men is time that, in theory, could be used for physical activity (NRC, 2005).

In fact, however, in the U.S. men, women, and children are spending less time being physically active (DHHS, 1996); more and more individuals are spending increasing amounts of time in sedentary behaviors, the most common of which are traveling to and from work (Brownson and Boehmer, 2004), sleeping (Cutler et al., 2003), and watching television (*ibid*). In their analysis of the National Personal Transportation Survey (NPTS) data, Pucher and Renne (2003) document the increase in the use of the automobile for work trips—from 67% of all work trips in 1960 to 88% in 2000—and they point to the decline in walk trips as a proportion of all daily travel, from more than 9% in 1977 to less than 6% in 1995. These transportation trends indicate that as part of achieving access Americans are reducing their levels of PA and increasing their participation in sedentary activities.

6.3 IT/instant access, and time use: Implications for physical activity

Very little is known about IT[6] and time use and what little exists on the subject pertains mostly to children. One exception is a review by DiMaggio et al. (2001) of the social implications of the Internet. They ask how the Internet has affected patterns of social interaction and note that with the advent of television, people substituted watching TV for listening to the radio, watching movies, and reading; that is, TV time displaced time devoted to those other activities. Has the Internet displaced offline social interaction? The answer appears to be that it has not: "Internet use [is] unrelated or positively associated with social interaction" (*ibid*, 315). These authors go on to say that in a time use study by Robinson et al. (2000) Internet users were found to be "no less active media users or offline socializers than nonusers, though they did do less housework, devote less time to family care, and sleep less" (*ibid*, 316).

A recent study, *Generation M: Media in the Lives of 8–18 Year Olds* (Roberts et al., 2005), documents the central role of IT in the lives of U.S. children. The study was a follow-up to a 1999 study, *Kids and Media at the New Millennium*, and the most revealing finding was just how much—and how quickly— children's media environment has changed: in the 5 years from 1999 to 2004 the proportion of 8–18 year olds with computers in their homes increased from 73% to 86%, and the proportion with Internet connections grew from 47% to 74%. Also, instant messaging, a computer activity that barely existed in 1999 has now become one of the most popular online activities. Table 6.1 shows that whereas in 1999 most children lived in homes with at least one of each of the various types of media, by 2004 much larger percentages of children were living in homes where three or more television sets was the norm (Roberts et al., 2005). Moreover, the proportion of young people's homes with video game consoles, computers, and cable or

Table 6.1 Percentage of children living in homes with 1 or more and 3 or more of each medium (Roberts et al., 2005, 10)

	1+		3+		Mean	
	2004	1999	2004	1999	2004	1999
TV	99%	99%	73%	70%	3.5‡	3.1
VCR	97	98	53‡	26	2.9‡	2.0
DVR	34		6		0.6	
Radio	97	98	63‡	73	3.3	3.4
CD/tape[1]	98	95	66		3.6	
Video game	83	81	31‡	24	2.1‡	1.7
Computer	86‡	73	15‡	8.0	1.5‡	1.1
Cable/satellite TV	82‡	74				
Premium channel	55‡	45				
Internet	74‡	47				
Instant messaging program	60					

[1] Differences in question format preclude comparisons of means and proportion with 3+ CD/tape players.

‡ For this and all following tables, a double dagger (‡) indicates that the difference in proportions for 1999 and 2004 is statistically reliable at p. < .05. For example, a significantly higher proportion of youths reported a household computer in 2004 than in 1999, and a significantly lower proportion reported 3+ radios.

satellite TV subscriptions has surpassed 80%. In short, as Roberts et al. (2005, 10) conclude, "a typical U.S. child between 8 and 18 years old is likely to live in a home equipped with three televisions, three video cassette recorders, three (VCRs), three radios, three CD/tape players, two video game consoles, and a personal computer."

How does this exposure to IT-based media affect children's use of time? Overall, Roberts et al. (2005) conclude that children spend nearly 6.2 hours per day using media, and during this time they are exposed to more than 8.3 hours of media messages (because for at least 1.5 hours per day children are using two or more media simultaneously). These numbers can be compared with the 2.2 hours children say they spend with their parents, the hour they spend on hobbies or clubs, or the 2.2 hours they say they spend with friends (the latter is for 7th–12th graders), and the 50 minutes they spend on homework (*ibid*). Translating these numbers into proportions reveals that the typical U.S. 8–18 year old spends 45% of his or her leisure media time with screen media (TV, videos, DVDs, and movies), 22% with audio media (radios, tapes, CDs and MP3s), 11% with print media (newspapers, magazines, and books), 11% with computers, and 9% with video games (*ibid*).

How this time spent with various forms of IT affects children's participation in PA, if at all, remains unknown. Ho and Lee (2001) offer some insight into the relationship between IT (in this case computer use) and PA among adolescents in Hong Kong. They wanted to determine if computer use is associated with less PA and less social support among adolescents. Of the 2,110 secondary students (mean age of 14.6 years) surveyed, more than 80% used the computer for an average of 2.5 hours per day (or 18 hours per week) for doing homework, surfing the Internet, playing computer games, and communicating with one another. Girls used the computer more for doing homework and communicating with others, whereas boys spent more time playing computer games. Ho and Lee concluded that computer usage does not compete for time spent on other social and recreational activities, nor does it affect the self-perceived social support of users. Although this study suggests that students are still engaged in PA in addition to the time they spend using the computer, the results are not conclusive because the authors do not specify how much PA is included in their category "other social and recreational activities."

Other studies suggest that children's IT use need not detract from their PA. In their study of adolescents in three Montreal schools, Feldman et al. (2003) found that for those adolescents who are physically active, PA is not inversely associated with watching television or playing video games. Similarly, Robinson et al. (2000) (cited in DiMaggio et al. (2001, 315)) reported that Internet users "watched *and played* more sports than comparable nonusers," suggesting that Internet use (amount of time not specified) does not necessarily imply less involvement in PA. Other scholars, however, have reported that time spent in sedentary behaviors is inversely correlated with PA, particularly among girls (Robinson et al., 1993; Bungum and Vincent, 1997). The link between IT use and PA remains poorly understood for both children and adults.

It is clear, however, that many Americans engage in very little PA. The U.S. Surgeon General recommends at least 30 minutes of moderate to vigorous exercise on most, but preferably all, days of the week (DHHS, 1996). Nearly one-third of high-school-aged teenagers report not meeting recommended levels of daily PA, and 10% classify themselves as inactive (NRC, 2005). The figures for U.S. adults are even worse from a public health perspective, with more than 50% saying they fail to meet the guideline for daily PA and approximately 25% saying they are completely inactive during their leisure time (CDC 2003). The negative consequences of sedentary living are well documented and include an increased risk of many chronic diseases such as cardiovascular disease, hypertension, diabetes and certain forms of cancer (Sparling et al., 2000).

Evidence of the connection between PA and health, together with concern over the generally low levels of PA in the U.S. population, have generated interest in the extent to which PA can be increased via participation in the activities of everyday life, including those associated with access, and via changes in geographic context, especially the nature of the built environment (BE). Health professionals, urban planners, transportation planners, and exercise scientists (inter alia) are now seeking an improved understanding of the relationship between PA and the BE (see for example: Sallis et al., 1998; Frank, 2001; Handy et al., 2002; Ewing et al., 2003; Brownson and Boehmer, 2004; Handy, 2004; NRC, 2005). The kinds of questions being asked, include: Does an abundance of sidewalks in the neighborhood, or dense and diverse land uses, for example, promote walking? Does the provision of bike lanes or bike trails mean that more people will make more of their trips by bicycle?

6.4 The built environment and physical activity

Previous research has defined the built environment (BE) in various ways. The recent report by the National Research Council examining the role of the BE as an important potential contributor to reduced levels of physical activity in the U.S. population defines the BE to include land use patterns, the transportation system, and design features that together generate the needs and provide opportunities for travel and physical activity (NRC, 2005).

The complexities involved in examining the BE-PA link are illustrated in the conceptual models built by the NRC committee charged with assessing the current state of knowledge of the BE-PA relationship (NRC, 2005). Figure 6.3 shows that the committee placed the individual at the center of the BE-PA relationship and saw the individual as embedded within the BE and the larger social environment. Figure 6.4 provides greater conceptual detail on the nature of BE-PA relationships. Note in particular that the BE is conceptualized at multiple scales and that four categories of physical activity are recognized: PA as part of leisure activities, transportation, home-based activities, or paid work. The framework recognizes that the BE-PA relationship depends on the individual's characteristics and preferences, which can, in turn, be affected by the nature of the built environment and by the person's participation in PA. Several characteristics of the BE at each scale are noted as potentially influential for the individual's propensity to be physically active.

Figure 6.4 illustrates the complexity of the causal chain from the individual to the BE to PA and includes multiple feedback loops including PA, which itself can reinforce the propensity of individuals to be physically active (NRC, 2005).

Finally, the NRC committee points out that the PA-BE relationship described in Figure 6.4 is mediated through a number of variables, such as the actual or perceived safety of an element of the BE as a site for PA; crime, traffic, and air pollution are all variables that can affect perceived safety in this regard.

Studies have documented some relationships between PA and the BE, but no causal link has been established. There is general agreement that built environments have the capacity to obstruct or to facilitate PA (Wendel-Vos et al., 2004; NRC, 2005), but there is currently no evidence that aspects of the BE are causally linked to levels of PA, i.e., that changes to the BE will result in increased PA (NRC, 2005). As examples, Cervero and Duncan (2003) report a positive relationship between walking and land use density, Greenwald and Boarnet (2001) find that walking is positively related to population density and to the density of retail facilities, and the Environmental Protection Agency (EPA) (2003) reports that walking to school is higher in areas with sidewalks. Still unresolved, however, is whether people who want to walk choose to live in places where they can readily do so. This endogeneity problem is currently the focus of on-going research.

Although research focusing on PA and the BE recognizes that increased PA can occur in the context of an individual's daily activity pattern (e.g., by substituting bike trips for car trips), missing from this line of research is any consideration of the role of IT/instant access. That is, how does IT/instant access figure into the relationship between physical activity and the BE? Those exploring the relationship between PA and the BE see that PA has been engineered out of daily life (via, inter alia, microwave ovens, garage door openers, elevators, riding lawn mowers, and the automobile), and they ask if, via manipulating the BE and geographic context, we can engineer PA back in. To answer this question, scholars must include the use of IT/instant access in their investigations of the relationship between the BE and PA. In the next section we consider how the NRC framework in Figure 6.4 might be expanded to include IT/instant access.

6.5 IT/instant access, the built environment, and physical activity

IT/instant access can interact with the concepts in each of the boxes in Figures 6.3 and 6.4; in this section we situate IT/instant access within the framework in Figure 6.4 and describe some examples of how IT/instant access might affect PA. Keeping in mind the recommended minimum of 30 minutes of moderate or

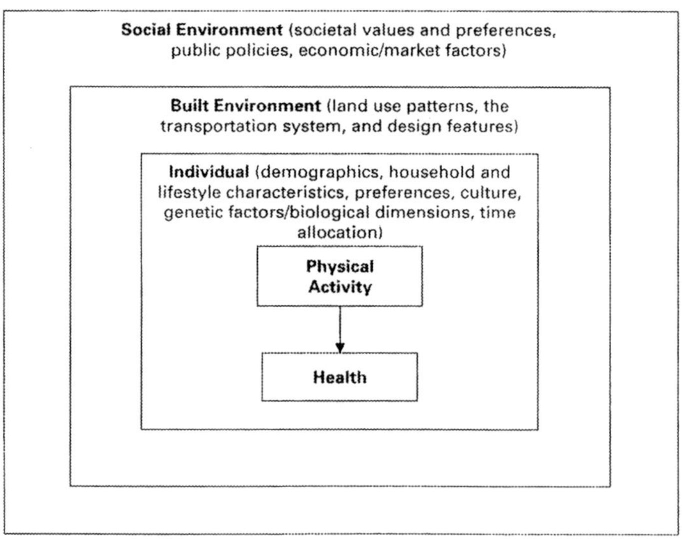

Fig. 6.3 Individuals are embedded in the built and social environment (NRC, 2005)

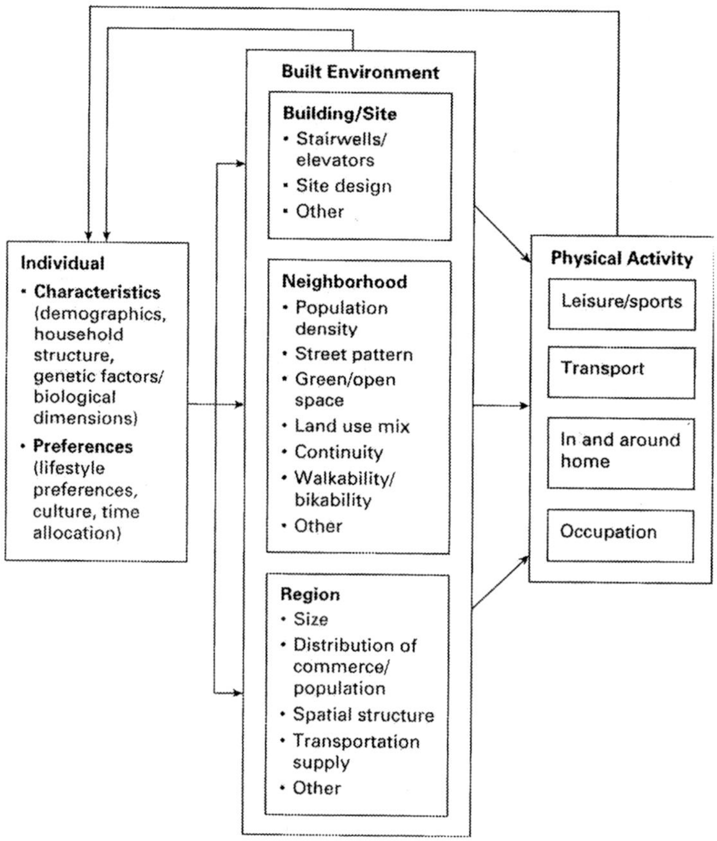

Fig. 6.4 Conceptualizing the built environment at multiple scales (NRC, 2005)

vigorous PA on most days of the week, we emphasize at the outset that it is the individual's *total* daily PA that is of concern. We also want to stress that we see the potential for people's use of IT/instant access to decrease *or increase* their PA. In discussing how IT, instant access, and the BE might affect PA, we adopt a time-space activity framework for thinking through some possible relationships. That is, we want to understand how IT/instant access and the BE affect how and where people allocate time to PA. Because existing work tends to emphasize IT as something that promotes sedentary behavior, we emphasize here the potential of IT to increase PA. Our thinking throughout this section situates IT/instant access as a new element, in conjunction with the BE, in understanding what facilitates or constrains individuals' participation in PA.

Individual characteristics and preferences (far left column of Figure 6.4) are known to affect PA (NRC, 2005) and access to IT (Hoffman et al., 1995; Hoffman and Novak, 1998). For example, PA is positively related to education, income, and to being white and male (NRC, 2005); access to and use of IT is similarly positively related to being white (Journal of Blacks in Higher Education, 1998; Bucy, 2000), having more education (Selwyn et al., 2001), and higher income (Bucy, 2000). It is plausible that IT can also affect individual preferences for allocating time to PA: i-pods and music during exercise, exercise videos, Internet discussion groups, and instant messaging could all encourage PA. Growing reliance on wireless networking (WiFi) and its propensity to encourage multi-tasking while at work could also encourage people to be more physically active. The addictive appeal of video games or email might also, however, precipitate a preference for sedentary behavior. Ho and Lee (2001) found that adolescents who showed a high preference for computer use (i.e., spent more time with computers) also spent more time socializing; to what extent this relationship also means that time spent using computers is positively related to time spent in social forms of PA (e.g., team sports) is unknown.

An example of IT's positive impact on preferences for PA is the video game Dance Dance Revolution, in which the player stands on a mat and moves to the beat of any one of more than 100 tunes as arrows on the TV screen tell the player where to place his or her feet on the mat. Testifying to the fad power of new forms of IT, the game has proven popular with American teenagers and is being promoted by health groups to improve cardiovascular fitness. Children who are not interested in other forms of PA have become devoted participants in Dance Dance Revolution (Barker, 2005; Kesten, 2005).

IT/instant access can likewise interact with the BE (middle column in Figure 6.4) to facilitate or constrain time spent in PA. It is well recognized that the availability of IT/instant access and the quality of that access varies substantially from place to place, even within urban areas; the nature of this variation, currently poorly understood, is important to know in conceptualizing how the BE together with IT/instant access might affect PA. For example, there is still some question as to how settlement density relates to the propensity to telecommute. Telecommuters do live farther on average from their main workplaces than do comparable non-telecommuters, suggesting but not demonstrating that they live in lower-density settings (Mokhtarian, personal communication). Ellen and Hempstead (2002), however, found that, compared to non-telecommuters, telecommuters were more likely to live in large urban areas and less likely to live in rural areas.

Research on IT-transportation tradeoffs[7] has explicitly focused on the substitution of IT/instant access for *motorized* transport (primarily autos) because of the overriding concern of policy makers to promote IT, as in telecommuting, as a way to reduce Vehicle Miles Traveled (VMT), air pollution, and energy consumption. A focus on PA suggests the need to examine the IT-transportation relationship as it pertains to the non-motorized modes, walking and biking. In view of how people's concerns about their own safety or the safety of their children can constrain walking or biking, IT/instant access has the potential to ameliorate those concerns so that people perceive walking or biking to be safer than it is without such IT. Cell phones or GPS units can increase the sense of safety (and perhaps actual safety), especially for parents of children walking or biking to school by allowing parents to track their children's locations. These technologies could potentially boost the proportion of children who walk (currently less than 15% of students aged 5–15) or bike (currently less than 1% of students aged 5–15) (Ewing et al., 2005) to school and might also have considerable impact on women's participation in outdoor PA as well.

IT/instant access can also influence PA as it relates to the BE by providing people with information about where in the BE it is possible or advisable to be physically active. The Internet is a source of information on bike routes, walking routes, gyms, golf courses, organized sports groups and the like and may be particularly important for newcomers to a place, including travelers passing through or newly arrived residents. For instance, Mountainproject.com[8] is one example of an online rock climbing database that is created and maintained by regular rock climbers to provide a valuable resource of information on climbing sites to visitors and experienced climbers alike in places such as Salt Lake City, Utah and Boulder, Colorado.

The Internet is also a source of information on the nature and importance of the link between PA and health, on the societal costs of automobility, on the environmental benefits of transit, or other subjects that might affect a person's propensity to be physically active.

IT can interact with the BE to affect PA in other ways, such as through surveillance cameras. Placed in stairwells, such surveillance might provide a heightened sense of safety and encourage people to abandon the elevator for the stairs. Surveillance might also, however, encourage people to avoid places they might otherwise have visited. The bicycle rights movement Critical Mass is an example of how use of the Internet can intersect with group action in the BE to create a social movement aimed at effecting changes in the BE to promote PA (Blickstein and Hanson, 2001). Critical Mass participants communicate via the Internet to set up mass bike rides in dense urban areas during the Friday afternoon rush hour, with the aim, inter alia, of demonstrating the effectiveness of the bicycle as an urban transportation mode.

In the same way that Critical Mass participants gather around a common interest, a contemporary fad is "flocking" or the en-masse movement of people who share a common interest to a designated location. The key to flocking is that it relies on people contacting one another using various IT in real time to organize the physical movement of individuals within a short time frame to a designated location. Flocking is one example of bringing together a group of individuals for a reason other than simply the incidence of proximity. By extension, it is conceivable that IT could lead to an increase in physical activity in the BE as individuals with a genuine interest in physical activity contact one another and arrange to congregate en masse at some designated location in the city (which in itself requires PA to get to the location) for PA such as walking or running.

Another related example that further highlights the utility of IT in promoting PA is geocaching. Labeled variously as "a high-tech treasure hunt" and a "great way to get outdoors" it is a game in which strangers hide objects in public places for others to find. The co-ordinates of the location of objects are posted online so that individuals can then go out using hand-held Global Positing Systems (GPS) to find the hidden "treasure."

Finally, IT/instant access can create efficiencies that can provide time in people's daily schedules for PA. Insofar as PA is now, for most people, a matter of choice rather than a necessary part of daily life, it has become a matter of time allocation. Like the automobile, IT/instant access can enable people to make more efficient use of time, thereby freeing up time for a walk in the park or a workout at the gym. Again, there are questions as to how IT/instant access intersects with the BE to enable or constrain PA.

With the proportion of the population working in low-activity occupations climbing and the near ubiquity of personal motorized transport, PA has become, for most Americans, a choice in the age of instant access. Given the importance of PA to health and the degree to which IT/instant access has saturated everyday life, especially for children, alarmingly little is known about how IT/instant access interacts with the BE to influence PA. In our view improved understanding of these relationships requires investigating the role of IT/instant access in everyday life. The study of time use and time-space activity patterns is one framework for pursuing this understanding.

6.6 Outlining a research agenda

There is no evidence that increased use of IT/instant access diminishes participation in PA, nor is there evidence that it increases PA. Certainly it has the potential to do both. Efforts to understand how IT/instant access interacts with the BE to affect PA must be sensitive to tradeoffs among activities (e.g., a sedentary, but faster, trip to work via auto or telecommuting can carve out time for PA later in the day), must focus on total daily PA, and must capture multi-tasking. There is no reason, for example, to equate screen time with sedentary behavior, as the video game Dance Dance Revolution demonstrates.

Data on individuals' time-space activity patterns could be used in conjunction with data on geographic context, including the IT context, to explore questions such as:

- How does IT/instant access affect overall travel time budgets? What happens to the "law of constant travel time" in an age of instant access? How does PA figure in this equation?
- How does IT/instant access interact with the BE to affect access via walking and bicycling?
- How does IT/instant access interact with the BE to affect the total daily PA of different groups of people defined by age, gender, ethnicity, and location?
- What quantitative and qualitative methodologies are likely to be most effective in building knowledge of interactions among IT/instant access, the BE, and PA?

- How can IT/instant access be used to build more PA into leisure/recreation, transport, home, and work - each of the four arenas of daily life outlined in Figure 6.4?

These are just some of the questions for a research agenda on this topic. So little is currently known about how IT/instant access interacts with the BE to affect PA that this field is wide open. We hope that at the very least people will stop assuming that the increased use of IT/instant access necessarily contributes to reduced levels of PA, that they will appreciate the complexities involved in investigating these relationships, and will see the importance of doing so.

Acknowledgements

This paper draws heavily on the work of the NRC Committee on Physical Activity, Health, Transportation, and Land Use which Susan chaired. She thanks the members and staff of that committee for many stimulating discussions. We also thank Pat Mokhtarian for her assistance with studies of telecommuting and participants in the November 2005 conference (Cities in an Age of Instant Access) for their ideas and suggestions.

Notes

[1] PA is broadly defined as "bodily movement produced by the contraction of skeletal muscle that increases energy expenditure above the basal level" (DHHS, 1996, 20).

[2] A plan is underway to geocode the National Health and Nutrition Examination Survey (NHANES).

[3] The American Time Use Survey, conducted every 5 years by the Bureau of Labor Statistics, has accommodated multi-tasking for certain activities such as child care, food preparation, and parent care (Shelley 2005).

[4] Considered the most reliable source on TV viewing

[5] Specifically, the percentage of the population involved in agricultural activity declined from 12% of the labor force in 1950 to 2% in 2000. Also, between 1950 and 2000 the number of employees in the manufacturing sector fell from 30% to 13%, while those in the service sector (with a higher fraction of less physically demanding white-collar jobs) grew from about two fifths to four fifths of civilian employment (Bureau of Labor Statistics 2004).

[6] IT is the use of electronic machines and programs for the processing, storage, transfer, and presentation of information (Bjork, 1999); it includes the television, computer (including the Internet and email), telephone and fax, computer games, compact discs (CDs), and various computer software programs.

[7] See Janelle 2004 for a review

[8] This online resource was previously hosted as a number of smaller sites, including, for example *http://climbingsaltlake.com*. To make information more easily available a number of these smaller, sister sites have now been consolidated into Mountainproject.com

References

Astrand, P. O. (1986) "Why Exercise? An Evolutionary Approach," *Acta Medica Scandinavica Supplementum*, 711, 241–242.

Barker, A. (2005) "Study uses video games to fight obesity," *USA Today*, 2 February.

Bjork, B-C. (1999) "Information Technology in construction: domain definition and research issues,"*International Journal of Computer Integrated Design and Construction*, 1, 3–16.

Blickstein, S. and Hanson, S. (2001) "Critical mass: forging a politics of sustainable mobility in the information age," *Transportation*, 28, 347–362.

Brownson, R. C. and T. K. Boehmer. (2004) "*Patterns and Trends in Physical Activity, Occupation, Transportation, Land Use, and Sedentary Behaviors*," Report prepared for the Transportation Research Board and the Institute of Medicine Committee on Physical Activity, Health, Transportation, and Land Use.

Bureau of Labor Statistics (BLS). (2004) "Time-Use Survey – First Results Announced by BLS and Technical Note," *News*, U.S. Department of Labor, Washington D.C., September 14.

Bucy, E. P. (2000) "Social Access to the Internet," *The Harvard International Journal of Press/Politics*, 5(1), 50–61.

Bungum, T. and Vincent, M. (1997) "Determinants of physical activity among female adolescents," *American Journal of Preventive Medicine*, 13, 115–122.

Centres for Disease Control and Prevention (CDC). (2003) "Prevalence of Physical Activity, Including Lifestyle Activities amongst Adults – United States, 2000–2001," *MMWR*, 52(32), 764–769.

Cervero, R. and Duncan, M. (2003) "Walking, Bicycling, and Urban Landscapes: Evidence from the San Francisco Bay Area,"*American Journal of Public Health*, 93, 1478–1483.

Cutler, D. M., Glaser, E. L. and Shapiro, J. M. (2003) "Why Have Americans Become More Obese?" *Journal of Economic Perspectives* 17(Summer), 93–118.

Di Maggio, P., Hargittai, E., Russel Neuman, W. and Robinson, J. P. (2001) "Social Implications of the Internet,"*Annual Review of Sociology*, 27, 307–36.

Ellen, I. G. and Hempstead, K. (2002) "Telecommuting and the demand for urban living: a preliminary look at white-collar workers," *Urban Studies*, 39(4), 749–766.

Environmental Protection Agency (EPA). (2003) "*Travel and Environmental Implications of School Siting*," EPA-231-R-03-004. Washington D. C.

Ewing, R., Schmid, T., Killingsworth, R., Zlot, A. and Raudenbush, S. (2003) "Relationship Between Urban Sprawl and Physical Activity, Obesity and Morbidity," *American Journal of Health Promotion*, 18, 47–57.

Ewing, R., Forinas, C. V. and Schroer, W. (2005) "Neighborhood Schools and Sidewalk Connections: What are the Impacts on Travel Mode Choice and Vehicle Emissions?" *TR News*, 237, 4–10.

Feldman, D. E., Barnett, T., Shrier, I., Rossignol, M., and Abenhaim, L. (2003) "Is physical activity differentially associated with different types of sedentary pursuits?" *Archives of Pediatrics and Adolescent Medicine*, 157(8), 797–802.

Frank, L. D. and Engelke, P. O. (2001) "The Built Environment and Human Activity Pattern: Exploring the Impacts of Urban Form on Public Health," *Journal of Planning Literature*, 16, 202–218.

Greenwald, M. J. and Boarnet, M. G. (2001) "Built Environment as Determinant of Walking Behavior: Analyzing Nonwork Pedestrian Travel in Portland," *Transportation Research Record*, 1780, 33–42.

Handy, S., Boarnet, M., Ewing, R. and Killingsworth, R. (2002) "How the Built Environment Affects Physical Activity: Views from Urban Planning," *American Journal of Preventive Medicine*, 23, 64–73.

Handy, S. (2004) "*Critical Assessment of the Literature on the Relationship Among Transportation, Land Use, and Physical Activity*," Paper prepared for the Transportation Research Board and the Institute of Medicine Committee on Physical Activity, Health, Transportation, and Land Use.

Ho, S. M. Y. and Lee, T. M. C. (2001) "Computer Usage and Its Relationship with Adolescent Lifestyle in Hong Kong," *Journal of Adolescent Health*, 29, 258–266.

Hoffman, D. L., Novak, T. P. and Chatterjee, P (1995) "Commercial Scenarios for the Web: Opportunities and Challenges," *Journal of Computer-Mediated Communication* 1(3), [http://jcmc.huji.ac.il/vol1/issue3/hoffman.html] (last accessed 15 September 2005)

Hoffman, D.L. and Novak, T. P. (1998) "Bridging the Racial Divide on the Internet," *Science*, 280, 390–391.

Janelle, D. G. (2004) "Impact of Information Technologies," in S. Hanson and G. Giuliano (eds.) *The Geography of Urban Transportation* (Third Edition), New York: The Guilford Press, 86–114.

Journal of Blacks in Higher Education. (1998) "African-Americans as the Have-Nots of the Information Age," *Journal of Blacks in Higher Education*, 21, 18–21.

Kesten, L. (2005) "Some video games force players off couch," *USA Today*, 7 September.

Mokhtarian, P. Personal Communication

National Research Council (NRC). (2005). "*Does the Built Environment Influence Physical Activity?"Examining the Evidence*. TRB Special Report 282. Washington D.C: The National Academies.

Okie, S. (2005). "Off the Couch and Away from the Screen," In S. Okie. *Fed Up: Winning the War Against Childhood Obesity*, Washington D. C.: Joseph Henry Press.

Powell, K. and Blair, S. (1994) "The Public Health Burden of Sedentary Living Habits: Theoretical But Realistic Estimates," *Medicine and Science in Sports and Exercise*, 26, 851–856.

Pucher, J. and Renne, J. (2003) "Socioeconomics of Urban Travel: Evidence from the 2001 NHTS," *Transportation Quarterly*, 57(3), 49–77.

Roberts, D. F., Foehr, U. G. and Rideout, V. (2005) "*Generation M: Media in the Lives of 8–18 Year-olds*," Washington D. C: Kaiser Family Foundation.

Robinson, T., Hammer, L., Killen, J., Kraemer, H., Wilson, D., Hayward, C., and Taylor, C. (1993) "Does television viewing increase obesity and reduce physical activity? Cross-sectional and longitudinal analyses among adolescent girls,". *Pediatrics*, 91, 273–280.

Robinson, J. P., Kestnbaum, M., Neustadtl, A. and Alvarez, A. (2000) "*IT, the Internet and Time Displacement*," Paper Presentation. Annual Meeting of the American Association of Public Opinion Research. Portland, OR. May 2000.

Sallis, J. F., Baumann, A., and Pratt, M. (1998) "Environmental and Policy Interventions to Promote Physical Activity," *American Journal of Preventive Medicine*, 15, 379–397.

Selwyn, N., Gorard, S. and Williams, S. (2001) "Digital Divide or Digital Opportunity? The Role of Technology in Overcoming Social Exclusion in U.S. Education," *Educational Policy*, 15(2), 258–277.

Sparling, P. B., Owen, N., Lambert, E. V. and Haskell, W. L. (2000) "Promoting Physical Activity: The New Imperative for Public Health," *Health Education Research*, 15, 367–376.

United States Department of Health and Human Services (DHHS). (1996) "*Physical Activity and Health: A Report of the Surgeon General*," Atlanta, GA: U.S. Department of Health and Human Services, Centers for Disease Control and Prevention, National Center for Chronic Disease Prevention and Health Promotion.

Utter, J., Neumark-Sztainer, D., Jeffery, R. and Story, M. (2003) "Couch potatoes or french fries: are sedentary behaviors associated with body mass index, physical activity, and dietary behaviors," *Journal of the American Dietary Association*, 103(10), 1298–1305.

Wendel-Vos, G, C, W., Schuit, A., De Niet, R., Boshuizen, H. C., Saris, W. and Kromhout, D. (2004) "Factors of the Physical Environment Associated with Walking and Bicycling," *Medicine and Science in Sports and Exercise*, 36, 725–730.

7 Revisiting Hägerstrand's time-geographic framework for individual activities in the age of instant access

Hongbo Yu[1] and Shih-Lung Shaw[2]
[1]*Department of Geography, Oklahoma State University, OK*
[2]*Department of Geography, University of Tennessee, TN*

7.1 Introduction

Our society recently experienced rapid developments of information and communication technologies (ICT), such as the Internet, cellular phones, and wireless-enabled personal digital assistants. These technologies enable a different space that can connect us electronically and transmit information more efficiently than the conventional physical space. This space has been named *virtual space* or *cyberspace* in the literature (Janelle and Hodge, 2000). With the help of ICT, people now enjoy new freedom in space and time to carry out activities and interact with others. For example, using cellular phones, we are no longer constrained to the fixed locations of landline phone services. We can now purchase air tickets or search for literature on the Internet regardless of the open hours of travel agencies and libraries. The new freedom can alter how people carry out their activities and how they interact with others, which may eventually lead to changes in the spatial and temporal distribution of travels. Therefore, it is crucial to gain a better understanding of spatio-temporal characteristics of human activities in the Age of Instant Access.

Hägerstrand (1970) proposed a time-geographic framework, which provides an effective approach of studying the relationships between various constraints and human activities in a space-time context. Researchers have frequently used the framework to study spatial and temporal characteristics of human activities in physical space (Parkes and Thrift, 1980; Carlstein, 1982; Ellegård, 1999). However, the original time-geographic framework focused mainly on human activities in physical space. Without accommodation of virtual space in the framework, time geography falls short of providing a complete view of human activities with their space-time constraints in the Age of Instant Access. Efforts are needed to extend the current time-geographic framework to deal with activities in both physical and virtual spaces. In this study, we will revisit the basic concepts of Hägerstrand's time geography and propose an

H.J. Miller (ed.), Societies and Cities in the Age of Instant Access, 103–118.

extended time-geographic framework to examine the spatial and temporal characteristics of human activities in both physical and virtual spaces.

7.2 Research backgrounds

7.2.1 Time geography

Originally proposed by Hägerstrand (1970), time geography was developed to study the relationships between human activities and various constraints in a space-time context (Golledge and Stimson, 1997). Hägerstrand and his colleagues argued that time should not be considered only as an external factor in activity study. Time, as essential as space, should be explicitly included in the investigation process. Considering the time dimension to be equivalent to that of space, the framework adopts a three-dimensional (3D) orthogonal coordinate system, with two spatial dimensions and one temporal dimension. The spatial dimensions are used to measure location changes of objects in a plane, while the temporal dimension is used to order the sequence of events and to synchronize human activities.

Time geography assumes that an individual's activities are limited by various constraints. Three types of constraints that can impact an individual's ability to conduct activities in space and time are defined in time geography: *capability constraints, authority constraints*, and *coupling constraints* (Golledge and Stimson, 1997). Physiological necessities (e.g., sleeping, eating) and available resources (e.g., auto ownership) that can constrain a person from participating in activities are recognized as *capability constraints. Authority constraints* reflect general rules or laws that limit a person's access to either spatial locations (e.g., a military base) or time periods (e.g., open hours of a library). *Coupling constraints* are spatial and temporal requirements that allow an individual to bundle with others to conduct certain activities (e.g., having a meeting at a conference center at 3pm). Among these three types of constraints, capability constraints and authority constraints focus on issues related to separate individuals while coupling constraints deal with interactions among multiple persons. Because people are social beings, interaction is an important element in most of their daily activities. Coupling constraints directly define the requirements in space and time that allow people to interact with one another. Meanwhile, capability constraints and authority constraints indirectly determine whether two individuals can couple; they limit the movements of each individual and control whether the individuals are able to be present at a certain location during a specified time period. These three types of constraints control the spatio-temporal patterns of people's movements (Golledge and Stimson, 1997).

Time geography developed two key concepts, *space-time path* and *space-time prism*, to portray human activities with their spatial and temporal characteristics in an integrated space-time system (Hägerstrand, 1970). A *space-time path* is the trajectory of an individual's movements in physical space over time and it

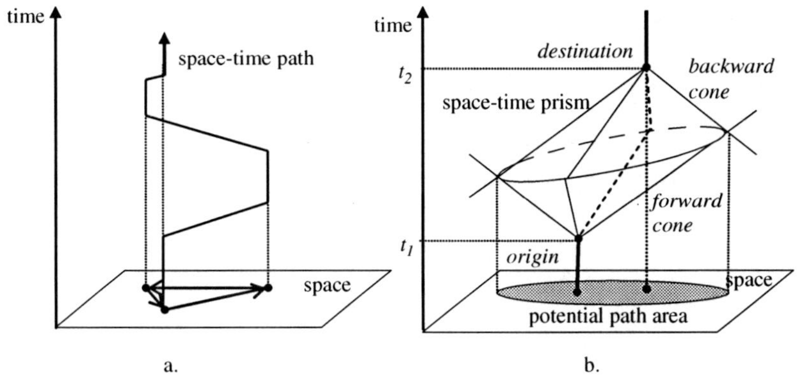

Fig. 7.1 Space-time path and space-time prism

is represented as a linear feature in the 3D space-time system (Figure 7.1a). It can be used to organize an individual's observed activities in their time sequence effectively and record spatial and temporal characteristics of activities integrally. A *space-time prism* depicts the extent in space and time that an individual can access under specific constraints (Lenntorp, 1976). It occupies a continuous space in time-geography's coordinate system (Figure 7.1b) and indicates where and when potential activities are available to an individual in physical space. Transportation serves as a means for people to trade time for space since movements in physical space take time (Miller and Shaw, 2001). Given a location and a time period, a person can stay at the location for the entire time duration. If the person wants to move to a new location, it will take time to make the physical movement, and the time available for activities at the new location is shortened accordingly. If we project a space-time prism onto a 2D plane, the result will be a region, which is known as *potential path area* (the shaded area in Figure 7.1b). The shape of a space-time path or a space-time prism is confined by the three types of constraints defined in the framework.

7.2.2 Virtual space, tele-presence, and virtual activities

With a virtual space enabled by ICT, activities available to an individual expand far beyond the physical proximity of the individual. The electronic linkages among people in virtual space can replace transportation in some situations to help people participation in activities remotely. This new participation mode has been named *tele-presence*, with comparison to the conventional participation mode – *physical presence*.

Janelle (1973) argued that human beings were agents who could sense their environment and interact with others. Without the help of any appliance, the sensing capability of a person is limited to the range of the person's physical proximity, which means that physical presence is required when an individual needs to participate in an activity or interact with others. ICT have greatly expanded the agency and sensation of human beings (Adams, 2000; Kwan, 2000). An individual now can reach out far beyond her/his physical proximity and participate in an activity remotely through tele-presence. This implies that space is not as important as before in constraining activities. In the case of tele-presence, as long as a person has access to virtual space, the person can complete activities while being physically separate from the activity locations and other participants. With tele-presence, the activity opportunities to an individual are greatly expanded. An individual certainly can conduct activities within her/his physical proximity, but s/he can also access and complete activities through tele-presence in virtual space. *Physical activities* and *virtual activities* will be used for activities conducted in the two spaces respectively in this study. With capabilities of conducting both physical and virtual activities, individuals start to observe different spatial and temporal constraints for their activities.

Researchers have studied different types of communication methods based on their spatial and temporal characteristics (Harvey and Macnab, 2000; Janelle 2004; Miller 2006). Four types of modes have been identified to classify communications with physical presence and tele-presence (Table 7.1). *Synchronous physical presence* (SP) requires coincidence in both space and time. *Asynchronous physical presence* (AP) requires coincidence in space, but not in time. With the use of ICT, people can be freed from physical presence to perform certain communications. *Synchronous tele-presence* (ST) only requires coincidence in time, but not in space. Finally, *Asynchronous tele-presence* (AT) is free from coincidence requirements in both space and time. This classification system can also be applied to describe different types of human activities and interactions based on their spatial and temporal requirements. Prior to the wide adoption of ICT, human activities used to be carried out mainly in physical space through either the SP or AP mode. With tele-presence enabled by ICT, the ST and AT modes have become increasingly used for interactions and they are changing the ways people interact with each other and altering their spatio-temporal activity patterns.

Table 7.1 Communication modes based on their spatial and temporal constraints (Adapted from Miller, 2006)

Temporal / Spatial	*Synchronous*	*Asynchronous*
Physical presence	**SP** Face to face (F2F) meetings	**AP** Post-it® notes Traditional hospital charts
Tele-presence	**ST** Instant messaging Teleconferences	**AT** E-mails Webpages

7.3
An extended time-geographical framework for physical and virtual activities

7.3.1
Relationships of physical space and virtual space

In today's society, people can conduct their activities in both physical and virtual spaces. A better understanding of the relationships between these two spaces can help us examine the changing spatial and temporal characteristics of human activities in the Age of Instance Access. First of all, physical space and virtual space have different characteristics. Negroponte (1995) views physical space as a material world made of atoms and virtual space as a world composed of bits of information. While physical space can work as a container for both physical materials and information, virtual space is specialized to carry the information flow efficiently. However, these two spaces are not exclusive and activities in the two spaces can influence each other (Salomon, 1986; Batty and Miller, 2000). Although we have seen that the significance of distance in our daily activities has been weakened since the emergence of virtual activities, speculations about "the death of distance" (Cairncross, 1997) are debatable because virtual space access channels are not available ubiquitously in physical space. Physical location is still important because it controls where individuals can access virtual space. Furthermore, an individual who has access to virtual space can retrieve information, and the information can affect the individual's travel decisions and behaviors, which in turn may change travel distribution in physical space and time.

This study proposes a conceptual model to portray the roles of the two spaces in containing human activities (Figure 7.2). Both physical and virtual spaces can be considered relatively independent because each has its own specific characteristics. Transportation can help people move around in physical space, while ICT are the means for people to navigate in virtual space. The two spaces also have intersections and they influence each other through the intersections. Two aspects of the intersections are addressed in this study. First, physical space provides access channels to virtual space because virtual space is built on information and communication infrastructures that reside in physical space. If a person wants to perform virtual activities, s/he needs to reach access channels in physical space. Consequently, movements in physical space may be required to help the person

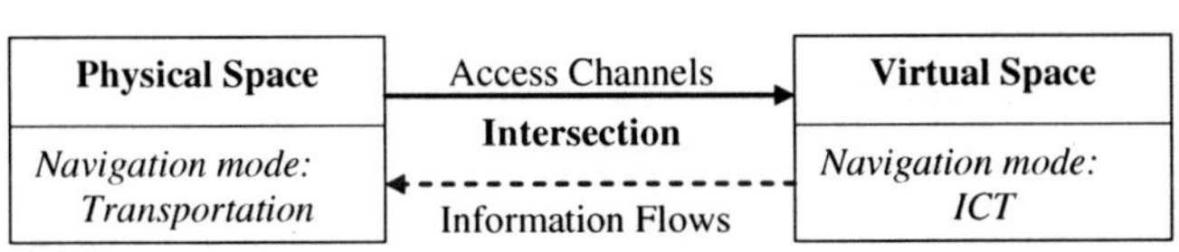

Fig. 7.2 A conceptual model of physical space and virtual space

find the access channels to virtual space. In the meantime, virtual space can feed information back to physical space. People can retrieve information from virtual space and use the information to help them make decisions, which may affect the activity patterns in physical space.

According to this proposed conceptual model, physical space plays two roles in supporting human activities: a *carrier of physical activities* and a *connector for virtual activities*. Hägerstrand's time geography provides an effective framework to address the role of physical space as a carrier of activities. However, the role of physical space as a connector for virtual space has not been clearly addressed yet. The time-geographic framework must be extended to handle the new role of physical space in supporting human activities. The three types of constraints (i.e., capability, authority, and coupling constraints) for human activities in the framework need to be re-visited and the key concepts (space-time path and prism) need to be adjusted to deal with the different roles of physical space.

7.3.2 Revisiting the three types of constraints for human activities

Researchers have noticed that different constraints apply to virtual activities. However, limited effort has been made to identify them in a systematic approach. Harvey and Macnab, (2000) studied new capability constraints and temporal coincidence of coupling constraints for personal communications through the Internet. Their study provides insight on examining constraints for virtual activities. Based on the proposed conceptual model of relationships between physical and virtual spaces, we revisit the three types of constraints for human activities in both physical and virtual spaces.

7.3.2.1 Capability constraints

We identify three subsets of capability constraints. First, physiological necessities, which are basic needs of human beings, can constrain an individual from conducting both physical and virtual activities. Personal capabilities, which are skills and resources owned by an individual, make up the second subset of this constraint type. Different capability constraints apply to physical activities and virtual activities. For physical activities, physical space acts as a carrier for the activities. Available resources (e.g., auto ownership) and skills (e.g., driving capability) can affect how efficiently an individual travels, which consequently determines the available activity opportunities in physical space to the person. Virtual activities need support from both physical and virtual spaces. Physical space works as a connector for virtual activities. Being able to reach virtual space access channels (e.g., fixed Internet ports or wireless phone service areas) in physical space and possessing appropriate devices (e.g., networked computers or cellular phones) become necessary conditions for individuals to conduct virtual activities. Without these capabilities and resources, an individual is excluded from the intersection of the two spaces and, consequently, from conducting virtual activities. In addition, capabilities of navigating in virtual space can also affect an individual's involvement in virtual activities. For example, an individual needs basic computer and Internet knowledge to surf the Internet, and an individual must know a particular language in order to browse a website developed in that language. Finally, features of infrastructures and facilities can also impact people's capability to conduct activities. Speed limit and capacity of roads can affect people's travel efficiency. Similarly, the bandwidth of an Internet connection (e.g., 56KB/s for a dial-up connection) and the coverage range of a cellular phone transmission tower can determine whether specific virtual activities can be performed. These characteristics of facilities can indirectly influence an individual's capability of conducting activities.

7.3.2.2 Authority constraints

Similar authority constraints apply to virtual activities as those to physical activities defined in the classic time geography. These constraints can be found in virtual space and in the intersections that provide access channels to virtual space. Virtual space also has rules that may limit the access of particular locations to selected people at specific time. Although most resources on the Internet are free and available around the clock, some websites do require membership and have limited operation hours. For example, a university's online class registration website requires students to provide correct user names and passwords before they log onto the website and browse their class registration information, and these functions may be available only during certain time periods due to administrative reasons. Virtual activities are also controlled by the intersections of the two spaces. Because physical locations that host connection facilities to virtual space can be affected by the same situations as discussed above, performance of virtual activities therefore is also limited by constraints in the intersections. When an individual does not have permission to visit a location hosting virtual space access channels or does not visit the location during its open hours, the individual cannot connect to virtual space and conduct virtual activities. Therefore, spatial and temporal authority constraints at these locations will restrict virtual activities.

7.3.2.3 Coupling constraints

The emergence of virtual space expands the coupling opportunities among people. With tele-presence, people now can communicate and interact with one another through any of the four interaction modes (i.e., SP, AP, ST, and AT). Therefore, people can be bundled in different spatial and temporal contexts rather than only in the situation of co-location in both space and time to conduct activities.

Both the SP and AP interaction modes require the physical presence of the participants. SP interactions such as face-to-face meetings, which have been the major focus in Hägerstrand's time geography for coupling situations, require participants to be physically present at the same location during the same time period. Consequently, participants share a *co-existence* relationship in space and time. An AP interaction, such as a bulletin board information exchange, requires participants to be physically present at the same location. The participants however do not have to be there at the same time. Therefore, individuals involved in an AP interaction share a relationship of *co-location in space*.

ST and AT interactions take advantage of tele-presence, which allows people join the same virtual activity remotely. ST interactions allow people from different locations to conduct the same activity, such as a videoconference or a session of instant messaging, but they require participants to access virtual space during the same time period. People bundled through this mode will have a *co-location in time* relationship. AT interactions relax constraints in both space and time. Participants of AT interactions need not to be at the same location or be present at the same time. E-mails and webpages are good examples for this interaction type. By conducting AT interactions, participants can have a relationship of *no co-location in either space or time*.

The existence of virtual space provides people more choices and greater freedom in space and time to conduct activities. However, researchers realize that the cliché that virtual space "enable[s] people to interact with anyone, anywhere, at any time and in any place" is a "crude vision" of the emerging phenomena rather than a precise description (Batty and Miller, 2000, p. 138). After revisiting the three types of constraints for human activities, we can see that physical location still plays an important role in confining human activities, even for those activities conducted in virtual space. Table 7.2 sums up the contents of the three types of constraints for physical and virtual activities. With the newly identified constraints for human activities in the Age of Instant Access, the key concepts of time geography need to be extended and adjusted accordingly.

7.3.3 Adjusted concepts of space-time path and prism for physical and virtual activities

7.3.3.1 Space-time paths with physical and virtual activities

ICT have strengthened the connections and interactions among individuals across the distance (Wiberg, 2005). The conventional space-time path concept, which represents only physical proximities around individuals, needs to be extended to reflect the changing characteristics of human activities. The extensible agent concept (Janelle, 1973), which treats human beings as agents whose sensation over space can be extended with appliances (e.g., cellular phones), provides an

Table 7.2 Constraints for human activities in physical and virtual spaces

Constraints	Contents
Capability constraints	Human capabilities and characteristics of infrastructures or facilities that can support the performance of human activities. • Physiological necessities: sleeping, eating, etc. • Individual capabilities: o In physical space: auto ownership, driving skills, etc. o For intersection: accesses to virtual space – wired accesses (e.g., Internet ports, fixed phone lines, etc) and wireless accesses (e.g., cellular phones, wireless Internet ports, etc). o In virtual space: computer skills, language ability to browse foreign websites, etc. • Characteristics of environment: types of roads, speed limit, band width of the Internet connections, etc.
Authority constraints	General rules or laws that limit the performance of activities at certain locations and/or time periods. o In physical space: military bases, open hours of a library, etc. o For intersection: student computer labs in a university, open hours of an Internet café, etc. o In virtual space: membership controlled websites, business hours for web services, etc.
Coupling constraints	Spatial and temporal requirements for people to interact with each other through either physical presence or tele-presence. • Co-existence (co-location in both space and time): synchronous physical presence (SP), e.g., face-to-face meeting, etc. • Co-location in space: asynchronous physical presence (AP), e.g., fridge note, bulletin board, etc. • Co-location in time: synchronous tele-presence (ST), e.g., instant messaging, videoconference, etc. • No co-location in either space or time: asynchronous tele-presence (AT), e.g., email, voice mail, etc.

effective approach of modeling the new characteristics of human activities, and it has been used in several studies (e.g., Adams, 2000; Kwan, 2000). We also adopt this concept to extend the space-time path concept with integrated representation of physical and virtual activities in this study.

A space-time path can be considered as a container for all activities performed by a person because all activities take place at certain locations and time periods, and each of them occupies a portion of the path. Activities can be located on a space-time path according to their time references. Virtual activities are different from physical activities in terms of constraints and action spaces. While physical activities impact only the physical proximity of a space-time path, virtual activities

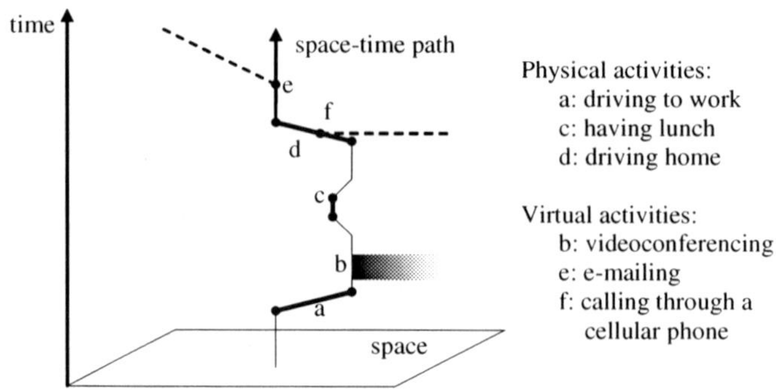

Fig. 7.3 An extended space-time path with physical and virtual activities

only take place at ICT-enabled locations and can extend to distant locations. Figure 7.3 shows the conceptual representation of an extended space-time path. Virtual activities, such as instant message and e-mail, are presented as extended links from space-time paths due to their extended action space. Because some virtual activities may experience delays in time, extended links may be tilted as the one for e-mail activity. Multi-tasking, which becomes more common due to tele-presence, also can be represented. An example of using a cellular phone while driving is given in Figure 7.3.

7.3.3.2
Adjusted space-time prisms for virtual activities

A conventional space-time prism delimits the extent in space and time that an individual can find potential physical activities under specific constraints. How can we define the confines for potential virtual activities to an individual and what is the prism for virtual activities? As previously discussed, virtual activities are constrained in both virtual space and the intersections of the two spaces. We use two steps to identify the potential virtual activities to an individual.

Physical space acts as a connector for virtual activities. A person needs to reach ICT-enabled locations (i.e., the intersections) before s/he can conduct virtual activities. Therefore, the first step of identifying potential virtual activities is to find reachable virtual space access channels. An *adjusted space-time prism for virtual activities* is defined in this study as a prism which delimits the space and time extent where an individual can reach virtual space access channels under specific constraints. Such a prism can be achieved by intersecting a conventional space-time prism with *space-time life paths* of virtual space access channels in physical space. In this study, the term *space-time life path* is used to describe the existence of a virtual space access channel in space and time. A space-time life path is attained by extruding a virtual space access channel along the time dimension according to its operational hours. It represents where and when a virtual space connection

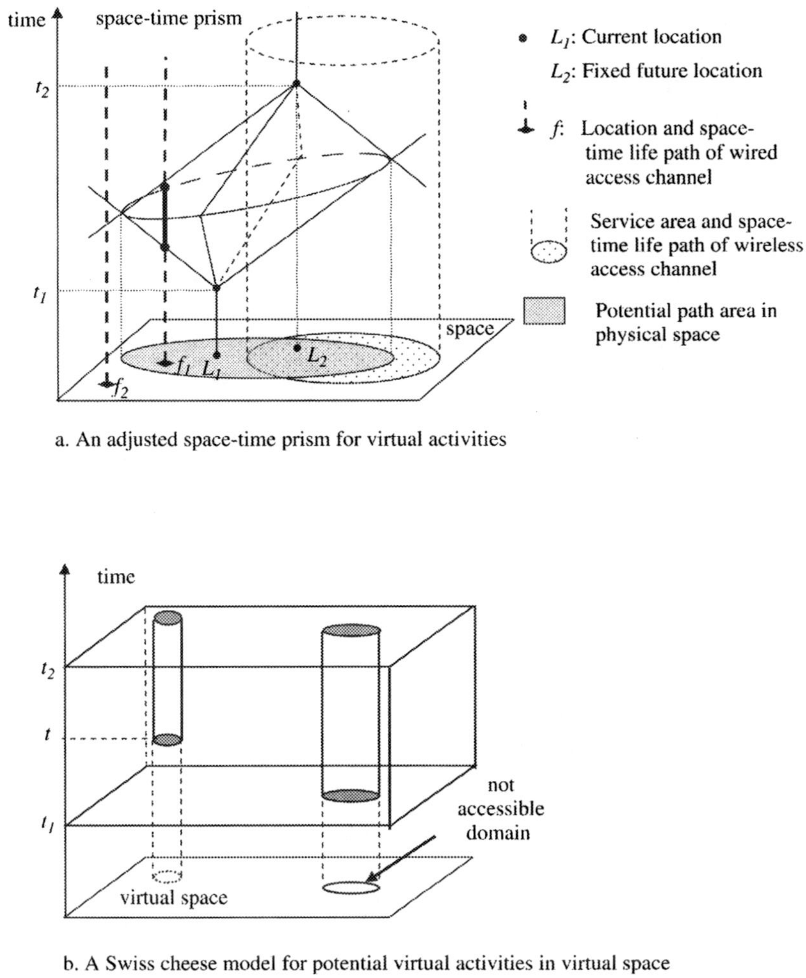

Fig. 7.4 Identifying potential virtual activities

service is available. Space-time life paths can take forms of lines or cylinders, according to the original shapes of virtual space access channels on the ground (e.g., points for wired connections such as fixed Internet ports and regions for wireless connections such as cellular phone coverage areas). Figure 7.4a illustrates the adjusted prism concept. The thick line segment and the portion of the prism falling inside of the cylinder in Figure 7.4a are the result of intersection, which indicates where and when an individual can access to virtual space.

Once an individual gets access to virtual space, constraints in virtual space take over to determine the available virtual activities to the individual. In virtual space, physical distance does not play a determinant factor in defining the activity opportunity space. Therefore, boundaries defining the opportunities in virtual space are no longer prisms in shape. In addition, it is unrealistic to assume ubiquitous opportunities in virtual space because virtual services are not available everywhere and all of the time. A *Swiss cheese model* is proposed in this study to define the opportunities in virtual space (see Figure 7.4b). The Swiss cheese model suggests that, once an individual gains access to virtual space, the opportunities exist across the entire virtual space except for some irregular holes in it. These holes represent the unavailable opportunities due to factors such as a service is down for maintenance between time t_1 and t_2 (i.e., an authority constraint) or an individual cannot use a virtual service developed in a foreign language (i.e., a capability constraint). The adjusted space-time prism and the Swiss cheese model together confine the potential opportunities of virtual activity.

7.4 Human interactions in the age of instant access

People, as social beings, cannot avoid interacting with one another. With virtual space, people can interact with other people over distance through tele-presence. The use of tele-presence is changing the spatial and temporal patterns of how people interact. Four interaction modes (i.e., SP, AP, ST, and AT) have been identified and participants involved in an interaction via each mode share a specific spatio-temporal relationship (i.e., co-existence, co-location in space, co-location in time, or no co-location in either space or time). These spatio-temporal relationships between individuals can be represented through their extended space-time paths. Also, identifying these spatio-temporal patterns between space-time prisms can help us explore opportunities for different types of interactions among people.

Figure 7.5 shows the four spatio-temporal relationships represented with extended space-time paths. In case 1, persons A and B had a relationship of

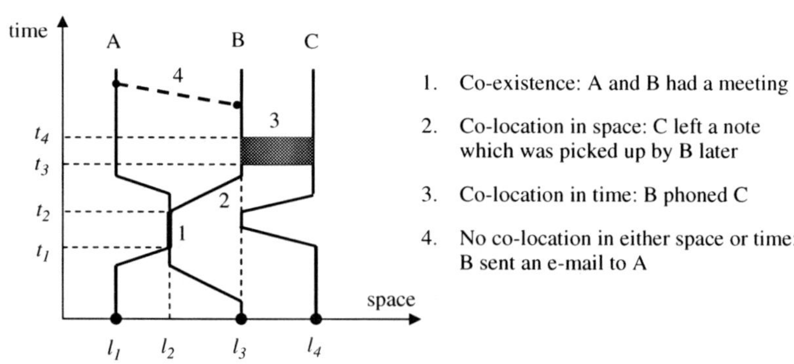

Fig. 7.5 Spatio-temporal relationships of human interactions

co-existence because they both stayed at the same location (l_1) during the same time period (from t_1 to t_2) to have a meeting. Case 2 describes a co-location in space relationship when both persons B and C visited the same location (l_3) at different times. In case 3, persons B and C had a co-location in time relationship because they stayed at different ICT-enabled locations and interacted through tele-presence during the same time period (from t_3 to t_4). Case 4 depicts a relationship of no co-location in either space or time in which B sent out an email and A received it at a different location at a later time.

When these spatio-temporal relationships applied to prisms, they can help explore the opportunities for potential interactions among people. If two conventional space-time prisms overlap as shown in Figure 7.6a, it indicates that the individuals can reach the same location during a common time window (i.e., co-existence); therefore, they can carry out potential SP interactions.

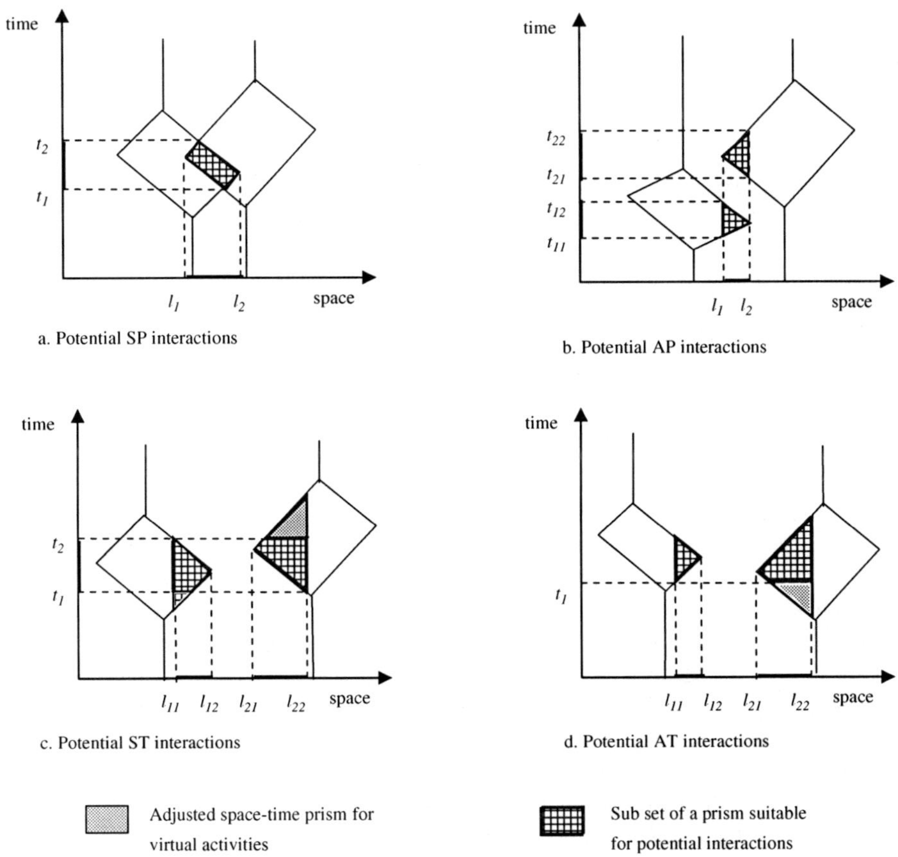

Fig. 7.6 Spatio-temporal relationships of prisms and potential interactions

Figure 7.6b shows that two individuals will be able to reach the same locations at different time periods (i.e., co-location in space). Such situations provide opportunities for these individuals to carry out AP interactions. Both ST and AT interactions involve activities in virtual space and adjusted space-time prisms are needed to help explore opportunities for these interactions. The gray areas of the space-time prisms in Figure 7.6c and 7.6d indicate the space-time extents that individuals can access virtual space. If two prisms for virtual activities overlap along the time dimension as shown in Figure 7.6c, the individuals will be able to share a co-location in time relationship and conduct ST interactions. AT interactions do not require coincidence in either space or time (i.e., no co-location in either space or time). As long as participants have access to virtual space and the receiver of an AT interaction has access to virtual space at a later time than the initiator, they will be able to conduct AT interactions. Figure 7.6d shows the situation for potential AT interactions. It indicates that the receiver's adjusted space-time prism needs to last beyond the earliest boundary of the initiator's prism along the time dimension so that the receiver will be able to pick up the incoming message. These spatio-temporal relationships of prisms provide an approach to exploring potential human interactions in both physical and virtual spaces.

7.5 Conclusions

The virtual space enabled by ICT enhances our capability of accessing information, conducting activities, and interacting with others. It is also changing the ways how people carry out their activities in space and time. In this study, we revisited Hägerstrand's time geography and proposed an extended time-geographic framework to help represent and examine spatial and temporal characteristics of individual activities in the Age of Instant Access. Based on a proposed conceptual model which depicts the relationships between physical space and virtual space, this study identifies the constraints applied to virtual activities and adjusts the concepts of space-time path and prism to accommodate both physical and virtual activities in the framework. Four spatio-temporal relationships (i.e., co-existence, co-location in space, co-location in time, and no co-location in either space or time) are recognized to portray different interaction modes among people (i.e., SP, AP, ST, and AT). With appropriate geographic information systems (GIS) designs, this proposed framework can be operationalized as a working system. Figure 7.7 shows a possible approach of implementing space-time paths and prisms in GIS. Efforts also have been made to develop GIS functions that can represent and analyze the four spatio-temporal relationships among people with extended space-time paths (see Yu, 2006). Operational systems of the proposed framework can offer useful tools to help researchers investigate the changing spatio-temporal patterns of human activities in the Age of Instant Access.

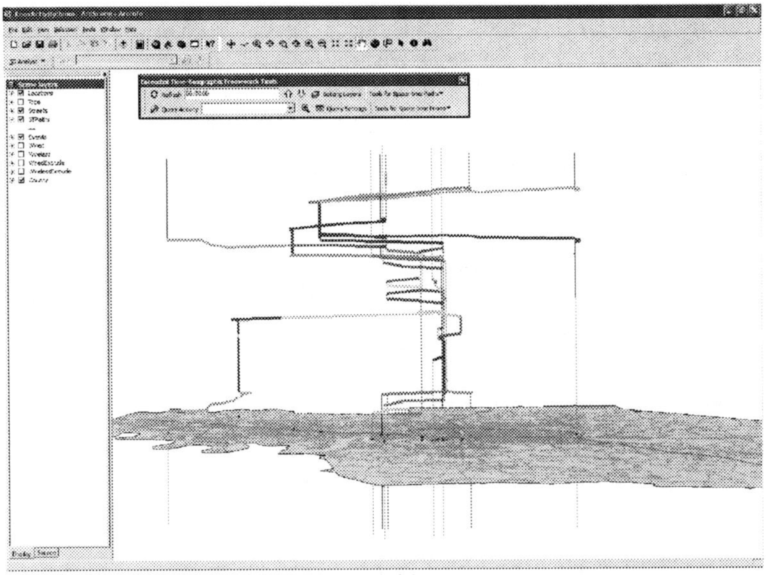

a. Space time paths and individual activities

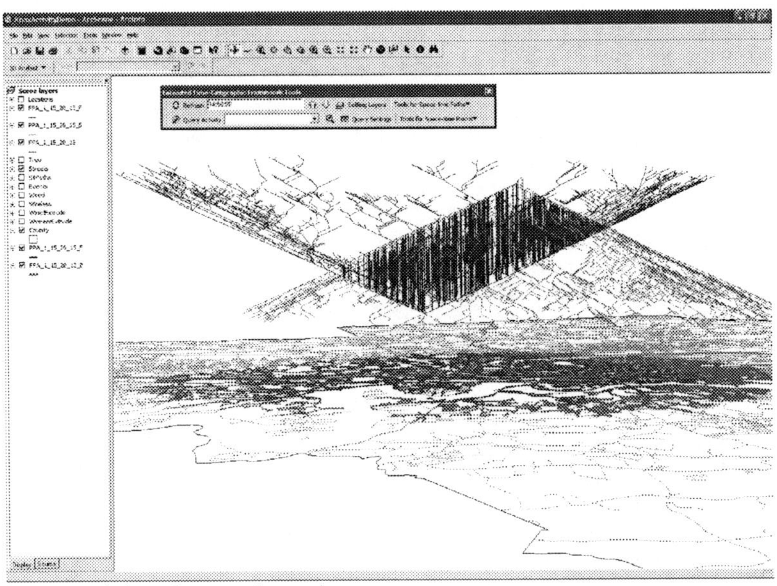

b. Network-based space-time prism

Fig. 7.7 Representation of adjusted space-time paths and prisms in GIS. a. Space-time paths and individual activities b. Network-based space-time prism

References

Adams, P. (2000). Application of a CAD-based accessibility model. In D. Janelle and D. Hodge (eds.) *Information, Place, and Cyberspace Issues in Accessibility*. Berlin: Springer, 217–239.

Batty, M. and H. Miller (2000). Representing and visualizing physical, virtual and hybrid information spaces. In D. Janelle and D. Hodge (eds.) *Information, Place, and Cyberspace Issues in Accessibility*. Berlin: Springer, 133–146.

Cairncross, F. (1997). The death of distance: How the communications revolution will change our lives. Boston: Harvard Business School Press.

Carlstein, T. (1982). *Time Resources, Society and Ecology*. London: George Allen and Unwin.

Ellegård, K. (1999). A time-geographical approach to the study of everyday life of individuals – a challenge of complexity. *GeoJournal*, **48**, 167–175.

Golledge, R. and R. Stimson. (1997). *Spatial Behavior: A Geographic Perspective*. New York: The Guilford Press.

Hägerstrand, T. (1970). What about people in regional science? *Papers of the Regional Science Association*, **24**, 1–12.

Harvey, A. and P. Macnab (2000). Who's up? Global interpersonal accessibility. In D. Janelle and D. Hodge (eds.) *Information, Place and Cyberspace: Issues in Accessibility*. Berlin: Springer, 147–170.

Janelle, D. (1973). Measuring human extensibility in a shrinking world. *Journal of Geography*, **72**(5),8–15.

Janelle, D. (2004). Impact of information Technologies. In S. Hanson and G. Giuliano (eds.) *The Geography of Urban Transportation (3rd Edtion)*. New York: The Guilford Press, 86–112.

Janelle, D. and D. Hodge (eds.) (2000). *Information, Place, and Cyberspace Issues in Accessibility*. Berlin: Springer.

Kwan, M.-P. (2000). Human extensibility and individual hybrid-accessibility in space-time: a multi-scale representation using GIS. In D. Janelle and D. Hodge (eds.) *Information, Place, and Cyberspace: Issues in Accessibility*. Berlin: Springer-Verlag, 241–256.

Lenntorp, B. (1976). *Paths in Space-Time Environments: A Time Geographic Study of Movement Possibilities of Individuals. Lund Studies in Geography B: Human Geography*. Lund: Gleerup.

Miller, H. (1991). Modeling accessibility using space-time prism concepts within geographical information systems. *International Journal of Geographical Information Systems*, **5**, 287–301.

Miller, H. (2006). Social exclusion in space and time. In K. Axhausen (ed.) *Moving through Nets: The Social and Physical Aspects of Travel*, Elsevier, 353–380

Miller, H. and S.-L. Shaw (2001). *Geographic Information Systems for Transportation: Principles and Applications*. New York: Oxford University Press.

Negroponte, N. (1995). *Being Digital*. New York: A. A. Knopf.

Parkes, D. and N. Thrift (1980). *Times, Spaces, and Places: A Chronogeographic Perspective*. New York: John Wiley.

Salomon, I. (1986). Telecommunications and travel relationships: a review. *Transportation Research A*, **20**, 223–238.

Wiberg, M. (2005). Introduction – the emerging interaction society. In M. Wiberg (ed.) *The Interaction Society: Practice, Theories and Supportive Technologies*. Hershey, PA: Information Science Publishing, 1–24.

Yu, H. (2006). Spatio-temporal GIS design for exploring interactions of human activities. *Cartography and Geographic Information Science*, **33**(1), 3–19.

8 Dynamic prisms and "instant access": linking opportunities in space to decision making in time

Pip Forer[1], Otto Huisman[2], Chris McDowall[1]
[1]*School of Geography and Environmental Science, University of Auckland, New Zealand*
[2]*SHORE, Massey University of the Manawatu, New Zealand*

8.1 Introduction

The ideal quotation to start this chapter would run roughly like this: "At the moment of birth we are nothing but prism, and at the moment of death nothing but timeline. Life is the inexorable destruction of space-time opportunity, and its translation into a lived experience". As far as I know no one has penned quite such a lurid and kitsch-mystic sentiment, so the quotation will have to be foregone, but if it existed, it would make a very valid point and remind us of the fundamental role of dynamic human decision-making in Hagerstrand's minimalist but essentially powerful view of space, time and geography.

The broad question posed by this book is how society will adopt, use and be changed by the ubiquitous high speed delivery of information through mobile technologies. This particular chapter attempts to examine these issues at the basic level at which innovations in communications operate: that of individuals and their everyday lives. Its aim is to describe an approach that helps in conceptualizing the avenues by which fast, mobile information delivery can impact on the exercise of individual choices in space-time, and the likely geographical variety of that response. In due course we might also expect to be able to speculate on the emergence of patterns of group activity that reflect the sum of individual actions, and thus gain insights on how the dynamic effects of new mobile ICTs play out in a spatial and social context.

There is enough in that statement of purpose to hint very strongly that this paper is going to look towards the work of Torsten Hagerstrand and Time Geography as its foundation. (Hagerstrand, 1970, 1978, 1995; Parkes and Thrift, 1980; Miller, 2005). That is largely true, but before we develop a specific perspective in this area we need to critically deconstruct the term 'age of instant access' in terms of the geographies it implies. Like many wide-screen metaphors that are used to describe ICT innovations this phrase encapsulates a broad phenomenon with a degree of panache, but communicates a misleadingly simplified world in the context of the details of the changes at work (e.g., Cairncross, 1997). Should we even think of an 'age of

H.J. Miller (ed.), Societies and Cities in the Age of Instant Access, 119–132.

instant access' as we discuss the way an individual might experience dramatically enhanced access to information? An 'age' implies an impact of landmark significance, usually with a dominant impact across a society. And 'instant access' suggests just that: immediate information, any mode, anywhere, anytime, anyone, right now.

Is this delusional? Certainly the majority of commentators enthusiastically acknowledge that there are good grounds for expecting that innovation in ubiquitous ICTs will drive broad-scale changes. However, there has to be caution in interpreting this as an undifferentiated phenomenon across populations, especially when strong voices and equally seductive metaphors (i.e. the digital divide) propose quite the reverse (Norris, 2001). Furthermore, the implication that the geography of information will be an undifferentiated one, an isosurface of instant gratification, is a similarly significant distraction from the truth. At almost every spatial scale, down to signal reception strength across a room, we should expect variations in both speed and access as services are rolled out and subjected to the usual dictates of capital cost and market demand. While empirical studies are few it is clear from commercial and research sources (Mitchell in press, Telecom 2005) that we are currently far from a position of ubiquitous spatial access. For the present, access to information (even within G3 service areas) is highly differentiated across both space and individuals. There are strong arguments that differentiation will always be the norm at various scales. Systematic constraints on infrastructure, education or the ability to pay for access mean that the complexities of geography and individual circumstance will persist indefinitely.

Having asserted that, we turn to Hagerstrand and his concern for the interrelationship between spatio-temporal geographies of information, individual decision making, and activity patterns in time and space. Firstly we explore the sometimes hidden nature of Hagerstrand's time geography relative to the current context. This involves a (re)affirmation of the active rather than passive perspective on a key concept: the *prism*. Normally seen as the volume of space time available to an individual between two fixed events (their choice space), this chapter focuses on the prism as an evolving construct that can be part of a recursive, information-constrained process that navigates towards an eventual timeline. These ideas are developed and then discussed in the context of two contrasting scenarios: the (infra)structured day of a tertiary student and the spontaneously structured travels of an itinerant tourist. The conclusions reflect on the degree to which the individual examples can illuminate the potential local impact of ICTs and be generalized into a better understanding of aggregate impacts, before speculating on a possible way to create a more realistic model of the prism for analysis.

8.2 Time geography for the age of instant access

Hagerstrand's work on time geography is frequently, and rightly, acknowledged as a critical work in human geography (Pred, 1980; Colbert, 2003). It is seen as elegant and powerful in its ability to provide a vision of the space-time environment of individuals in respect of their everyday lives, and in its capacity to illustrate how the performance of such tasks forms social and spatial networks of considerable importance (Hagerstrand, 1970, Parkes and Thrift *op.cit.*).

As other contributors note, time geography is essentially about people, their physical mobility, their activity needs and their constraints. (see Chapters 1 and 7 in this volume). For our purposes three key concepts dominate the framework: the *timeline* (the locational trace of an individual through space and time); the *marker episode* or *event* (a high priority and locationaly constrained activity within a person's day, typically work or school attendance but could be leisure activities); and the *prism* (the volume of space-time which can be accessed by an individual for discretionary activities between marker episodes). These are identified in Figure 8.1a in the classic 'aquarium' model, adapted from What *about people in Regional Science?*, the paper that for many people is still the founding manifesto of Time Geography (Hagerstrand, 1970). Certainly it is a work that set the theoretical tone for a flush of time geography research extending to the mid 1980s.

The alternative rendition of an Auckland student's daily prism (Figure 8.1b), reflects the greater engagement with computation and real spaces that characterizes the resurgence of research from the mid 1990s, work which endeavors to situate theory within a stronger empiric context (Miller, 1991, 2005). Much work has focused on two areas: the description and analysis of timelines, increasingly based on new sensor technologies (Mountain, 2001; Laube et al., 2005), and prisms as definitions of access (Miller, 1991, 2000; Kwan, 1998, 1999; Forer, 2000). The majority of this work is reflective, either analyzing given movement patterns in static information geography or using pre-defined marker episodes to identify resultant prisms/options as access indicators. Naturally any current timeline data will capture revised patterns of movement that reflect the decisions of individuals using ICT in various ways, but particularly those capitalizing on the ability to virtualize tasks and unlink certain activities from their traditional locations (Couclelis

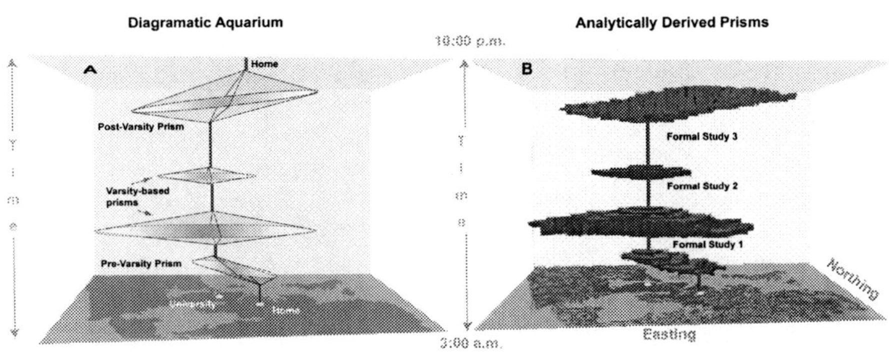

Fig. 8.1 A simple diagramatic timeline and prism and an analytically generated prism illustrating a student's day

and Getis, 2000). However, in the absence of longitudinal data sets these insights are somewhat tangential to the process of changing information use.

What is needed for this is a move away from the passive acceptance of timelines and marker episodes to a dynamic view of prisms and the process by which timelines emerge from prisms; one following Hagerstrand's example and steering clear of many of the difficult issues surrounding human agency. This may provide a focus on the enhanced decision options that better access to information confers, rather than on the way a specific individual might make a particular decision.

8.3 The creation of marker episodes, assignations and prisms

As noted, much analysis with prisms is based on measurement of, or assumption of, pre-defined (binding) marker episodes. The resultant prisms are a valuable but derivative artifact based on a binary classification of location-constrained activities as either vital or not. In turn the derived prism is a similar binary: space-time split into volumes that an individual can corporeally access, or not. Thus the traditional prism prescribes, for any period of uncommitted time, a local universe in which physical presence can be manifest, and in which an individual's corporeally-based activity can occur, duration and necessary resources permitting.

In the 1970s such a polarized view mirrored a simpler society. There was a largely 9-5 work-force, a much stronger differentiation by gender, a more limited choice of recreation and mobility options, and activities that were more strongly linked to specific sites. Institutionally defined marker episodes dominated: school terms, the work day, child care, perhaps church. Such markers endured over long periods, imposing distinct constraints on when, where and what discretionary activities were possible in an individual's remaining time. Marker episodes were either compulsory, negating the need for decision making as such, or had some degree of initial choice. In either case information was needed to allow the activity to be planned and executed. Such information included personal knowledge (carried with the individual or family), or knowledge gained from timetables, books, maps, phone calls to friends, newspapers, encyclopedias and atlases: in brief a universe of knowledge constrained by the geographies of the public and private phone, family members, the opening time and location of libraries, the presence of a newsagent and, of course, the physical mobility of the individual.

To summarize, individual mobility and the nature and location of local services/resources determined major constraints on what activities could be sustained. However, choice was also constrained by adequate information, and information itself had its own geography of access, which in turn acted as a constraint on choice or practicality of options. New marker episodes and their related project tended to be predictable, with a long lead time, and so in the 1970s could be based on information assembled over a relatively long period. By nature 'discretionary' activities were more transient, usually of shorter duration and planned at significantly shorter notice than the markers. It seems fair to assume that the time geography of information access was a more significant issue for the non-routine trips, and in many cases a constraint.

Three decades, on the societal situation has changed. Institutional markers are much weakened: part time employment, flexitime and various lifestyle choices reflect in much more complex personal movement patterns and role sharing. Various factors contributing to space-time compression and wider consumer choice and expectation have generated more complex and dynamic lives in western economies, a process which started prior to any manifestation of 'instant access'. For households that juggle part-time work, childcare, and socializing is a constant adjustment in their daily patterns. Similar adjustments can be seen appearing in other contexts such as students moonlighting to cover fees (Boswell, 1995) or free independent tourists building their itineraries on the run (Forer, 2005).

In this situation the concept of the structural marker episode loses ground, and discretionary, or more accurately transitory, activities dominate, generating a growing need for information on how to identify and execute choices. Also the decisions that individuals need to make are qualitatively different from the previous epoch. In a Western city in the current era the marker/discretionary dichotomy is increasingly challenged as mobile individuals trade-off multiple, competing demands on time, and where the certainty of access to various locations is complicated by chaotic traffic and occasionally delinquent public transport.

At this point it may be helpful to outline a perspective on prism formation which is more dynamic and can be directly related to the impact of instant access. At the heart of this is the renaming of unrealized (future) marker episodes as 'assignations', a move designed to distinguish the existence of these sections of pre-dedicated time lines and their virtual episodes from earlier usage. The nature of assignations is that they are in the future, they represent individual intent, contract or aspiration, that they exhibit varying degrees of priority, and that enjoy varying levels of discretion. At any point in time an individual has a series of prioritized assignations in mind, as visualized in Figure 8.2. This specific structure defines in its turn a series of prisms which have yet to be utilized for any specific purpose, but naturally can house a new 'assignation' if the individual spots an opportunity to fulfill a need. Equally and fittingly, existing assignations may be canceled, creating new prisms of freed space-time.

This notation is not intended as a model of individual decision-making so much as a framework for examining how an individual's options might adjust through time, going forward, particularly under different models of communication and information provision. It allows us to identify specific purposes and contexts, and to experiment with what opportunities might be identified under different time constraints and information provisions. To do this it is useful to identify the kinds of conditions needed for an activity option to be selected, and the kind of information an individual needs to satisfy themselves as to whether, where and when the conditions can be met.

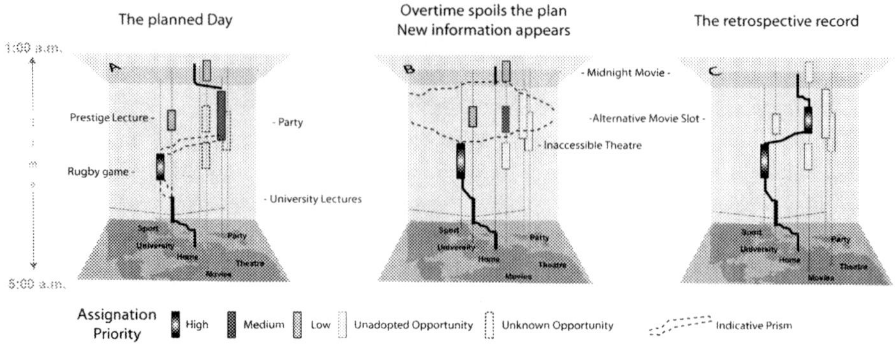

Fig. 8.2 Markers, assignations and prisms in a daily frame

Time geography identifies a number of constraints that can be relevant in determining if a specific activity can occur effectively. All relevant criteria need to be simultaneously met. Significant criteria are: i) *access* (the ability to be in a location within the time window available); ii) *duration* (the ability to be present for an adequate time); iii) *capability* (the presence of any necessary fixed resources at a location); iv) *coupling* (the ability to coincide with a mobile entity such as a person or a traveling resource), and; v) *authority* (the right to occupy that part of space-time). The basic prisms of Figure 8.1 essentially define the volume which meets the Access criteria. Meeting other criteria will reduce this volume in accord with the needs of any particular purpose, for instance to meet a friend (coupling) for 60 minutes (duration) to shop (capability), when shops are open (authority), Any proposed or adopted assignation will lie within the remaining opportunity space.

One way to visualize the dynamic process of prism formation and the adoption of assignations is suggested in Figure 8.3. The driver is *desire formation*, the expression of an activity goal to be achieved, often in conjunctions with a wider project. The outcome is some form of *action*, which is based on *decision making*, and the critical precursor to this is *information gathering* (and analysis).

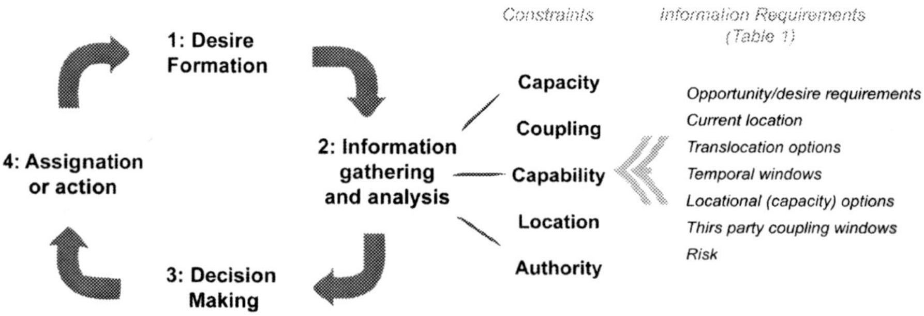

Fig. 8.3 Decision making and the information-gathering loop

Critically, this step needs to assemble the necessary information to express the key constraints, or put more positively to identify the conditions where all constraints are met. Working optimally, and with a rational, omniscient decision maker, the data would provide a perfect illustration of available opportunities that met the individual's needs. They might also suggest totally new purposes/opportunities for the individual to pursue. Conversely, the more information that is omitted then the more incomplete or risk prone the opportunity set might be.

The commitment to any assignation will lie somewhere between compulsion and whimsy. An assignation with the dentist may be close to the former, and a last-minute visit to the art gallery more the latter. These assignations may be in place months ahead, or decided minutes before they are actioned. Decisions may be made with a closure at day's end, or over a period of days. There may be recursive decision making, establishing key decisions at one temporal scale (to visit Ireland for a week) followed by decisions over activities on specific dates and at specific places (which castle to visit).

Figure 8.3 acknowledges that assignations are essentially the reflection of choices, possibly to submit to a given regime of work, accept attendance at a social event or to be at home when a child leaves or returns. Some choices are cyclic and repeat every week day, some are socially driven, some driven by the availability of a needed resource, some spontaneous. All are made in a particular information context, which is both a result of the activity pattern and a conditioning of it.

Just what information might the individual require? Table 8.1 identifies some

Table 8.1 Example Information Needs for Identifying/Validating a Possible Activity Option

Opportunity/Purpose: which may be more or less precisely known, but in some cases can be quite dynamic or a serendipitous product of information retrieval.

Present Location of the Individual: a factor which is often assumed in time-geography studies (presumably since it must be known to draw the aquarium) but which is in fact not a safe assumption at all in some circumstances

Translocation Opportunities: knowledge of which is a fundamental need. The rather awkward phrase identifies the key requirement for the individual to know where they can get to, and how quickly they can get there.

Temporal Window(s): the unallocated time period(s), the plural accepting that there are potentially multiple time-related constraints, for instance availability of equipment or opening times of retailers.

Locational (Capacity) Options: locations where the capacity constraints for an activity can be met i.e. necessary resources are available.

Coupling Window(s): knowledge of the availability of and location of people or artifacts needed to be assembled for the activity, whether static or mobile

Risk and Change: factors not frequently considered in a retrospective view of prisms as indicators for activity options, and where possible presence or absence in a volume of time-space is seen as a binary. In reality for many of us timing is an approximate science, wayfinding hazardous and transportation systems exhibit chaotic tendencies, as do doctors' queues.

important data which are prerequisites for assessing the degree to which the space-time constraints can be met, and which can be a deployed for answers in a number of space-time analysis techniques. The degree to which such information is available will be a significant factor in determining the envelope within which options, or new possibilities, can be established. From a research perspective, existing spatial data infrastructure can already respond to many of the more mundane information needs for the purpose of such analysis.

8.4 Information and assignations

The rationale for working within a time-geographic framework now becomes clear. The dynamics of decision making for assignations provide the process by which activities are scheduled and eventually consummated. The wider framework for this allows a consideration of the geography of information relative to the needs of the specific purpose/activity as well as for a specific ICT milieu.

Any decision is made with access to a quantum of information. The nature of that quantum will depend on the specific activity desired. If *all* required information is held in the decision-makers mind that ensures that a full and balanced decision is possible, but of course not guaranteed. However, (almost) instant information access is assured.

If information is not immediately available it will take time to assemble and/or the decision will be made without it, but with the risks of error and omission. In assembling the information the time available is governed by the lead time to the desired activity. If certain data is simply not available, *for* this location or *from* the current location, the search will wither or the decision be made at higher risk.

Available information feeds into the planning loop from various sources, and that feed is itself dependent on the local context of the individual. In a harmonious hamlet, with oral knowledge and everyone within shouting distance, the feed is simple. In large urban centers formal data sources become central in identifying a full range of opportunities. Printed sources, contact with individuals, codified information sources or that general infrastructure of civilization, which Stewart and Cohen (2001) have termed extelligence, all come into the picture. Certainly the widening scope of available data allows a greater power and range of queries (spatial and otherwise) related to constraints and opportunities.

If we look over time we can begin to identify some of the changes that evolving mobile ICTs have made or will make to the decision loop. If we acknowledge the evolving geographic variations in availability, service level and the cost of different ICT -mediated information services, then we can begin to construct a picture of how opportunity searches (and their resultant outcomes) may vary over space as a result of differential information access. Finally, if we speculate on new ICT services we can perhaps debate how they might work with the individual or community to further modify the bounds within which opportunity and choice are identified.

What is proposed here is not that decision making for an individual can be modeled using time geography, but that three steps provide a means to explore how mobile ICT innovation might structurally influence activity outcomes. These steps are: *identifying* the information needs to make a decision on a specific activity in a certain context; *specifying* the ICT environment (mobile or not) and the consequent information availability; and *analysing* whether the activity is possible in that context, and if so what options the available information can identify. This is at least a framework in which we can explore the underlying influence of mobile ICTs on the range and quality of the decisions that can be formulated within a certain spatial context. There is no room in this chapter to work through multiple examples of this procedure, but two contexts in which related research is under way are examined.

8.5 Opening game: assignations and ICT in two domains

Operating in an environment that is well known and heavily structured poses less demands for new, external, instant information, and often provides dedicated channels to acquire what is needed. Visiting a new area and being heavily dependent on serendipity to fill the day is quite the reverse. The highly structured world of the University (of Auckland) student during term time represents one example here, and a tourist traversing the West Coast of the South Island of New Zealand the other.

Student lives are interesting because of the relatively complex set of prioritized daily assignations timetabled into their future for the duration of each term. These are the numerous labs and lectures tied to specialist facilities of various sorts, and involving compulsive coupling for much of the day with groups of various sizes. How do mobile ICTs and fast access to data assist students? Flexible learning approaches have certainly been used to moderate the timetable's rigidity through virtual technologies that allow substitution of on-line experiences for corporeal presence, but with that in place the key issue for students remains finding ways to balance their non-study activities, often across a home location, a commuting experience, a casual paid work location and several social locations. The earlier Figure 8.1 showed a typical set of daily prisms for a student, where prisms identified opportunities outside of home, travel and study. If that is their blueprint for tomorrow, what information do they need in order to identify assignations that will establish structure within their as yet uncommitted time?

Perhaps text messaging will suffice, and for the established student more sophisticated and ubiquitous information may be of limited added value. Perhaps there is a desire to generate a spontaneous study group, and remote e-mail chat is too weak an environment. Certainly there are social issues to pursue. In both cases the ability to exchange information at short notice on mutual meeting times will be crucial. However, probably most of such meetings will constitute 'virtual teams', where the spatial impact may be minimal and or at least engagement is likely to be in an established environment at a place of residence. For a really casual part-time job, or a job requiring travel to new parts of the city, travel information may be important. Maybe e-mail on the run helps with some space-time substitution. But, in something as slow moving as a University and its timetable, with an environment permeated by broadband web access and large screens, it may

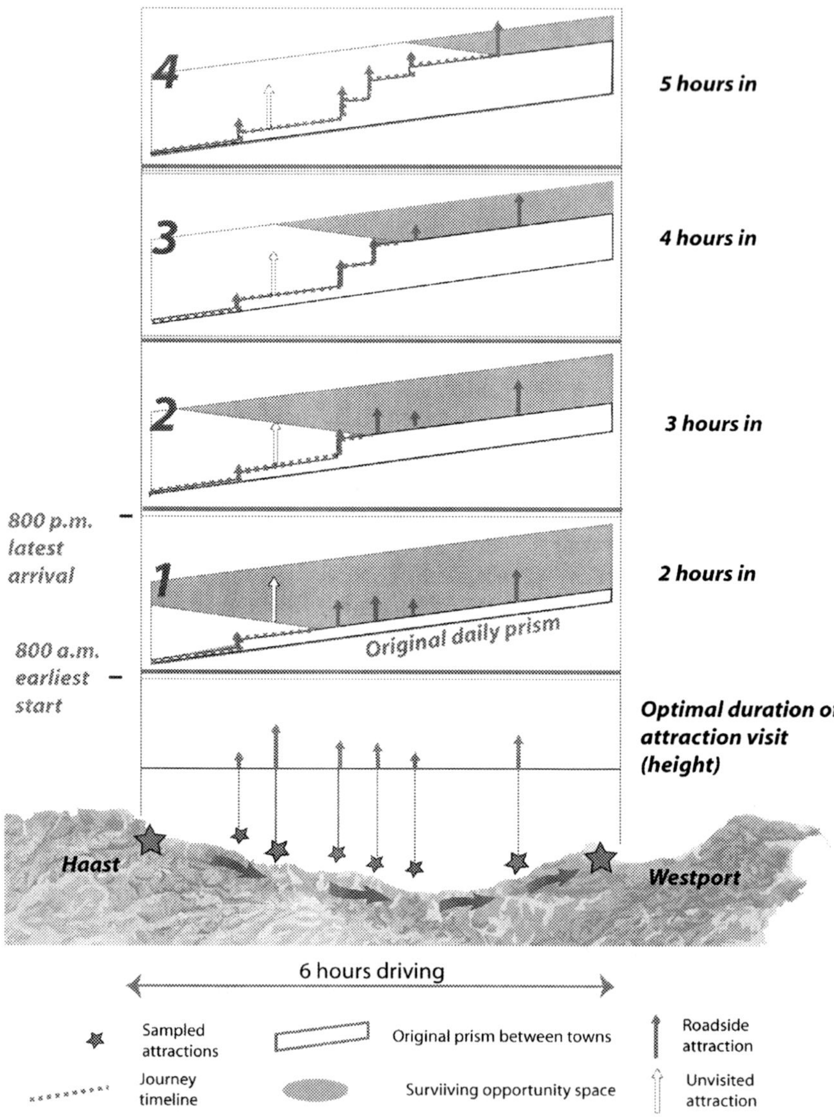

Fig. 8.4 A hypothetical sequence illustrating the evolving prisms of opportunity during a tourist day

be that further instant, ubiquitous access will prove a marginal benefit, with only peripheral positive effects compared to what is offered by existing domestic and institutional portal infrastructure.

Tourists are a different issue. They possess far less internalized knowledge for their holiday environment, even after reading *Lonely Planet*. Consequently they are far more open to the real-time discovery of opportunity when they encounter an unknown but delightful attraction. Decisions are often made at short notice, often on the basis of unpredictable factors such as the weather. Their information deficit is considerable, as is witnessed by the popularity of information centers and internet cafes. The same phenomena also illustrate the degree of information overload experienced at times. This is the opportunity and the challenge for mobile information services.

Figure 8.4 illustrates a tourist's decision outcomes at the daily level. A higher level of planning has already created two assignations as part of an itinerary: namely sleeping at Haast and then at Westport. In fact overnighting choices (assignations) have already partitioned the next seven days into seven daily prisms The diagram explores the sequential destruction of prisms on the one day Hasst-Greymouth leg. There is an inescapable logic as the party drive their selected route and attractions are chosen to be visited, bye-passed, or are lost due to commitments to earlier stops, perhaps commitments made in ignorance.

Tourists movements are very interesting but seldom studied in time geography. They are far from random as they evolve they are information needy. They feature individuals who are usually relaxed and receptive to mobile communications; and (certain in New Zealand) they operate in areas with diverse geographies of communication. If you look at the information needs in Table 8.1 the University student context might light up only a few of the information demands as having an enhanced provision from mobilized access, and thus a significant impact on patterns of activity and decision making. If you do the same for the itinerant tourist, quite the reverse is true. That is of course why so much interest focuses on LBS and tourists (Dias et al., 2004; Brimicombe and Li, 2006), and why instant access can be expected to generate a number of ongoing space-time adjustments..

8.6 Conclusions

ICTs, now in a mobile phase, continue to impact not just on cities, but on society and all its geographic manifestations. However, multiple questions emerge from this picture regarding the degree to which we are seeing a really significant reorganization of the nature of many everyday experiences, or whether the change is more minor and qualitative and in many cases ephemeral. There is no doubt that text messaging is a major phenomenon in areas supporting cellular phone communications: that is easily and incontestably established. It is in your face, frequently disrupting local conversation for distant communication and apparently re-prioritizing the distant over the local. But the degree to which the text messaging creates enduring new social networks with new expressions of social and spatial activity is far harder to establish. The fact that a PDA can technically access much of the wealth of the web (and increasingly in a readable and usable way) is known, but the nature of usage can only be roughly gauged, and volume remains below expectations. The actual use of the technology in decision making is even less well known.

This chapter has tried to explore 'instant access' at a localized level, but in a context grounded in geography and space-time. It makes no claim to understand the social or personal forces which compel certain activity choices or lifestyles, but it does investigate the changing context in which decisions can be made, and the implications of this for realized choice.

While such a framework makes no pretense to solve the issue of choice and agency it does provide a tool for examining, in a number of contexts, the workings of a decision-making environment in which prisms of opportunity are identified and then annihilated, often with different levels of extant information and different levels of fast access.

Several questions are raised. Predictably, the neo-tech narrative should run cleanly through to the delivery of true instant access, envisioning a transformation as one by one the information needs identified in Table 8.1 get ticked off as deliverable, and instantly deliverable at that (although even the neo-tech voice would probably apostrophise). This chapter has argued against such a comprehensive development, at least in the short term, but two lesser issues are raised in closing. One is the imprecise nature of the future and the other that of mobile *spatial decision support systems* (SDSS).

The vagaries of traffic, human timekeeping and prescience are such that in whatever tense one uses the concept a precise prism can not exist. Nor can a precise assignation. The suggestion that innovative GIS be used to assess the opportunities that can be identified from a given information quantum may have merit, but it is handicapped if it can not describe such future options as vague imprecise entities. For a better description it needs to be possible to represent prisms as dynamic, vague objects, where the prism's space-time volume represents the probability of being able to be present, or the likelihood of meeting the various constraints. There are possible ways to represent this for analysis (Mcdowall, 2006), and work is ongoing to use this approach to enable the better modeling of the interactions between different individuals' time geographies.

This links to the issue of instant access and the ability of an individual to manage a substantial amount of information at once. Accessing, arranging and interpreting large quantities of information takes time and often invites confusion, particularly where information sources are providing contesting stories. To date the most effective uses of mobile ICTs appear to involve well-defined simple decisions, for instance providing informal locational information to friends for a social gathering or meeting within a sports stadium. Assembling complex information takes time and a truly mobile device is often not well suited to the task. Thrift and French (2002) have noted the growth of mediating software in society, and even a basic location-based service (LBS) could classify as such. An SDSS for day planning would be even more in that vein. From one perspective it is just what is needed, providing less but more potent information and more manageable choice. From another it is intrusive mediation and possibly covert manipulation (tainted information). Thus the chapter ends as it started: deconstructing the concept of 'instant information'.

References

Brimicombe A. and Li Y. (2006): "Mobile Space-Time Envelopes for Location-Based Services". Transactions in GIS 10(1): 5–23

Boswell, S. (1995): "The splintering of student geographies: Restructuring of the tertiary education sector in New Zealand." Unpublished M.A. Thesis. Department of Geography, University of Auckland, New Zealand.

Cairncross F. (1997): "The Death of Distance". Harvard Business School Press, 302pp

Corbett J (2002): "Torsten Hagerstrand: Time Geography", Centre for Spatially Informed Social Science, University of California Santa Barabara, http://www.csiss.org/classics/content/29

Couclelis H. and Getis A. (2000): "Conceptualizing and measuring accessibility within physical and virtual spaces," in D. G. Janelle and D. C. Hodge (eds.) Information, Place and Cyberspace: Issues in Accessibility, Berlin: Springer, 15–20.

Dias, E., Beinat, E., Rhin, C. and Scholten, H. (2004): "Location Aware ICT in Addressing Protected Areas' Goals". Research on computing Science vol. 11 (special edition on e-Environment). Edited by Prastacos, P. and Murillo, M. Centre for Computing Research at IPN. Mexico City. p.273–289.

Forer, P.C. (2005): 'Tourism Flows and Dynamic Geographies: Applying GI Science to Understanding Tourist Movements and Impact.' In Simmons D. and Fairweather J. (editors) 'Understanding the Tourist Host-Guest Encounter in New Zealand', EOS Publishers, Christchurch, New Zealand, 22–56

Forer, P.C. and O. Huisman (2000): "Space, Time and Sequencing: Substitution at the Physical/Virtual Interface". In Janelle D.G. and Hodge D.C. (eds), Information, Place and Cyberspace: Issues in Accessibility, Springer-Verlag, Berlin, 73–90.

Hägerstrand, T. (1970): "What about people in regional science?" Papers of the Regional Science Association 24: 7–21.

Hägerstrand, T. (1978): "Survival and arena: On the life–history of individuals in relation to their geographical their Geographical Environment (1975). Monadnock Vol. 49: 9–29. Clark University

Hägerstrand, T (1995): "Action in the physical everyday world" In A.D. Cliff, P.R. Gould, A.G. Hoare and N.J. Thrift (eds). Diffusing Geography: Essays for Peter Haggett. Blackwell,Oxford, pp. 35–45

Kwan, M.-P. (1998): "Space-time and integral measures of accessibility: A comparative analysis using a point-based framework", Geographical Analysis, 30, 191–216.

Kwan, M.-P. (1999): "Gender and individual access to urban opportunities: A study using space-time measures", Professional Geographer, 51, 210–227.

Laube P, Imfeld S. and Weibel R. (2005): "Discovering relative motion patterns in groups of moving point objects", International Journal of Geographical Information Science, Vol. 19, No. 6, July 2005, 639–668

Mcdowall C. (2006): Modelling vague dynamic entities within a GIS. Thesis presented for the degree of Doctor of Philosophy, University of Auckland

Miller, H.J. (2005): "What about people in geographic information science?" In Re–Presenting Geographical Information Systems. P. Fisher and D. Unwin.(editors), John Wiley, 215–242.

Miller, H.J. (1991): "Modelling accessibility using space–time prism concepts within geographical information systems". International Journal of Geographical Information Systems 5(3): 287–301.

Miller H.J, Wu, Y. (2000): "GIS software for measuring space–time accessibility in transportation planning and analysis". GeoInformatica 4:141–159

Mitchell, P (in press): "ICTs and Emerging Timespaces: The Role of Local Government in the Auckland ICT Environment", Social and Cultural Geography special issue on Spatial Technologies/Technological Spaces.

Mountain D. and Raper J. (2001): "Modelling human spatio-temporal behaviour: A challenge for location-based services": Proceedings of the 6th International Conference on GeoComputation, University of Queensland. "GeoComputation CD-ROM" Publisher David V. Pullar,, or at http://www.soi.city.ac.uk/~dmm/research/pubs/B05.3.pdf

Norris P. (2001): "Digital Divide: Civic Engagement, Information Poverty, and the Internet Worldwide. Cambridge University Press, Cambridge 320 pp

Parkes, D. N. and N. J. Thrift (1980): "Times, Spaces, and Places: A Chronogeographic Perspective"., John Wiley & Sons, Chichester, 524 pp.

Pred, A.,(ed.) (1981): "Space and time in geography: Essays dedicated to Torsten Hagestrand". Gleerup, Lund

Karlqvist A. Lundqvist L. and Snickars F. (eds) (1975): "Dynamic Allocation of Urban Space". Saxon House, Westmead, 280 pp

Stewart I. and Cohen J. (1999): "Figments of Reality: The Evolution of the Curious Mind", Cambridge University Press, 340 pp

Telecom New Zealand 2005: "Mapping of national and local coverage areas for mobile phone and data communications". July 2005. Telecom New Zealand, Wellington, 4pp

Thrift N. and French S. (2002): "The automatic production of space", Transactions of the Institute of British Geographers, Volume 27, Number 3, 309–335(27)

9 Where do you want to go today [in attribute space]?

André Skupin
Department of Geography, San Diego State University, San Diego, CA

9.1 Introduction

As we move across geographic space, aren't we simultaneously moving through a high-dimensional attribute space in which the geographic entities are located that we encounter along the way? Building on this premise, imagine that one has captured several paths that either traverse the same region or are even contained in wholly disjoint geographic regions. If these paths are now passing through areas that exhibit similar attributes, they are coming close to each other in *n*-dimensional space. This fact, as well as such characteristics as *n*-dimensional direction, looping, and so forth may aid our understanding of the paths and of the entities traveling on them, depending on how meaningful the set of measured attributes is.

The human mind can not easily cognize such high-dimensional movement without first implementing means for dimensionality reduction. The key to achieving this is to conceive of a combination of previously separate technologies. The convergence proposed in this paper involves location determination (e.g., through GPS) and spatialization. The latter is here understood as the process of transforming high-dimensional data into a visual form via low-dimensional geometry (Skupin and Buttenfield, 1997), which has in the last few years been implemented for diverse data types from text documents to population census tables (Skupin, 2004; Skupin and Hagelman, 2005).

Attribute-time paths (ATPs) may be encountered in a number of principal circumstances. One would be the case of a geographic object remaining stationary but changing its non-spatial characteristics, thus moving in attribute space. Such a conceptualization of geographic change and its application to area-based census attributes has been described elsewhere (Skupin and Hagelman, 2005). In this paper, a different situation is addressed, where individuals move across geographic space and along the way encounter geographic objects with different characteristics. Moving from one object to the next thus becomes movement through attribute space.

H.J. Miller (ed.), Societies and Cities in the Age of Instant Access, 133–149.

The idea of an ATP derived from the movement of individuals across geographic space constitutes a natural extension of space-time paths (STPs). These have been most prominently associated with the work of Torsten Hägerstrand (Hägerstrand, 1970; Pred, 1977). The last decade has seen a resurgence of work in the general area of time geography, notably under the influence of improved methods for position measurement and geospatial databases and modeling. These efforts have included the further development of major concepts, such as space-time prisms (Miller, 1991). Early investigations of time geography tended to rely on origin-destination pairs (e.g., places of work and residence), but the capture and analysis of individual space-time paths with high geometric details has now become common place (Kwan, 2002; Mountain and Raper, 2001).

The approach put forward here addresses two issues with existing methods for analysis of space-time paths. One problem is that the existing visualization approaches are too restrictive with regard to geographic space providing x-y coordinates in the display space. This is the case both for 2-D maps of space-time paths and for 3-D space-time cubes, where the third dimension conveys the progression of time. Both approaches are useful for data sets containing paths from a single geographic study area. However, they do not accommodate visualization of paths traversing multiple, disconnected geographic areas, such as different metropolitan areas.

Another issue relates to attempts to categorize geographic objects based on attribute similarity. Particularly prominent have been examples using socio-economic data, such as ESRI's *Community Tapestry* or similar efforts in Great Britain (Rees et al., 2002; Webber and Longley, 2003). The resulting categories can be mapped onto geographic space and geographic patterns inspected. However, classes are typically treated as discrete *n*-dimensional locales, without much ability to *see* transitory, field-like variation, nor nuanced relationships among categories that in reality exist in *n*-dimensional space.

The approach put forward here addresses both of those concerns. First, by not considering explicit geographic coordinates in the display of space-time paths it becomes possible to compare paths from different geographic areas. Second, major topological structures existing in the *n*-dimensional input space are preserved in the two-dimensional display space. Therefore, one can not only distinguish regions in the *n*-dimensional space, but develop some understanding of their relative location.

9.2 Integrating spatialization with geospatial data and technology

Central to the notion of spatialized attribute-time paths is the idea that in order to learn more about certain geographic phenomena we may have to [temporarily] ignore traditional geographic coordinates during visualization. Some existing methods, like parallel coordinate plots (PCPs), and scatter plot matrices, follow a similar approach as they transform the geometric representation of geographic objects on the basis of their descriptive attributes. However, those methods tend to also lose some fundamental affordances of traditional geographic geometry, most notably the ability to support observation of complex geographic distributions and relationships in holistic two-dimensional or three-dimensional form. Axes in individual scatter plots correspond to single variables, which makes it difficult to *see* complex high-dimensional relationships, even when larger scatter plot matrices are constructed. In a PCP, multiple axes are placed in parallel, each associated with a particular variable. Through a side-by-side display of these axes and connections made between attribute values for each object, the PCP transforms each geographic object into a line consisting of multiple segments, with strict norms on line geometry (e.g., no looping).

These information visualization methods also tend to be highly interactive, and this interactivity often includes manipulation of geometric and topological relations, far beyond what would be considered permissible when dealing with geographic coordinate space. As a result, users of such methods are dealing with an ever-changing display, where little is fixed and few aspects can be taken for granted.

Compare this to the notion of the base map in traditional cartography, where the relationship is established and fixed between the world coordinate system (e.g., in latitude and longitude) and the projected or display coordinate system. A fixed geographic feature located at a certain latitude and longitude will occupy the same map position today as it did yesterday and as it will tomorrow. The known distance between two fixed positions can likewise be trusted to remain such for some time. Having established geographic location as a relatively fixed aspect of a geographic map, we are then free to utilize the remaining set of visual variables for most mapping tasks. If an object indeed changes its position in the visualization, then this is detected against the back-drop of an otherwise stable base map and thus interpreted as movement of the object across geographic space.

The method proposed here attempts to extend the power of maps in enabling humans to detect complex relationships towards n-dimensional attribute space. Instead of using a very limited number of variables, as in scatter plots, or of axes with distinct, predefined meaning, as with PCPs, we propose to create map-like information visualizations (Skupin, 2002b), in which output dimensions are based on transforming a large number of input variables into a low-dimensional output space. Once a geographic object is represented with zero-dimensional, point coordinates in a low-dimensional space, that location remains fixed unless the object's attributes have changed. Thus, a stable base map is created on top of which other features can be displayed. A number of methods could be used to perform dimensionality transformation/spatial layout, including multi-dimensional scaling (MDS), spring layout, and the self-organizing map (SOM). In the experiments presented in this paper SOMs are used due to their scalability to very large data sets.

Meanwhile, space-time paths (STPs) are geographically overlaid with the same geographic objects from which the point spatialization was derived (see Figure 9.1). Each STP is conceptualized as moving from object to object,

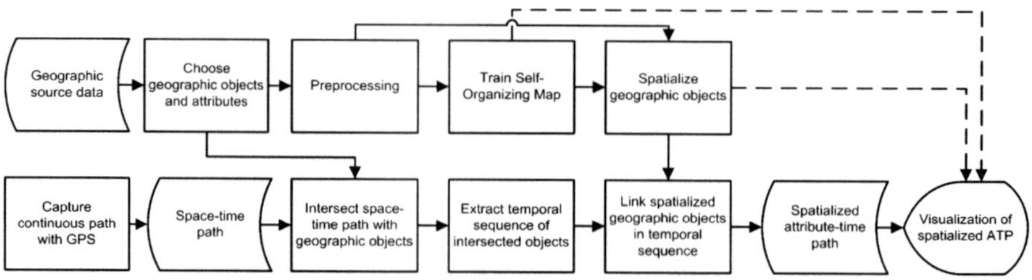

Fig. 9.1 Creation of Spatialized Attribute-Time Path (SATP) through combination of GPS and SOM

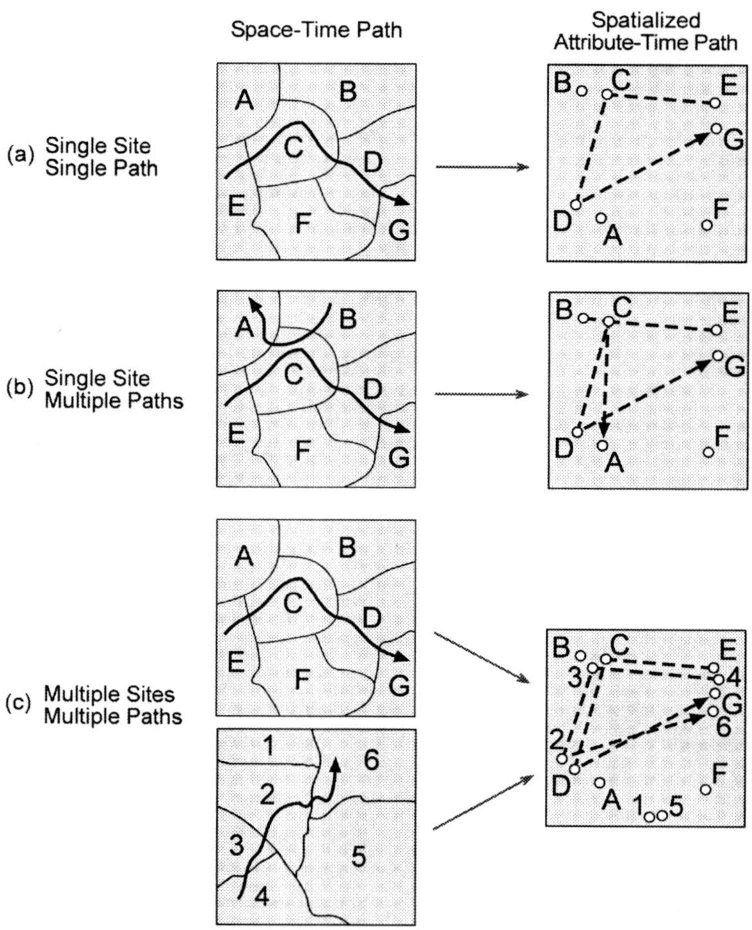

Fig. 9.2 Different application modes for the SATP concept

forming a topological sequence of geographic objects. When dealing with polygon objects, STP vertices are matched with polygons through point-in-polygon overlay. This is the case for all experiments discussed in this paper. Point or line objects will require certain distance-based transformations before becoming associated with defined STP portions. With geographic objects becoming nodes in a high-dimensional, directed graph (i.e., an ATP) one can finally derive the low-dimensional position of each node from the point spatialization in order to create the SATP (Figure 9.1).

One can imagine several principal modes for using SATPs (Figure 9.2). First, one can construct a point spatialization from geographic objects of a single study area and observe the transformation of an individual STP as it becomes represented in the spatialization. A geographically straight path may thus actually represent a circuitous route that eventually returns to its origin in attribute space (Figure 9.2a). For an application example, consider spatializing geographic features in a city and then tracing the path of a criminal suspect to see whether the path leads toward places that are measurably similar to the location of crimes for which that person was previously convicted, possibly indicating an impending infraction.

Second, multiple STPs could be mapped onto a spatialization for an individual study area (Figure 9.2b). For example, the known paths of multiple suspects may be mapped simultaneously to find out whether any of them have patterns of movement through suspiciously similar locations compared to the locations of previously unsolved crimes.

Third, one could capture STPs within multiple, disjoint sites and spatialize them using a base map containing objects from all sites (Figure 9.2c). This may prove useful in uncovering broader, generalizable patterns, for example regarding the movement of criminals in multiple cities. Notice in Figure 9.2c how two geographically completely disjoint STPs turn out to be extremely similar SATPs.

9.3 Traveling in attribute space: Applications and examples

This section presents a number of examples for SATPs, each intended to convey one particular application mode, as introduced in the previous section. Each example is accompanied by a discussion of the data sources, differences in processing, and other choices made during analysis. In each of these examples the n-dimensional source data consist of population census attributes attached to polygon objects.

Two different data sets are used. The first study is based on population data for Vienna, Austria combined with a GPS transect of that city. The remaining examples are all derived from population data for the United States, at the block group level, from which a detailed base map spatialization is derived. Various GPS tracks are transformed into SATPs and presented in different scenarios.

9.3.1
Single path across a single geographic area

The first example is included here because its scope, source data, and the geographic structures encountered lend themselves well to demonstrating some important features of the SATP concept. It is an example for a single path traversing a single study area (Figure 9.2a). The source data for constructing the base map consist of 1353 polygons with 158 associated population attributes for the Austrian capital, Vienna, and immediately surrounding areas. These attributes fall into two broad categories:

(1) population attributes (age structure, sex, household size, citizenship, educational attainment, income)
(2) buildings and land use (building purpose, types of heating and bathroom fixtures, land use types, size of places of employment)

A path was captured by GPS while navigating public roads starting in the *Donaustadt* section, which lies at the north-east periphery of the city (Figure 9.3). After

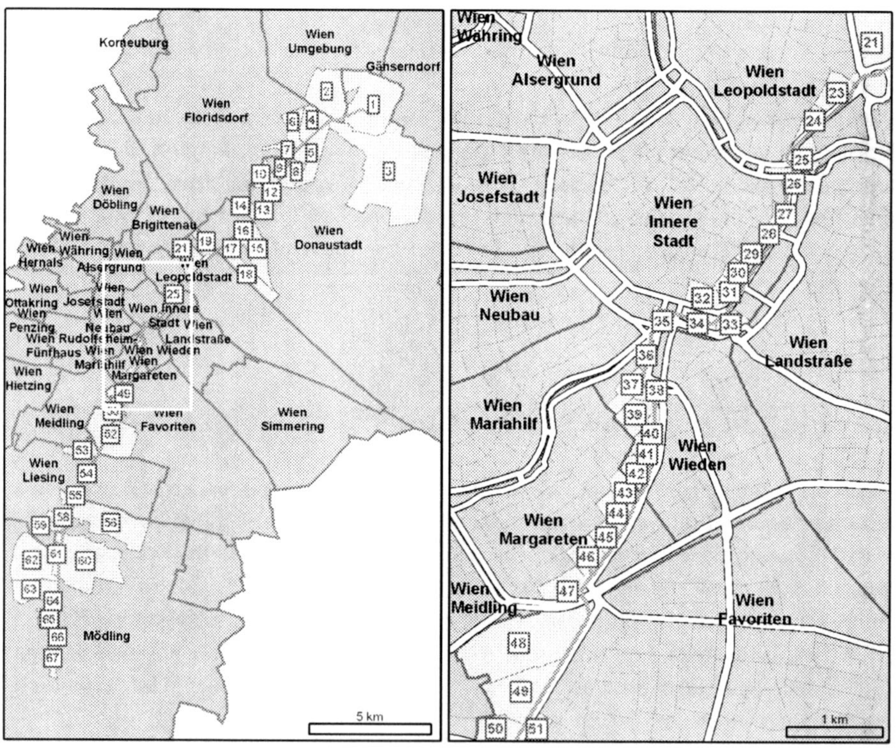

Fig. 9.3 A drive through Vienna, Austria. Polygon objects numbered in order of traversal

passing through the city center (*Innere Stadt*), the geographic path continues in a south-west direction leading to *Mödling*, outside of Vienna proper. The path crosses 67 different polygon objects.

Attributes for the complete set of 1353 polygons are used to train a self-organizing map consisting of 5625 neurons (75 × 75). Then, instead of visualizing the neural network model itself, the closest matching neuron vector is found for each input vector and two-dimensional coordinates are accordingly derived (see point symbols in Figure 9.4) by distributing each input object randomly near the best-matching neuron (Skupin, 2002a). Finally, the sequence of 67 geographically traversed polygons is retraced in the spatialization (see node and line symbols in Figure 9.4).

The trajectory roughly describes a horseshoe shape with the ends located in the lower right and left corners. This shape is traversed twice, from the lower right to the lower left and then proceeding to revert back to the origin. How should this be interpreted? In brief, the SATP indicates the existence of ring-like structures in geographic space that are transected by the geographic path. More detailed analysis of the SATP supports this. The lower right tip of the horseshoe corresponds to the outermost geographic areas in *Donaustadt* and *Mödling*. In other words, arriving in *Mödling* in some sense means returning to *Donaustadt*! Meanwhile, the lower left tip of the horseshoe is geographically located in the center of the city, aptly named *Innere Stadt*. Once there, while continuing to move towards the south-west,

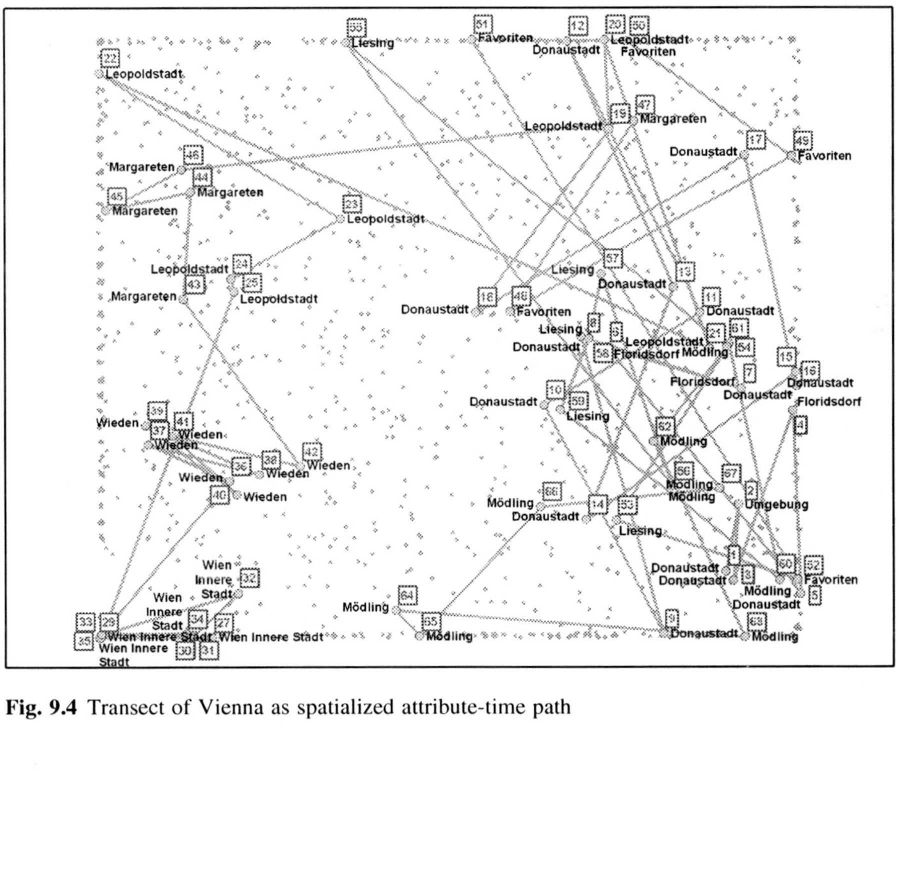

Fig. 9.4 Transect of Vienna as spatialized attribute-time path

in attribute space we are actually beginning the return journey to *Donaustadt* via all the other regions previously traversed.

This mirroring effect can be observed even at finer levels *within* regions, as apparent for the *Leopoldstadt* and *Margareten*. Notice that the outermost polygons in these regions (19 and 47) are neighbors in the spatialization (Figure 9.4). Geographic movement toward *Leopoldstadt* and away from *Margareten* corresponds to parallel paths, as indicated by the pairing of the previous/next polygons in *Donaustadt* and *Favoriten* (18/48 and 17/49). Within *Leopoldstadt* and *Margareten* one moves towards *Innere Stadt* (polygon sequence 22–23–24–25 in *Leopoldstadt*) or away from it (43–44–45–46 in *Margareten*). Discontinuities in the ring-like structure are also occurring. Notice how *Wieden* is what separates *Margareten* from *Innere Stadt* both in geographic and attribute space, while no such separation exists in geographic space between *Leopoldstadt* and *Innere Stadt*.

In further exploring the specific relationship between STP and SATP the next step would be to look into the distribution of attributes across the spatialization. In the interest of brevity this is not done for the Vienna example, but examples are contained in the following sections.

9.3.2 Multiple paths across a single geographic area

The previous section described a spatialization based on a relatively small number of geographic objects that cover a limited geographic area. The remaining examples in this paper are all based on a significantly larger data set consisting of all 200,000+ U.S. census block groups and 31 attributes (Table 9.1).

Like in the previous example, the SOM method is used to create a neural network model, which in this case consists of 250,000 neurons (500×500). Since this model is fairly intricate and constitutes the base map for all further examples, it is useful to first explore the model itself. Visual exploration of a self-organizing map is best begun with a side-by-side view of various component planes. In a SOM, every neuron is associated with an n-dimensional vector of the same dimensionality as the input vector (here: $n = 31$). With a two-dimensional lattice of neurons, one can conceptualize each variable as a field that is sampled at the neuron locations. Component plane visualization depicts each of these fields individually (Figure 9.5).

This side-by-side comparison of component planes allows a first glimpse at relationships between variables, for example based on correlated patterns of local maxima and minima. This can include simultaneously elevated values, as in the case of the variables *female households with children* (i.e., single mothers) and *black* population percentage. Negative relationships are also visible, as when the peak of persons *over the age of 64* is matched with a low percentage of *households of married persons with children*. An important advantage of self-organizing maps is that they allow observing local variations, as opposed to remaining at the level of global correlations. For example, while the *age over 64* and *married household*

Table 9.1 Variables for 200, 000+ U.S. census block groups used as input to SOM training

	Variable	Normalized by
1	Population size	Area
2	White population	Population size
3	Black	Population size
4	American Indian / Eskimo	Population size
5	Asian	Population size
6	Hawaiian / Pacific Islander	Population size
7	Other	Population size
8	Multi-race	Population size
9	Hispanic	Population size
10	Males	Population size
11	Females	Population size
12	Age < 5	Population size
13	Age 5–17	Population size
14	Age 18–21	Population size
15	Age 22–29	Population size
16	Age 30–39	Population size
17	Age 40–49	Population size
18	Age 50–64	Population size
19	Age $>= 65$	Population size
20	Median age	n/a
21	Average household size	n/a
22	Households w 1 male	Households
23	Households w 1 female	Households
24	Households married w/ children	Households
25	Households married w/o children	Households
26	Male head of household w/ children	Households
27	Female head of household w/ children	Households
28	Average family size	n/a
29	Vacant housing units	Housing units
30	Owner-occupied housing unit	Housing units
31	Renter-occupied housing unit	Housing units

with children variables show plenty of negative correlation, there is a region along the right edge of the SOM, where both have low values, together with a high percentage of persons *age 22 to 29* and *households consisting of single males*.

In the remainder of the paper, this highly detailed self-organizing map is used to create various SATPs. First, the case of multiple paths traversing a single study area is demonstrated. The study area consists of the city of New Orleans, Louisiana. The choice of paths in this example is informed by the argument that differences in income may result in a different experience of geographic space. Much of this may be due to different modes of transportation being available, i.e., private vehicle versus public transport. Accordingly, two paths were captured using GPS. The author's previous residence in the *Mid-City* section of New Orleans served as the

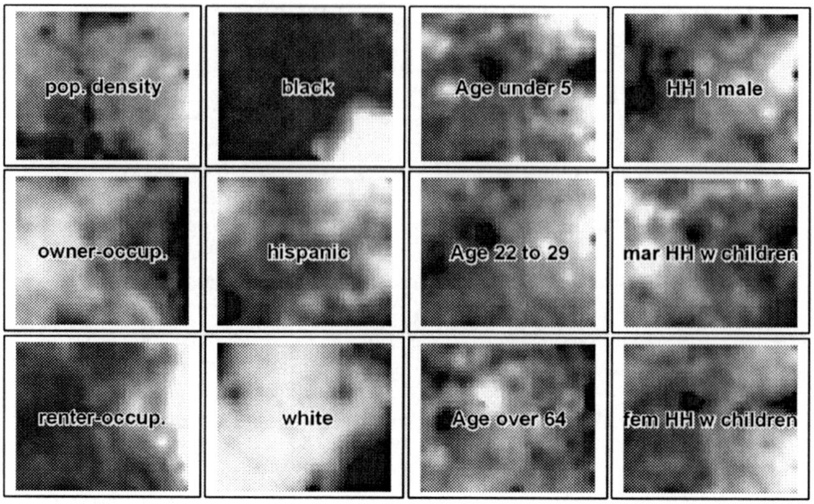

Fig. 9.5 Some component planes of a self-organizing map trained with all 200, 000+ U.S. census blockgroups. Lighter shading indicates higher values

origin and the place of employment at the University of New Orleans was the destination for both paths. *Mid-City* is a racially and economically very diverse area, so the assumption of income-driven alternative modes of transport to reach the city's main public university is quite realistic. The first path was captured while traveling on buses of the Regional Transit Authority (RTA), while the second path involved taking a private vehicle on a typical commuting journey using the quickest path (Figure 9.6a). The sequence of traversed block groups is the basis for showing the two SATPs on the SOM (Figures 9.7 and 9.8). Block groups are again numbered in sequence (Figure 9.8), but a new number is assigned every time a block group boundary is crossed, so that multiple entries into the same block group (i.e., a crisscross path) lead to multiple numbers for that block group, for example (e.g., labels 14, 16, 18 along the public transport path).

Apart from the common start and end point, the two paths mostly cross very different portions of attribute space. Initially, as they make their way out of *Mid-City*, both paths move upwards in the SOM, but then the RTA path moves back towards the origin (notice proximity of vertices 1 and 11) and continues towards the extreme bottom right of the SOM. This is in fact an extreme region, even at the national scale, with the census data indicating 100% of the population being black. As the central business district (CBD) region is entered, the RTA path suddenly bridges a gigantic distance in attribute space, arriving close to the *French Quarter* in both attribute and geographic space. As the RTA path leaves the *French Quarter* and enters the *Faubourg-Marigny*, it temporarily meets up with the path taken by private vehicle, while the latter traverses the *Bayou St. John* area. From there, the two paths diverge again, as the RTA path moves toward a region that is predominantly black and with high percentage of rental property

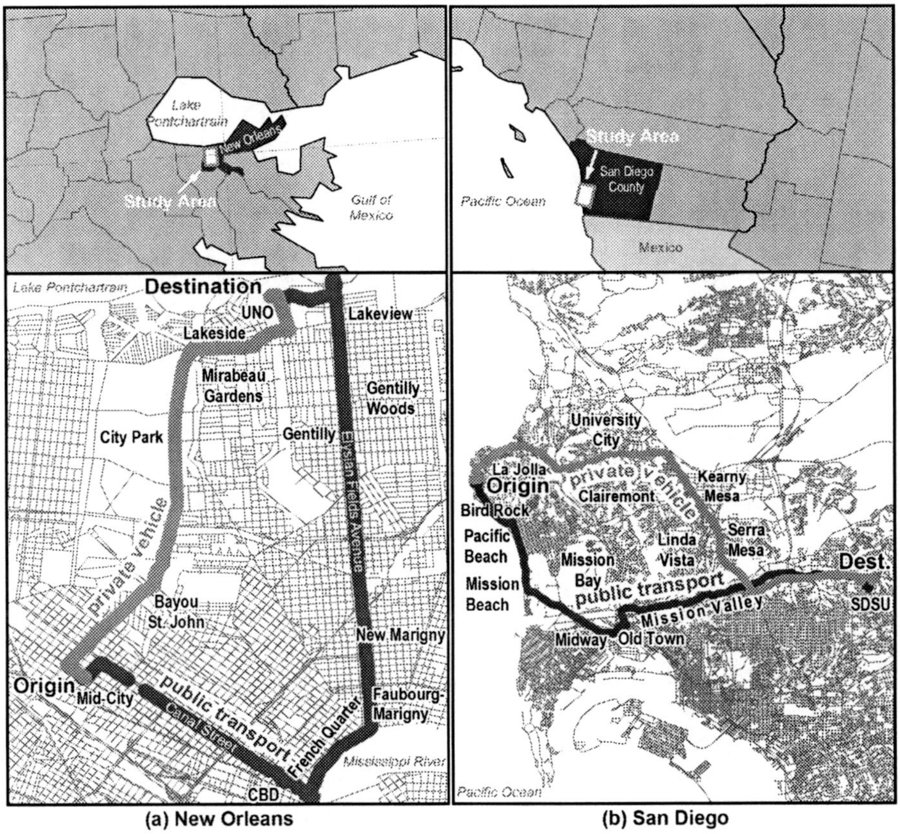

Fig. 9.6 Routes through two different study areas - geographic overview

(*New Marigny*) and the private vehicle travels through areas with larger percentage of white population and owner-occupied housing (across from *City Park* and in *Mirabeau Gardens*) (Figure 9.7).

Overall, the two SATPs traverse distinct *neighborhoods* that are compact in both geographic and attribute space, reflecting the common history of block groups in a neighborhood. Traveling between neighborhoods can involve bridging tremendous gaps in n-dimensional space, either directly (e.g., entering the CBD) or via bridge vertices (e.g., node 23).

9.3.3 Comparing paths within multiple geographic areas

One of the driving motivations behind the development of the SATP concept was the desire to compare space-time paths captured in disjoint geographic areas.

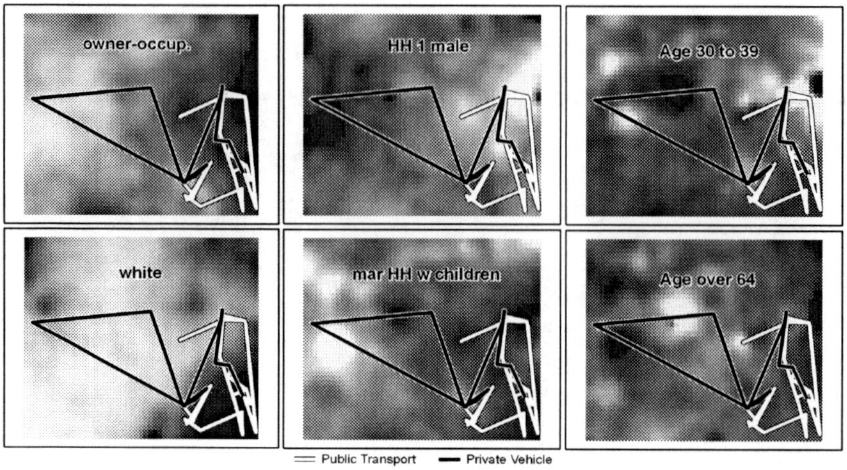

Fig. 9.7 Single site traversed by multiple paths. Example of commuting in New Orleans

This is demonstrated in the final example in this paper. The commuting paths captured in New Orleans are here combined with those captured in San Diego, under the same assumption of having either private or public transport options at one's disposal (Figure 9.6). Origin of San Diego commuting paths is the author's residence in La Jolla and destination is San Diego State University.

Comparison of paths from different cities may first simply involve observing the degree of diversity of locations along the path according to spatial patterns of links and vertices in the spatialization (Figure 9.9). In the case of New Orleans, one observes traversal of named neighborhoods as distinct regions, with clear separation

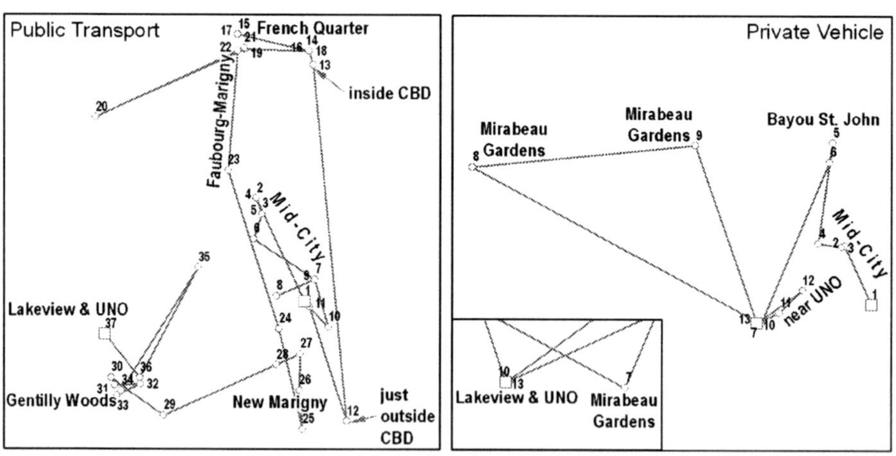

Fig. 9.8 Commuting in New Orleans – overview of traversed neighborhoods in SOM space

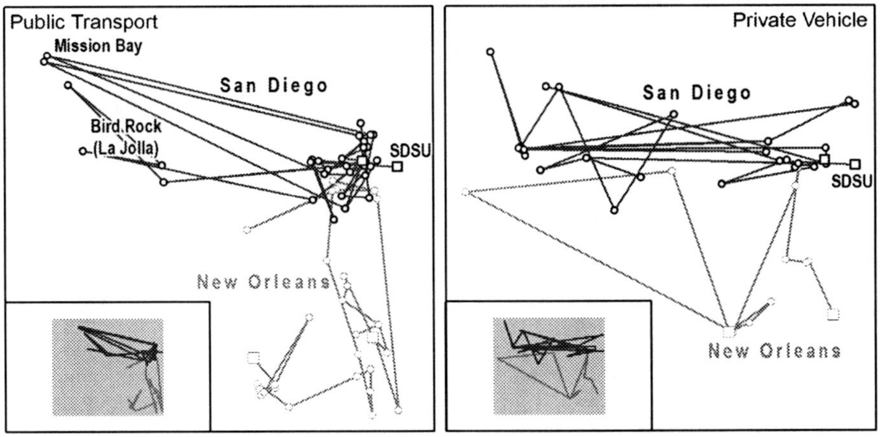

Fig. 9.9 Multiple sites traversed by multiple paths – comparison of commuting in New Orleans and San Diego

along an SATP (see also Figure 9.8). Compared to this, neighborhoods along San Diego paths appear less well organized and distinct. The private vehicle path shows plenty of crisscrossing between sometimes distant vertices (right portion of Figure 9.9). Meanwhile, the vast majority of vertices along the public transport path are closely clustered together (left portion of Figure 9.9), indicating relative uniformity in attribute space, when compared to New Orleans. There are basically only two major breaks from that uniformity. One corresponds to the *Bird Rock* community on the southern edge of La Jolla. The other stems from the path barely entering two block groups along *Mission Bay*.

A major advantage of the SATP approach is that the common base map allows overlays of various paths. When this is done for the public transport paths, one observes that the San Diego and New Orleans paths have fairly little in common (left portion of Figure 9.9). For the most part, they do not enter the same regions in attribute space. One major exception is seen where the upper portion of the New Orleans path intersects with the lower portion of the San Diego path. In other words, when traveling by public transport from La Jolla to SDSU, the closest parts of the New Orleans public transport path that one will encounter are the *French Quarter* and *Faubourg-Marigny* (see public transport portions in Figures 9.8 and 9.9).

In further analyzing the specific causes for relative similarity/dissimilarity one may now want to examine which variables are most related to the observed patterns. There are two principal choices in doing this: (a) exploring the distribution of variables within the finished computational model; or (b) exploring the distribution of original input values across the set of transected objects. The former is shown in Figure 9.7, where the New Orleans paths are combined with SOM component planes. The other approach is illustrated in Figures 9.10 and 9.11, where multivariate point symbols represent a total of 21 variables in four logical groupings.

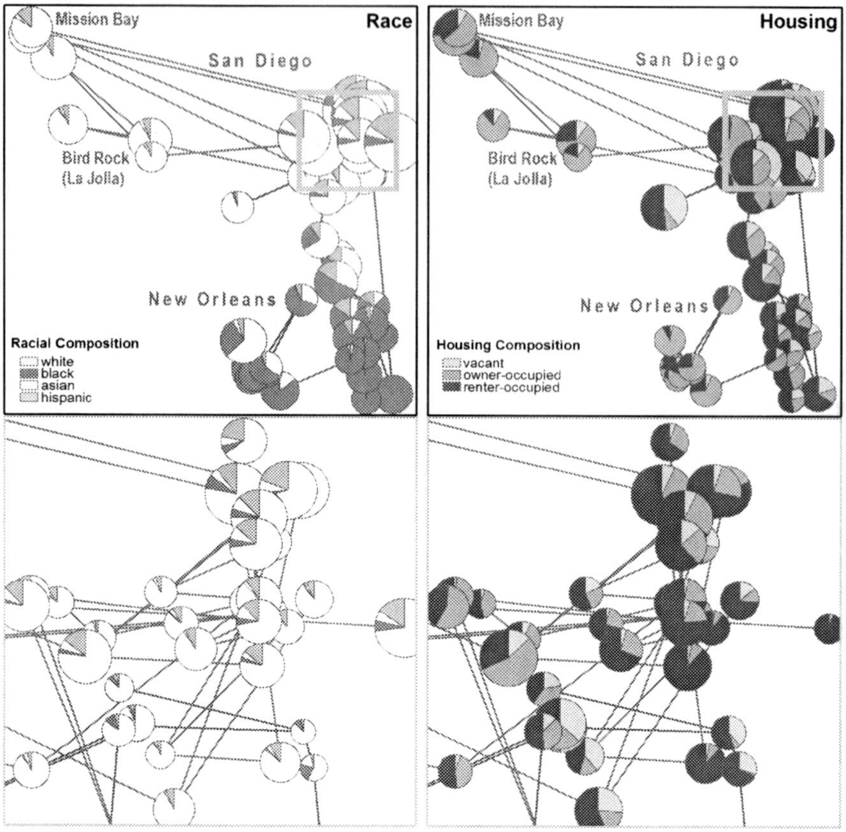

Fig. 9.10 Commuting via Public Transport in New Orleans and San Diego – Race and Housing Variables. Overlap Zone Shown in Detail

The public transport paths for New Orleans and San Diego are visualized side-by-side, with inset maps showing more detail for the dense cluster of block groups that includes most of the San Diego path and the *French Quarter/Faubourg-Marigny* portion of the New Orleans path.

One can now observe the attributes of the *Bird Rock* and *Mission Bay* portions of the San Diego public transport path, which set these block groups apart from the rest of the San Diego and New Orleans paths alike, most notably the large proportion of *white population* and *owner-occupied housing*. Finally, the overlap area between the public transport paths in the two cities can be investigated. Here one will find the most striking similarity not in the racial and housing variables, but in the age and household variables. Specifically, the overlap area consists of block groups with a large proportion of persons *between 22 and 39 years* old and a high percentage of *single male households* and, to a lesser extent, *single female*

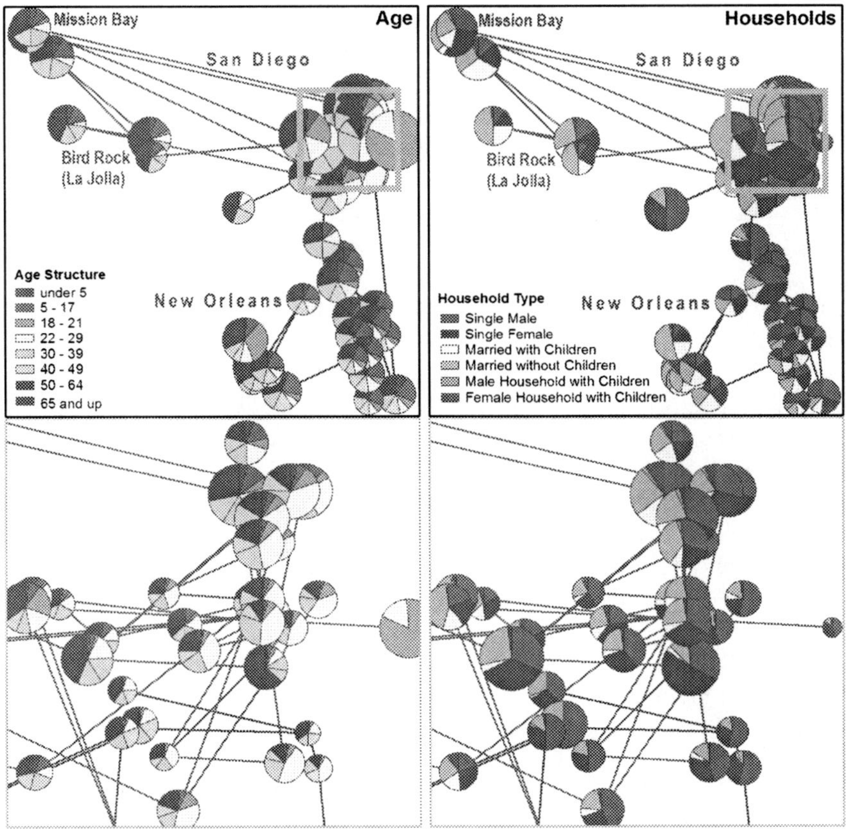

Fig. 9.11 Commuting via Public Transport in New Orleans and San Diego - Age and Household Variables. Overlap Zone Shown in Detail

households. Those then are the variables pulling the *French Quarter/Faubourg-Marigny* and a portion of the *Mission Valley* towards each other.

9.4 Conclusions

This paper introduces a conceptualization of geographic travel as a series of discrete geographic objects being encountered in n-dimensional attribute space. Path continuity is here solely derived from objects' topological relationships in geographic coordinate space. Alternatively, a geographic path could be conceptualized as simultaneously traversing n continuous fields of attributes of varying smoothness, like elevation, temperature, or land cover. This will lead to an SATP with higher geometric detail and more differentiation among paths, similar to the effects previously observed after insertion of temporally interpolated vertices into multi-temporal attribute trajectories of spatially fixed objects (Skupin and Hagelman, 2005).

When dealing with census data, one will frequently find that the street segments on which people travel coincide with area boundaries of census enumeration units. Combined with the peculiarities of a particular data capture technique (here: GPS from within a moving vehicle), this can lead to a somewhat erratic crisscrossing of boundaries, which becomes accordingly represented in the spatialization. To address this, one may want to recognize such razor's edge travel through incorporation of attributes from both of the adjacent objects and map out the SATP accordingly.

One of the ideas informing this research is that people's trajectories through geographic space exist in a complex relationship with their personal histories, attitudes, and perspectives. A number of recent research efforts are founded on this notion, included co-called feminist visualization (Kwan, 2000; Kwan, 2002). This paper asserts that additional insight could be gained through visualization of geographic paths derived from their location in *n*-dimensional attribute space. Defining a spatialization methodology is a necessary first step that this paper focuses on. In the short term, some questions have to be answered regarding the specific technical solutions proposed here. This includes algorithmic choices among spatial layout techniques (here: SOM) and respective parameters (such as the model's granularity). Other questions relate to cognitive plausibility. For example, there may be an inherent conflict between the spatial continuity of an SATP and the apparent crossing of potentially diverse regions in a spatialization. What one observes here is more akin to wormhole jumping or commercial air travel than to the continuity encountered during road-based travel.

The work described here is driven by an aspiration to develop explicit visual manifestations of *n*-dimensional patterns and structures within and among the paths of people though geographic space. Use of the easily measured attributes of geographic space, as captured during a population census or through satellite remote sensing, is a natural first step. However, a long-term goal is to more directly incorporate visual impressions that are experienced by and are influencing people along their path, most notably along roadways (Appleyard et al., 1964). This ultimately connects with the need to develop visualizations supporting better understanding of the relationship that people have with geographic reality, in particular their *sense of place* (Relph, 1976). To that end, impressions based on concrete, ground-level sensory input, including sight and sound, will have to be incorporated as the SATP notion is developed further.

Acknowledgements

The assistance of Josef Benedikt in capturing the GPS track through Vienna, Austria, is gratefully acknowledged. Austrian geographic base data and census data were published by ArcData Österreich and made available through TU Wien, Institut für Soziologie in der Raumplanung und Architektur.

References

Appleyard, D., Lynch, K. & Myer, J. R. (1964) *The View from the Road*, Cambridge, MA, M.I.T. Press.

Hägerstrand, T. (1970) What About People in Regional Science? *Papers of the Regional Science Association*, 24.

Kwan, M.-P. (2000) Gender Differences in Space-Time Constraints. *Area*, 32, 145–156.

Kwan, M.-P. (2002) Feminist Visualization: Re-envisioning GIS as a Method in Feminist Geographic Research. *Annals of the Association of American Geographers*, 92, 645–661.

Miller, H. (1991) Modelling Accessibility Using Space-Time Prism Concepts within Geographic Information Systems. *International Journal of Geographical Information Systems*, 5, 287–303.

Mountain, D. & Raper, J. (2001) Modelling Human Spatio-Temporal Behaviour: A Challenge for Location-Based Services. *GeoComputation 2001*. Brisbane, Australia.

Pred, A. (1977) The Choreography of Existence: Comments on Hägerstrand's Time Geography and Its Usefulness. *Economic Geography*, 53, 207–221.

Rees, P., Denham, C., Charlton, J., Openshaw, S., Blake, M. & See, L. (2002) ONS Classifications and GB Profiles: Census Typologies for Researchers. In Rees, P., Martin, D. & Williamson, P. (Eds.) *The Census Data System*. Chichester, Wiley.

Relph, E. (1976) *Place and Placelessness*, London, Pion Limited.

Skupin, A. (2002a) A Cartographic Approach to Visualizing Conference Abstracts. *IEEE Computer Graphics and Applications*, 22, 50–58.

Skupin, A. (2002b) On Geometry and Transformation in Map-like Information Visualization. In Börner, K. & Chen, C. (Eds.) *Visual Interfaces to Digital Libraries (Lecture Notes in Computer Science 2539)*. Berlin, Germany, Springer-Verlag.

Skupin, A. (2004) The World of Geography: Visualizing a knowledge domain with cartographic means. *Proceedings of the National Academy of Sciences*, 101, 5274–5278.

Skupin, A. & Buttenfield, B. P. (1997) Spatial Metaphors for Visualizing Information Spaces. *ACSM/ASPRS Annual Convention and Exhibition*. Seattle, WA, Bethesda: ASPRS/ACSM.

Skupin, A. & Hagelman, R. (2005) Visualizing Demographic Trajectories with Self-Organizing Maps. *GeoInformatica*, 9, 159–179.

Webber, R. & Longley, P. (2003) Geodemographic analysis of similarity and proximity: their roles in understanding the geography of need. In Longley, P. & Batty, M. (Eds.) *Advanced Spatial Analysis*. Redlands, CA, ESRI Press.

Transportation

10 Reexamining ICT impact on travel using the 2001 NHTS data for baltimore metropolitan area

Feng Zhang[1], Kelly J. Clifton[2], Qing Shen[1]
[1] *School of Architecture, Planning and Preservation, University of Maryland, College Park, MD*
[2] *National Center for Smart Growth, University of Maryland, College Park, MD*

10.1 Introduction

Information and communications technology (ICT) creates rich opportunities for individuals to reduce economic and social costs associated with their spatial interaction with others. The potential effects on travel behavior, however small individually, can aggregate substantially because ICT is rapidly permeating all aspects of our daily lives (Golob and Regan, 2001). Given that ICT usage necessarily involves reallocation of time and resources, it is conceivable that public policy could play an important role in facilitating desirable changes in travel outcomes by influencing the spatial and temporal reorganization of individuals' activities (Shen, 2004).

The last decade has witnessed a growing number of empirical studies on the relationship between ICT and travel (see, for example, Handy and Yantis, 1997; Johansson, 1999; Mokhtarian and Meenakshisundaram, 1999; Harvey and Taylor, 2000; Hjorthol, 2002; Farag et al., 2003; Fadare and Salami, 2004; Nobis and Lenz, 2004; Srinivasan and Athuru, 2004). These research efforts have undoubtedly broadened and deepened our understanding of the multidimensional, multidirectional, and complex ICT-travel connection. However, due to data and methodological limitations, such as small sample size, inadequate measurement of ICT usage, and lack of experimental design, there are many gaps in the existing knowledge about the actual and potential impacts of ICT on travel. The knowledge base must be strengthened in order to effectively support transportation planning and policy making for achieving the full benefits of ICT.

This empirical study aims to gain additional insights into the relationship between ICT and travel using the 2001 National Household Travel Survey (NHTS) data, which contain a large sample of households and individual members and

H.J. Miller (ed.), Societies and Cities in the Age of Instant Access, 153–166.

serves as an important data source for travel behavior research. The specific research objective is to examine the impact of several indicators of ICT on three measures of travel. The ICT indicators include the frequency of Internet use, the number of mobile phones, and the presence of a telephone at home for business purposes. The travel outcomes examined are vehicle miles traveled (VMT), total daily trips, and daily walking trips. These ICT indicators and travel outcomes are all measured at the individual level. We ask two basic research questions. First, what are the individual effects of the ICT indicators on the travel outcomes? And second, what is the overall relationship between ICT and travel as seen in the empirical data?

The findings contribute to the growing body of work in the area of ICT and travel behavior by demonstrating the differential effect that the various indicators of ICT have on the range of travel outcomes examined. Furthermore, because the NHTS is expected to be a continual data collection effort, studies such as ours can help identify desirable modifications to the existing questionnaire and, subsequently, improvements for future surveys. As Golob and Regan (2001) point out, household activity and travel surveys should and can be improved to collect richer data that are essential for better understanding the profound changes induced by ICT.

10.2 Literature review

Scholars (de Sola Pool, 1979; Salomon, 1986 and Mokhtarian, 2002) have theorized that the interaction between telecommunication and travel can take several different forms. These include substitution (alternatives replacing each other), complementarity (use of one generating additional use of the other), modification (use of one changing the way in which the other is used), and neutrality (use of one having no effect on the use of the other). This general framework has provided a useful conceptual basis for empirical investigations.

Mokhtarian and Salomon (2002) reviewed a considerable number of empirical studies on the ICT-travel relationship. They concluded that, although short-term studies that focus on a single activity (such as telecommuting) have often found substitution effects of ICT, in more comprehensive, long-term empirical studies, complementarity effects are often observed. The reader is cautioned, however, that the question remains as to how much the simultaneous increases observed in telecommunications and travel reflect true causal complementarity, and how much they are due to spurious third-party correlation with other variables (Mokhtarian, 2002).

An example of the studies that found a positive association between ICT and travel was a report by Johansson (1999). Using data from the 1997 Swedish National Communication Survey in which a sample of about 2,000 respondents completed a diary to record information on all trips and contacts, Johansson observed that the number of trips is positively correlated with the number of telecommunication contacts. However, the data analysis did not control for some obvious confounding factors, such as income. Therefore, the results told us little about causality.

Similarly, using data collected from a survey of 3,500 individuals in Germany, Nobis and Lenz (2004) showed that the heavy ICT users also tend to be heavy car users. Again, while the data analysis showed a clear pattern of positive statistical association, it did not use multivariate methods to explore possible causality. Such an effort will likely be undertaken in the future by these researchers, however, because the survey, conducted in 2003, was intended to be only the first data collection effort of a long-term panel study of ICT applications and daily activities.

Stronger evidence of complementarity was provided by Hjorthol (2002). Based on the Norwegian national personal travel survey 1997/98 and a connected mail back survey of the use of ICT at home, Hjorthol examined the relationship between transportation mobility and use of home computer. In this study, use of home computer was divided into four categories: (1) private tasks without need of an Internet connection, (2) private tasks with need of an Internet connection, (3) work tasks without need of an Internet connection, and (4) work tasks with need of an Internet connection. Linear regression models were estimated for a sample of almost 800 individuals. The models showed that—after controlling for gender, age, household income, and number of cars—using a home computer for work with or without Internet connection both have a small but statistically significant positive effect on daily distance traveled.

Understanding the complementarity effect of ICT and recognizing the substantial empirical evidence of an overall net travel generation outcome is certainly important, especially from the standpoint of forecasting short-term change in travel demand. However, it is equally important to recognize that recent empirical studies have also identified the variety of ways in which ICT generates effects on travel, revealed the dynamic nature of ICT-travel interactions, and suggested potential roles of public policy in influencing the outcomes of these interactions.

In a study of the impacts of ICT on nonwork travel, Handy and Yantis (1997) examined the relationship between in-home and out-of-home versions of the same activities. Three specific activities were selected, and the sets of potentially substitutable versions of those activities were examined: movies (theater vs. VCR vs. television), shopping (store vs. catalog vs. television), and banking (bank vs. ATM vs. phone vs. on-line). A household survey was conducted to characterize the use of the different versions of the three case study activities and explore the trade-offs between them. The results suggested that the degree to which in-home versions substitute for out-of-home versions of an activity depends on the nature of the activity—with in-home entertainment generating additional travel, whereas online maintenance activities reduced travel—and the characteristics of the individuals.

Similar insights about the complex and dynamic nature of ICT-travel interactions have been obtained by other researchers who undertook empirical research in different contexts. In a study of travel behavior implications of ICT in Puget Sound region, Viswanathan and Goulias (2001) found that Internet use was correlated negatively, whereas mobile phone use was correlated positively, with time spent on travel. In the aforementioned study by Hjorthol (2002), it was found that users of home computer tend to make somewhat less work trips but more chauffeuring and total trips after the effects of income, gender, age, and car availability have been controlled for.

Srinivasan and Athuru (2004) conducted probably the most extensive statistical analysis of interaction between Internet communication and travel activities to date. Using a series of statistical models, they investigated three dimensions of the interaction: (1) factors affecting the propensity to perform virtual activities using the Internet, (2) the relationship between activity participation and ICT use, and (3) interaction between ICT use, activity attributes, and observed travel behavior. The models were estimated based on the 2000 activity-diary data from the San Francisco Bay Area. The results indicated the presence of both substitution and generation of trips due to Internet use. Furthermore, they showed the influence of a wide range of technological and socioeconomic factors on the interactions between ICT and travel.

Several studies focused on the impact of telephone on travel. One early study was reported by Claisse and Rowe (1993), who surveyed the residential telephone use of 663 respondents in the Lyon, France metropolitan area in 1984. The respondents recorded a one-week diary of all calls made and received at home. Claisse and Rowe found that residential phone use generated trips 3–5% of the time, and replaced trips 21–27% of the time, for a net substitution of 17–22%. Contrary evidence was presented in a more recent study, however (Fadare and Salami, 2004). Focusing on the effect of telephone use on travel behavior in Nigeria, the researchers found that telephone usage tends to increase the number of trips. In particular, most business-related calls tend to induce travel rather than substitute for it.

Sociological studies of the impact of the Internet on individuals' time use and social participation also suggest great complexity in the ICT-travel relationship. Sociologists (Nie and Hillygus, 2002; Robinson et al., 2002) tend to agree that the Internet has changed people's use of time and consequently relationships with various communities (including the residential community), but they disagree on the nature and extent of the substitution effect of the technology. Therefore, impact on social interaction and travel is expected, but the direction of the impact is unclear.

Taken together, these empirical studies not only have painted a general picture of the ICT-travel relationship in which complementarity appears most prominent, but also have uncovered aspects of the dynamics and complexity of this relationship. As the outcome is not always certain and seems to be influenced by a wide range of factors, it begs the question of what potential roles can public policy play in exerting desirable influence on the outcome (Shen, 2004).

The existing knowledge is insufficient for answering the question raised above. The empirical basis needs to be expanded and strengthened, with causal-effective mechanisms more clearly understood and their connections with policy instruments established. This will require a great volume of empirical research that employs appropriate data and methodology.

10.3 Research methodology

This research is intended to gain additional insights about the impact of ICT, in the forms of Internet and conventional and mobile telephones, on individuals' travel behavior. The research design is a cross-sectional study that tests the associations of various indicators of ICT usage with three different measures of travel. Using the 2001 National Household Travel Survey (NHTS) Add On for the Baltimore metropolitan area, complemented with local land use data, three statistical models of travel are estimated at the individual level. The NHTS data were collected through a household-based, random digit–dialed telephone survey, administered under the guidance of the Baltimore Regional Transportation Board (BRTB), the Metropolitan Planning Organization (MPO) for the region. The data include information about individuals and their households, and detail information about all trips made in one day for a sample of residents in the six-county region surrounding Baltimore City. The sample of this survey consisted of 7,825 respondents from 3,519 households residing in the study area. Because respondents aged under 16 appropriately skipped the question about "the frequency of Internet usage", the sample used in this study was restricted to the adult population aged 16 years and older, resulting in a total of 5,429 individuals in the models.

The dependent or outcome variables capturing individual travel patterns include the vehicle miles traveled (VMT) in the travel day by personal vehicle, the total number of trips made in the travel day, and number of walking trips made in the travel day. For the VMT, the ordinary least squares (OLS) is employed to estimate the coefficients of the linear regression model. In this model, the continuous measure of vehicle miles traveled by the individual is regressed against several explanatory variables, including: characteristics of the individual and their household, measures of residential land use, and the policy variables of interest – measures of information and communications technology usage.

For the total number of trips and the number of walking trips made by the individual in the given travel day, Poisson regression is used for the model estimation because these two dependent variables are measured as integer counts. In these cases, the model outcome is the probability of occurrence of the number

Table 10.1 Descriptive Statistics of the Variables

	N	Min	Max	Mean	S.D.
Travel Behavior (dependent variables)					
VMT	5,429	0	3001	34.35	68.390
Number of trips	5,429	0	15	3.67	2.463
Number of walk trips in travel day	5,429	0	12	0.43	1.128
ICT usage					
Web use frequency: High	5,429	0	1	0.42	0.493
Web use frequency: Medium	5,429	0	1	0.19	0.395
Web use frequency: Low	5,429	0	1	0.11	0.307
Cell numbers per HH adult	5,429	0.00	1.00	0.54	0.420
HH has telephone exclusive for business/fax	5,429	0	1	0.16	0.371
Socioeconomic Characteristics					
Age	5,429	16	95	47.57	17.143
Age squared	5,429	256	9025	2557	1718
Sex: male	5,429	0	1	0.45	0.498
Race: white	5,429	0	1	0.77	0.422
Foreign born	5,429	0	1	0.06	0.231
Driver status	5,429	0	1	0.85	0.359
Worker status	5,429	0	1	0.66	0.473
Occupation: sale	5,429	0	1	0.16	0.368
Occupation: clerical	5,429	0	1	0.08	0.273
Occupation: manufacturing/construction/farming	5,429	0	1	0.07	0.257
Occupation: professional	5,429	0	1	0.33	0.470
Income per HH member (thousand dollars)	5,429	1	100	25.12	16.922
Vehicles per HH driver	5,429	0.00	9.00	0.97	0.516
Number of children in HH	5,429	0	8	0.59	0.990
Tenure of housing unit: owner	5,429	0	1	0.78	0.415
Land Use Characteristics					
Population density at HH location (track level)	5,429	0.05	30	7.82	8.257
Employment density at HH location (track level)	5,429	11.04	16.99	3.50	3.867
Accessibility to commercial activities (TAZ level)	5,429	0	19.77	10.63	6.683
Land-use mix entropy	5,429	0	1.00	0.551	0.318
One-way distance to work (miles)	4,910	0	210	9.26	14.200

of trips and is estimated using the same explanatory variables mentioned above. Descriptive statistics of all dependent and independent variables are illustrated in Table 10.1.

10.3.1 The three measures of ICT usage

The first consists of categorical measures of the frequency of Internet usage ("Web use frequency: High", "Web use frequency: Medium", and "Web use frequency: Low"). There are four groups of self reported Internet users - high, medium, and low frequency users, and people have no access to Internet. High frequency stands for "almost every day"; medium frequency is equivalent to "several times a week" and "once a week"; and low frequency corresponds to "once a month" or less. The first three groups have access to Internet at home and/or work. The fourth level of Internet use is "no access to Internet", which means that one has no access to Internet neither at home nor at work.

The second measure of ICT is the number of mobile phones per adult member of household, which proxies for both access to and use of the cell phone. If the value is between 0 and 1, it represents the probability of sharing a cellular phone with other household members. If the value is greater than 1, which means that an individual has more than one mobile phone, it will be assigned a value of 1 because the extra phones serve no practical purpose.

The third measure of ICT is the presence of a land line exclusive for business/fax purposes, which can to a certain extent capture the probability that one conducts business at home.

One limitation of these aggregate indicators of ICT use is that there is no information about the specific activities performed on the Internet or the purpose of accessing the web. Another limitation is that the ownership of mobile or conventional phones provides no indication of the extent or purpose of their use. Nonetheless, this information allows us to distinguish between several types of ICT technologies and estimate their separate effects on travel.

10.3.2 Socioeconomic characteristics

Age, gender, race, foreign born, driver status, worker status, and occupation are included in models as individual-level covariates. Because age often does not have a linear relation to travel outcomes, both age and its square are included to fit the curvilinear relationship. The reference groups for socioeconomic variables are female, non-white, those who were born in US, non-driver, non-worker, and people whose occupations are other than the four listed in Table 10.1. Variables that describe household characteristics are annual income per household member, vehicles per household driver, number of children in household, and tenure of housing unit.

10.3.3 Land use characteristics

To control for the influence that the urban environment may have on travel behavior (Dieleman et al., 2002), several land use variables are incorporated into the models. These include population density (persons/square mile), employment density (jobs/square mile), regional accessibility to commercial activities, and entropy of land use balance, all measured from the residential location of the individual. Population density and employment density are calculated at the census tract level. Regional accessibility is defined in terms of the accessibility to retailing activities at traffic analysis zone (TAZ) level. In this study the measure of regional accessibility is calculated based on the cumulative opportunities, measured by the number of retailing jobs, located within 30 minutes of travel from home. The land use balance entropy is computed using data on the relative proportions of land use types. In this study, residential, commercial, industrial, and institutional uses were considered. The entropy index for individual household is computed for a 800-meter radius circle around each residential location.

Finally, one-way distance to work is included in the VMT model to control for the residential and regular workplace locations, which are considered exogenous in this study. ICT increases location flexibility, which may cause households to move further away from workplace in the long term (Janelle, 1995; Shen, 2000). However, recent empirical research found little evidence that the increased location flexibility has led to changes in the residential patterns of households (Ellen and Hempstead, 2002).

10.4 Results

The specification and estimation results for the three models are illustrated in Table 10.2.

10.4.1 VMT

As has been found in previous research, VMT is greater for white than other races/ethnicities, for drivers than non-drivers. The age and square of age are both significantly associated with VMT in different directions which represent a parabola curve. Two household variables have significantly positive effect on VMT. The result can be interpreted that a person in a household which has more private vehicles and ownership of housing unit will generates more VMT.

Interestingly, after controlling for the effect of one-way distance to work, none of the land use variables has a statistically significant association with VMT even though the coefficients are all negative. It seems to show that, consistent with

Table 10.2 Estimated Models of VMT, Walking Trips, and Total Trips

	VMT		Walking Trips		Trips	
	Coefficient	t	Coefficient	z	Coefficient	z
Constant	−7.451	−1.187	**−1.530**	**−7.80**	**0.397**	**6.17**
ICT usage						
Web use frequency: High	**6.489**	**2.741**	0.107	1.55	**0.116**	**4.92**
Web use frequency: Medium	4.896	1.918	−0.031	−0.39	**0.127**	**5.06**
Web use frequency: Low	1.914	0.697	−0.101	−1.14	−0.020	−0.68
Cell numbers per HH adult	**4.740**	**2.377**	**−0.380**	**−6.64**	−0.028	−1.43
HH has telephone exclusive for business/fax	1.606	0.763	−0.081	−1.30	**−0.049**	**−2.46**
Socioeconomic Characteristics						
Age	**0.469**	**1.970**	**0.016**	**2.17**	**0.015**	**6.08**
Age squared	**−0.005**	**−2.090**	**−.0003**	**−4.15**	**−.0002**	**−6.07**
Sex: male	2.414	1.573	−0.032	−0.72	**−0.029**	**−1.94**
Race: white	**5.779**	**2.877**	**0.386**	**6.96**	**0.056**	**2.74**
Foreign born	−0.475	−0.151	0.098	1.19	−0.034	−1.08
Driver status	**9.321**	**3.662**	**−0.347**	**−5.44**	**0.306**	**10.78**
Worker status	−6.583	−1.576	−0.091	−1.06	−0.033	−1.08
Occupation: sale	2.870	0.664	**−0.287**	**−3.01**	**0.090**	**2.75**
Occupation: clerical	3.746	0.771	−0.109	−0.98	0.057	1.51
Occupation: manufacturing/construction/farming	5.857	1.213	**−0.404**	**−3.16**	0.013	0.33
Occupation: professional	1.428	0.329	0.143	1.59	**0.084**	**2.66**
Income per HH member (thousand dollars)	0.055	0.962	**0.008**	**6.01**	**0.001**	**2.57**
Vehicles per HH driver	**4.959**	**2.856**	**−0.550**	**−9.90**	0.024	1.41
Number of children in HH	1.305	1.468	0.017	0.66	**0.044**	**5.19**
Tenure of housing unit: owner	**4.900**	**2.460**	−0.070	−1.39	**0.046**	**2.28**
Land Use Characteristics						
Population density at HH location (track level)	−0.052	−0.219	0.004	0.72	−0.0002	−0.07
Employment density at HH location (track level)	−0.484	−0.999	**0.042**	**4.06**	0.006	1.13
Accessibility to commercial activities (TAZ level)	−0.063	−0.466	**0.062**	**9.68**	0.002	1.34
Land-use mix entropy	−3.510	−1.331	**0.214**	**2.55**	**0.083**	**3.19**
One-way distance to work	**0.882**	**14.443**				
Number of observations	4910		5429		5429	
R square or Pseudo R square	0.130		0.123		0.028	

previous findings, the local population or employment density, land use mixture, and regional accessibility have very weak, if any, effect on individual VMT. The one-way distance to work variable has a highly significantly positive effect on personal VMT.

Controlling for these covariates, the high frequency of Internet use is significantly and positively associated with VMT. All else equal, persons with relatively higher frequency of Internet usage will travel about 5 or 6 more miles than people with low frequency or no access to Internet. The standardized coefficient (not shown in Table 10.2) is 0.059 for high frequency. In comparison with other coefficients in the model, this one is not small, which means that the magnitude of Internet effect is not negligible. Although not conclusive, these results indicate an overall complementarity effect of Internet usage on vehicle miles traveled.

Use of mobile phones is also positively associated with VMT, and the levels of statistical and practical significance of the association are similar to those characterized the relationship between VMT and high frequency Internet use. Again, the result also suggests an overall complementarity effect—between mobile phone usage and travel.

10.4.2 Number of trips

Similar to what is found in the VMT models, the number of trips is higher for females than males, for white than other races, and for drivers than non-drivers. And people whose occupations are sale/service or professional generate more trips than others. Likewise, the curvilinear relationship between age and number of trips can be clearly demonstrated through the signs and significance of age and its square.

Three household variables – household income per household member, number of children in household, and tenure of housing unit – are positively associated with number of trips the individual in the household makes. The variable "vehicle per household driver" shows insignificant effect on number of trips. The problem of multicollinearity is suspected to function here.

Only one of the four urban environment variables, land use mix entropy, is found significantly related to number of trips. It suggests that a person with more mixed land use around his or her household tends to generate more daily trips, regardless of the mode of travel. Another finding is that the regional accessibility variable is highly correlated with land use mix measure (correlation = 0.505). If we estimate a model without the land use mix measure, the accessibility variable shows a significantly positive effect on dependent variable. A problem of multicollinearity between these two measures of urban environment may be concluded here. Generally speaking, higher land use mixture and regional accessibility tend to be positively associated with increased number of trips.

After controlling for these covariates, the high and medium frequency of Internet use are positively related to total number of trips at a highly significant.

The marginal effect of high frequency of Internet use is 0.426, suggesting that high frequency of Internet use will increase the expected number of trips generated by the individual by approximately 0.426, other things being equal. In other words, the difference between high frequency Internet users and people with no access to Internet is approximately half a trip. Likewise, medium frequency will increase expected number of trips by approximately 0.466.

This evidence indicates that frequent Internet users generates a greater number of total trips than less frequent users or people who have no access to Internet, all else equal. Coupled with the results from VMT model, it is clear that frequent Internet usage is associated with higher number of trips as well as higher VMT, which suggests a complementarity relationship between Internet use and travel.

Use of mobile phones does not show a statistically significantly relationship with number of trips made. The business telephone variable is negatively associated with the number of trips at the 0.05 level of significance. This suggests a small but significant substitution effect of business telephone use at home on travel behavior.

10.4.3
Number of walking trips

Consistent with common sense and expectation, the number of walking trips is higher for white than other races, for non-drivers than drivers. Also, people whose occupations are sale/service or manufacturing/construction/farming have less walking trips in comparison with others.

Vehicles per household member show a significantly negative effect on number of walking trips, which makes sense because usually people with more personal vehicles walk less. Interestingly, household income per household member is positively related to number of walking trips at a highly significant level. It seems to suggest that higher income people walk more.

The track level employment density at household place is significantly positively related to number of walking trips. It is consistent with findings in previous pedestrian behavior research. A higher employment density represents higher intensity of various activities in an area, and possibly attracts local residents who may reach the place through walking mostly. The regional accessibility to retailing activities is also found to be positively related to number of walking trips. In addition, the entropy index of local land use balance has significantly positive effect on number of walking trips. These evidences seem to reiterate recent research findings that desirable local urban environment characteristics will stimulate people's walking activities.

The only ICT variable statistically linked to number of walking trips is the number of cellular phones per household adult. The result suggests that an additional cellular phone per household adult will decrease the number of daily walking trips made by each adult in this household by approximately 0.205. Evidences show that persons with access to cellular phones are less likely to walk, all else equal. This may suggest a substitution effect of mobile phone usage for personal walks. However, the impact of this relationship cannot be discerned from this analysis. The three Internet usage variables don't have statistically significant effect on number of walking trips.

As Hanson and Fontaine (2005) points out, ICT could interact with the built environment and hence facilitate or constrain individuals' participation in physical activity. The above empirical finding shows that ICT, in this case mobile phone, may have an influence on people's walking activity, an important component of physical activity.

10.5 Conclusion

This study reveals that Internet and telephone usage are significantly associated with some forms of individual travel behavior. After controlling for covariates including socioeconomic characteristics of individuals, household characteristics, and urban environment, high and medium frequency of Internet use have significant association with both number of trips and VMT. These findings seem to attest to an overall complementarity impact of telecommunication on travel, which has been found in most of comprehensive empirical studies at disaggregate level.

Although extremely important, Internet is but only one component of ICT. Our research demonstrates that other measures of ICT usage cannot be overlooked. In particular, the empirical results indicate that presence of telephone for conducting business at home is associated with decreased number of total daily trips. Only by including multiple measures of ICT could we have identified the complementarity and substitution effects simultaneously.

Note that the magnitude of the complementarity effect of Internet usage observed in this study is relatively large in comparison with that of the substitution effect of business telephone. In fact, the substitution effect is observed in the model of total daily trips but not in the more fundamentally important model of daily VMT. The results, taken as a whole, seem to suggest that the net effect of ICT on travel is to generate additional travel. But before we take this conclusion too far, we must qualify the empirical results in light of several limitations of this study that have become clear through the research process:

First, because this study is cross-sectional in nature, the relationships between ICT usages and travel it has identified are essentially statistical associations, not causations. The empirical results basically indicate that certain ICT usages are associated with certain travel behavior changes, most of which are in the direction of complementarity. Research incorporating more rigorous experimental controls will be required to tackle the more difficult problem of establishing causality.

Second, the models are simplistic in that they ignore some likely simultaneous relationships among variables. For example, our exploratory data analyses revealed that some forms of ICT usage may be simultaneously determined with some outcomes of travel. More sophisticated statistical tools need to be explored to sort out these complexities.

Third, our measures of ICT usage are still too aggregate and do not allow for separated indicators of how and why Internet and telephones are used. More detailed information about ICT use can potentially shed additional light on the ways in which various technologies create different effects on travel. A survey that collects information on the purpose for accessing the Internet or making a telephone call would help to understand the impacts on travel. For example, was the Internet used to access information about congestion in order to better time a trip or select a route, suggesting a modification effect on travel? Or was it used to place an order for goods or services, substituting for a trip? In the case of complementarily, perhaps an individual seeks out information about various restaurants before choosing which one to visit. Traditional travel diary data is not adequate to meet the ICT research needs. Time use and activity surveys provide more detailed information but if they are not designed with ICT in mind, they cannot advance our understanding of these complex relationships.

Finally, the estimated models have rather limited explanatory power, and we have little knowledge of the nature of the large amount of unexplained variation. Are there explanatory variables that show more powerful influence on the dependent variables in certain contexts? Are the differences, if exist, in some way connected with public policy?

These limitations provide important hints for desirable improvements in the future, which include serious efforts in data collection, research design, statistical modeling, and policy analysis. None of these efforts will be trivial.

Acknowledgement

This research was partially supported by two research entities at the University of Maryland, College Park: the Maryland Transportation Initiative (MTI) and the National Center for Smart Growth Research and Education.

References

Claisse, G. and F. Rowe (1993). Domestic telephone habits and daily mobility. *Transportation Research A* 27(4): 277–290.

de Sola Pool, I. (1979). The communications/transportation tradeoff. In *Current Issues in Transportation Policy*, edited by A. Altshuler. Lexington, Massachusetts: Lexington Books.

Dieleman, F. M., M. Dijst and G. Burghouwt (2002). Urban form and travel behaviour: Micro-level household attributes and residential context. *Urban Studies* 39:507–527.

Ellen, I. G. and K. Hempstead (2002). Telecommuting and the demand for urban living: A preliminary look at white-collar workers. *Urban Studies* 39(4): 749–766.

Fadare, O. and B. T. Salami (2004). Telephone uses and the travel behaviour of residents in Osogbo, Nigeria: An empirical analysis. *Journal of Transport Geography* 12: 159–64.

Farag, S., M. Dijst and M. Lanzendorf (2003), Exploring the use of e-shopping and its impact on personal travel behaviour in the Netherlands. *Transportation Research Record* 1858: 47–54.

Golob, T.F. and A.C. Regan (2001). Impacts of information technology on personal travel and commercial vehicle operations: Research challenges and opportunities. *Transportation Research C* 9: 87–121.

Handy, S. and T. Yantis (1997). *The Impacts of Telecommunications Technologies on Nonwork Travel Behavior.* Southwest Region University Transportation Center, Center for Transportation Research, The University of Texas at Austin.

Hanson, S. and D. Fontaine (2005). Making Time for Physical Activity: A Choice in the Age of Instant Access. Symposium on Societies and Cities in the Age of Instant Access, University of Utah, Salt Lake City, Utah USA, 2005.

Harvey, A. S. and M. E. Taylor (2000). Activities settings and travel behaviour: A social contact perspective. *Transportation* 27(1): 53–73.

Hjorthol, R. J. (2002). The relation between daily travel and use of the home computer. *Transportation Research A* 36: 437–452.

Janelle, D. G. 1995. Metropolitan expansion, telecommuting, and transportation. In *The Geography of Urban Transportation*, edited by S. Hanson. New York, NY: Guilford.

Johnasson, A. (1999). *Transport in an Era of Communication.* Stockholm, Sweden, Swedish Institute for Transport and Communication Analysis.

Liao, T. F. (1994). Interpreting Probability Models: Logit, Probit, and Other Generalized Linear Models. Thousand Oaks, CA: Sage Publications.

Mokhtarian, P. L. (2002). Telecommunications and travel: The case for complementarity. *Journal of Industrial Ecology* 6(2): 43–57.

Mokhtarian, P. L. and R. Meenakshisundaram (1999). Beyond tele-substitution: Disaggregate longitudinal structural equations modeling of communication impacts. *Transportation Research C* 7(1): 33–52.

Mokhtarian, P. L. and I. Salomon (2002). Emerging travel patterns: Do telecommunications make a difference? In *Perpetual Motion: Travel Behavior Research Opportunities and Application Challeges*, edited by H. S. Mahmassani. Oxford, UK: Pergamon Press/Elsevier.

Nie, N. H. and D. S. Hillygus (2002). The impact of internet use on sociability: Time-diary findings. *IT & Society* 1: 1–20. *http://www.ITandSociety.org*

Nobis, C. and B. Lenz (2004): Changes in transport behavior by the fragmentation of activities. *Transportation Research Record* 1894: 249–257.

Robinson, J. P., M. Kestnbaum, A. Neustadtl and A. Alvarez (2002). Information technology and functional time displacement. *IT & Society* 1: 21–36. *http://www.ITandSociety.org*

Salomon, I. (1986). Telecommunications and travel relationships: A review. *Transportation Research A* 20(3): 223–238.

Shen, Q. (2004). ICT, local community characteristics, and the spatial-temporal reorganization of human activities. *STELLA Focus Group 2 Meeting (ICT, Innovation and the Transport System)*, Budapest, Hungary.

Shen, Q. (2000). New telecommunications and residential location flexibility. *Environment and Planning A* 32: 1445–1463.

Srinivasan, K. and S. R. Athuru (2004). Modeling interaction between internet communication and travel activities: Evidence from Bay Area, California, Travel Survey 2000. *Transportation Research Record* 1894: 230–240.

Viswanathan, K. and K. G. Goulias (2001). Travel behavior implications of information and communications technology (ICT) in Puget Sound region. *Transportation Research Record* 1752: 157–165.

11 Influence of mobility information services on travel behavior

Verena Franken, Barbara Lenz
German Aerospace Center (DLR) in the Helmholtz Association, Institute of Transport Research (IVF), Berlin

11.1 Introduction

The intention of the so-called "mobility information services" is to support individuals before and on travels. In general, mobility information services provide information for the user concerning traffic. A 'service' is created if the information is provided regularly and in a standardized form with a given level of quality. Road users use mobility information services, because they are able to make decisions which lead to an optimized travel behavior (Hirschhausen et al., 2001). In the context of this contribution mobility information services are defined as any active information concerning the everyday traffic. "Use of mobility information" is understood as the active procurement of information.

In an advanced version, mobility information services shall not only assist the users in their way finding but also enable them choosing the optimal way in terms of time or distance or individual preferences. In an ideal case services do not only provide information on the "best" route but also on alternatives for routes, departure times and transport modes. By the use of dynamic data services can also be offered en route. This, however, requires mobile access to the service provider by mobile phone, internet or PDA.

The travel optimization on the individual level is supposed to offer new options for improvement on the macro-level, too, in particular for traffic management, if the information that is transmitted to the user is altered into dynamic information. The mutual and ongoing feedback between traffic management and transport demand could then lead to a better utilization of the existing infrastructure. While network equilibrium will provide a smoother flow of traffic for all road users, the access to options for an individual optimization of travels is still restricted to particular user groups – this is at least the European experience.[1]

The diffusion rates of individual mobility information services are very low in Europe. The only service that is used frequently by a larger number of users is

H.J. Miller (ed.), Societies and Cities in the Age of Instant Access, 167–178.

traffic information broadcasted by almost each radio station. As this information is for free, the low use of other information services is mostly explained simply by the financial aspect. The demand for mobility information services and the underlying demand for transport, however, are not only dependent from financial resources, but also from mobility needs and from skills that enable a person to get access to mobility services (Franken Lenz, 2004).

Most research efforts concern technical aspects of the provision and transmission of travel related information. Only few studies give insight into the use of "mobility information services". Major knowledge gaps exist concerning the profiles of current and potential new users, their expectations and requests towards mobility information services and their [potential] behavioral change supported by the use of the services. This knowledge, however, is essential for both the diffusion of services on the individual level and the realization of a better utilization of infrastructures on the transport system level. The study performed by the DLR Institute of Transport Research makes a contribution to fill this gap. The article will use empirical data to find answers to the following questions:

- Who are the current users or user groups of mobility information services?
- How do people assess the [potential] use of mobility information?
- Which influence does the use of mobility information services have on travel behavior?

In the following paragraphs, we will start by discussing the conceptual problem of linking acceptance of mobility information services, travel behavior and information and communication technologies (ICT) use and by explaining our approach. In the empirical section we will explore the relationship between mobility information services, travel behavior and ICT use. The assessment of the potential impact of mobility information services on people's travel behavior is based on a cluster analysis. The final conclusions will also touch on what the disparate access to mobility information means with respect to social exclusion.

11.2
Acceptance of mobility information and travel behavior

Mobility information aims at improving the user's information level to ease his or her decisions for an adaptive behavior. The basic problem about the use of mobility information services concerns their implementation into the existing travel behavior of individuals. The quite simple assumption that is often related to the provision of mobility information says that more information will support the decision process of the road user: As long as detailed and on-time information is missing, road users have to rely on their experience. But access to the relevant information makes them take their decisions adapted to the real traffic situation. The argumentation is based on a concept of rational choice that identifies information as the prerequisite of rational action (Kirsch, 1998). In the case of travel behavior it means that the road user considers all potential transport modes and routes to move from point A to point B. The rational consideration of cost and utility ends up with an optimal choice (Kenyon and Lyons, 2003).

This assumption, however, is not in line with the understanding of travel behavior as it is currently found by researchers applying a social sciences perspective in transport research. In the everyday context people often act as they did before in the same or similar situations. They (re)consider the way they act only if situations are completely new or unknown so that previous behavioral patterns do not fit. This leads to a "real" moment of decision-making that has to be distinguished from habitual behavior. Additionally recent studies show that even decision-making itself may not be understood as something purely rational but that emotions are often linked to it (Schnabel, 2005).

The discussion about "routines" in travel behavior leads in a similar direction. Routines seem to dominate the individual's everyday travel behavior. They are understood as the outcome of a sequence of actions that ended up with a satisfying result so that the way the action is performed has become "automated" (Harms, 2004). Yet, routines are not an automatic reflex, but are based on target-oriented or "strategic" behavior (Bargh et al., 2001). This means that at least in the beginning there was a decision-making about how to perform an activity, e.g. by using a particular mode of transport. If taking the train to go to the workplace turns out to be "successful" several times, it becomes an automated action ever more. Several studies (Kenyon and Lyons, 2003; Aarts et al., 1997; Verplanken et al., 1997) have shown that routines characterize both the choice of transport modes and the choice of routes. Inertia, then, is particularly high. As a consequence, travel behavior often consists in suboptimal behavior which is not modified even under changing conditions or in situations which differ from everyday procedure. It is assumed that one major reason for not modifying routines is the lack of information supporting the behavioral change. This means that providing information in the case of changing conditions or different situations might influence the experienced travel behavior. Usually, the use of mobility information services will first occur in situations of particular 'urgency' e.g. congestion, trips under time pressure for extrinsic reasons or in unknown areas. At the same time it can be expected that intrinsic motivations, in particular the affinity to technical devices like computer and mobile phone, trigger the use of mobility information services.

The acceptance of mobility information must be assessed against this background. From a general point of view, current models of acceptance presume that the discernable and discerned usefulness of any innovation are the key to its acceptance. An innovation is labeled to be "useful" if

- the potential user can profitably use the functions of a service for the tasks in his [everyday] life context
- the configuration of the system fulfils the requirements of the user concerning both operability and functionality.

Usefulness in the context of travel behavior is primarily noticeable in situations which do not have the usual course or which are new (Lenz and Franken, 2005).

A particular problem then is the transmission of the information: How can the road user be informed? For this problem, particular attention is given today to ICT based devices like mobile phone or internet. Considering their rapid diffusion and their capacity to deal with high data volumes they seem to be the appropriate means for transmission. As very often for technical devices, availability has been mistaken for use. Studies about the use of mobile phones, for instance, have revealed that mobile phone owners are primarily interested in the use of the original function of the phone which is voice communication (Nobis and Lenz, 2005). On the other hand, empirical research has shown that the more an individual is used to use ICT the more he or she tends to assess ICT applications positively. From there, the assumption originates that mobility information services which are transmitted via ICT are particularly quickly accepted and adopted by people with affinity to ICT.

Recapitulated, the approach – applied in this research activity – is characterized by following assumptions:

- travel behavior as suboptimal behavior (routines)
- extrinsic motivation, inexperienced situations as stimulus to use mobility information (services)
- intrinsic motivation/necessity e.g. general affinity to ICT as stimulus to use mobility information (services)
- general affinity to ICT as pre-condition of a easier access to ICT based mobility information (services)

11.3 Data basis

The following analysis is based on the study "Acceptance and use of mobility information" performed by the DLR – Institute of Transport Research (IVF). This representative telephone survey was held in 2004, by INFAS, Institute for Applied Social Science GmbH, Bonn. The sample size is 2,200 German speaking persons from 16 years and older. The survey about user needs and user affinities as to mobility information, mobility and use of ICT explores the following aspects:

- knowledge about and use of possibilities to get mobility information (frequency, use depending on trip purpose etc.)
- demand of mobility information depending on trip purpose
- previous satisfaction and potential of support of mobility information
- attitudes towards mobility information
- technical and content features by mobility information services as well as potential additional services
- availability and use of navigation systems

In relation to the particular context of the interviewed person (e.g. employed or non-employed) we asked for individual routines and decisions in specified traffic situations, for attitudes in the field of mobility and ICT use and, in addition, the respondent's socio-demographic characteristics.

11.4 Awareness and use of mobility information (services)

A closer look at the awareness and the use of mobility information transmitted by different media identifies a strong discrepancy between knowing about and using mobility information. This is true with respect to all media, in particular to internet and mobile phone (figure 11.1).

The gap between knowledge and use might be interpreted as a lack of both the need for information and its appreciation. On the one hand, the missing active access to up-to-date mobility information indicates that people do not need the information as they do not intend to take any decision concerning their travel behavior. On the other hand there might be too little confidence in the accuracy and freshness of the mobility information so that people are not ready to search or even pay for it.

The assessment that the respondents give about the importance of mobility information confirms the general assumption that the more situations are unique or unusual the more additional information is requested. Among those who have

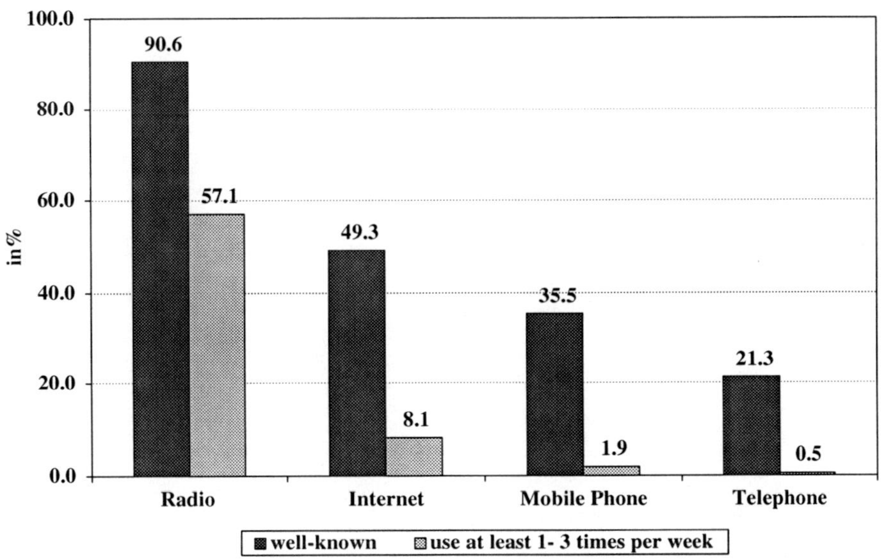

Fig. 11.1 Awareness about and use of mobility information services via different information and communication modes (n = 2154)

ever used mobility information via telephone, mobile phone or internet and who confirm a generally positive attitude towards mobility information services (which is altogether 37.2% of the entire sample), the most important request for information occurs for holiday and leisure trips (44.7% and 19.7%) while information for trips to work or school/university scores only at 14.4%. Business travels are another important trip purpose in relation to information gathering. 45.2% of those who make business travels (N=431) say they need and use special information for almost each travel.

As almost one half of the respondents think that, in general, "mobility information" represents something useful in their everyday life, there cannot be any doubt

Table 11.1 Attitudes toward mobility information (services)

	I agree completely/ rather	I agree partly	I agree rather not/I don't agree
1) A mobility information service would not affect my daily travel behavior. (n = 1229)*	43.0	25.7	30.2
2) By the use of such a new service I would save time. (n = 1229)*	40.9	23.8	32.9
3) If I got the necessary travel information at any time and everywhere my daily travels would be more comfortable. (n = 1229)*	49.0	21.7	28.0
4) I don't think about the choice of means of transport. (n = 1229)*	60.9	17.8	20.8
5) I always choose the means of transport which appears to be the most appropriate at that very moment. (n = 1229)*	70.1	11.0	18.1
6) For travels which I do rather rarely I consider exactly the best route. (n = 1229)*	82.3	7.8	9.0
7) I know too little about local public transport. Therefore I don't use it. (n = 727)*	28.2	19.9	49.8
8) If I were better informed I would leave the car more often and use public transport. (n = 1103)*	20.0	16.7	61.4

* Only a part of the persons, according to certain criteria, were be asked to their attitudes towards different statements.

about people's positive expectations. However, almost the same number of people estimates that mobility information services would not affect their travel behavior. If asked for a self-estimation about the impact of mobility information on their travel behavior, the respondents do not expect major changes (table 11.1).

While approximately half of them (45.5%) suppose that, generally speaking, mobility information represents a useful support on their everyday trips, people do not assume any influence on their usual travel behavior from this (statement 1).[2] The latter findings have to be considered against the background that 60.9% say they do not reflect about their choice of transport modes, but always use the same one (statement 4). At the same time they indicate to use the most suitable mode of transport for the specific purpose of any trip. This can be interpreted along the lines of the arguments given by AARTS et al.(8) who point out the persistence of routines even under changing conditions. Another argument for the interpretation refers to the existence of "strategic" decisions during the first phase in the development of routines which later "legitimate" actions by habit (see chapter 2).

What is the motivation for using mobility information if not the support of decisions for travel alternatives? It is interesting to see that increase in comfort and time savings are the particular outcomes for a greater number of people. 40.9% agree to the statement (statement 3) that they expect their daily trips to become more comfortable if mobility information is accessible anytime and anywhere. Another 49.0% think mobility information will help them saving time (statement 2). This leads us to the conclusion that the potential of mobility information to influence travel behavior is relatively small. If the use of mobility information happens, it is primarily for supporting existing behavioral patterns and not changing them.

11.5
Influencing travel behavior by mobility information?

In parallel to the self-estimations, the study used "virtual sets" to find out about the potential of mobility information to influence travel behavior. For this purpose people should imagine to be involved in typical traffic situations and they should describe their reaction if they got specific mobility information:

> Imagine you are on the way to work with your car and you get traffic information every half an hour by the radio. If they report congestion on your way and give recommendations for deviation so that you do not loose too much time – how would you react? (figure 11.2)

In all, 77.4% of the respondents would react actively to this information. 39.8% would even opt for the recommended detour. Any shift to other transport modes or even abandonment of the trip cannot be expected – only 0.1% would use public transport instead of the private car, only 0.2% would give up the intended trip.

In a next step subgroups of the entire sample (e.g. car drivers, public transport users) were set in different virtual situations to which they could react in various ways. Given the relevant information, 77.6% of the public transport users (N=82)

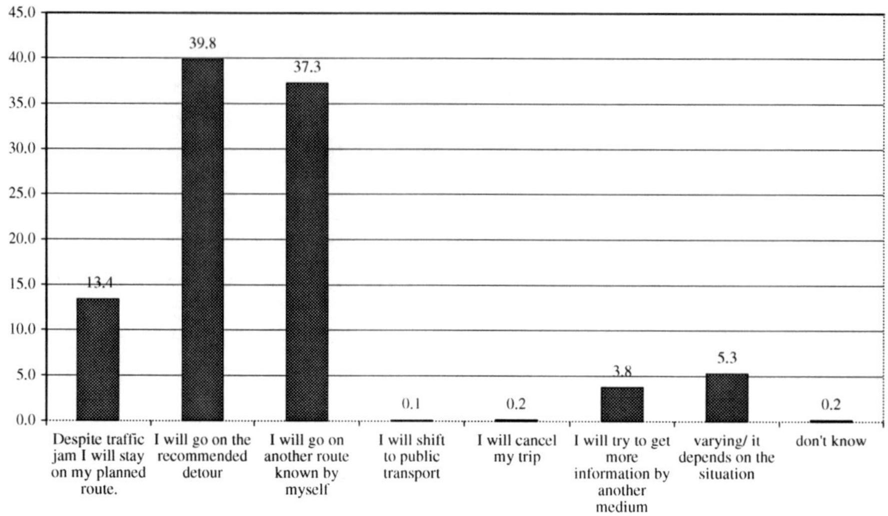

Fig. 11.2 Reaction to traffic advices on the radio (n = 1510)

would be ready to change to another public transport mode if they had to react to a breakdown of the transport system on their trip to work. Similarly, 87.0% of the car users (N=212) say they would choose another route if they had information about it. Much lower is the degree of influence for travels that are done for leisure or for shopping (tables 11.2 and 11.3). Apparently, the lack of time pressure reduces people's wish for alternative options.

To go further than just register the potential impact we wanted to find out how much the influence of mobility information depends on the "type" of person. For this purpose we used a cluster analysis. It was based on the approach – as outlined earlier – that the acceptance and use of mobility information respectively mobility information services is based on attitudes as well as on intrinsic and/or extrinsic motivation. Accordingly the clusters were created with the items "attitudes towards mobility information" (table 11.1) and "attitudes towards information and communication technology". The cluster analysis identified clearly three clusters which

Table 11.2 Reaction to virtual situations occurring for public transport trips

Change of the route by use of another public transport, which is recommended by the service	
Commuter traffic (n = 82)*	77.6%
Leisure traffic (n = 66)*	66.4%

* According to certain criteria, a part of the persons were be faced with a certain virtual situation.

Table 11.3 Reaction to virtual situations occurring for trips with the private car

Compliance of the route recommended by the service without changing the means of transport (car)	
Commuter traffic (n = 212)*	87.0%
Shopping traffic (n = 453)*	67.5%

* According to certain criteria, a part of the persons were be faced with a certain virtual situation.

are also distinct by their socio-demographic characteristics. We used the standardized method of hierarchical cluster analysis (clustering method: Ward-Linkage, measure: quadrupled Euclidean distance). One advantage of the hierarchical clustering method is the possibility to define the optimal number of clusters by using the graphical representation of dendrogram.

As the dendrogram in figure 11.3 shows, both a three and a four cluster solution would have fit the clustering objective. To assess people's reactivity to mobility information, however, we chose the three cluster solution as there were only very slight differences between clusters 1a and 1b. 1a and 1b are now included in cluster 1 which represents 27.5% of the sample. Cluster 2 stands for 44.5%, cluster 3 for 28.0%.

The clusters can be characterized as follows:[3]

- Cluster 1 is the "sceptical-negatory" cluster as its members do not appreciate either mobility information or ICT use. The cluster is characterized by a large share of elderly people with low income and low levels of formal education. Most cluster members are not employed or retired. Independently of the trip

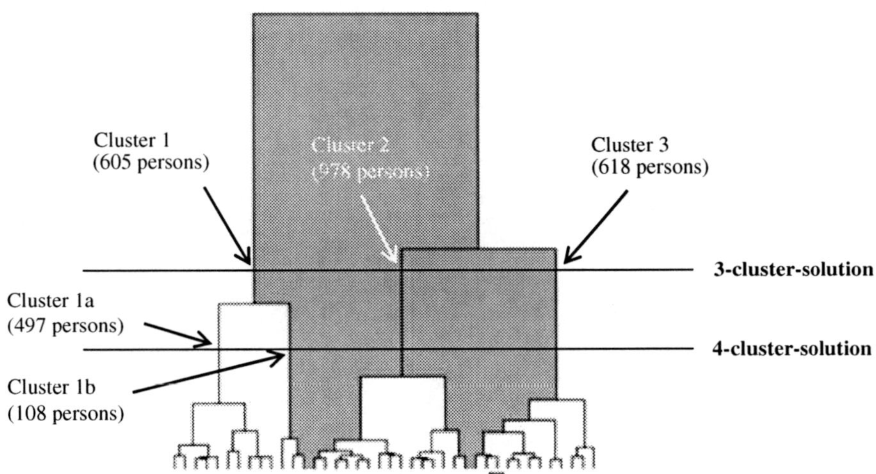

Fig. 11.3 Dendrogram of the WARD-algorithm

purpose or the transport mode, members of cluster 1 are in principle less mobile compared with the entire sample. This matters primarily the use of transport modes like car or public transport. 58.2% of all persons without driving license belong to cluster 1. However, if they are on the road they mostly use the car. The public transport use is below average in this cluster.

- Cluster 2 is the "conservative-indifferent" cluster. Cluster members show only a weak tendency to appreciate both mobility information and ICT. This cluster is made up of the large group of car users who are quite inflexible in terms of mode choice. As to its socio-demographic characteristics, the cluster has a large number of employed people and a small overhang of male persons.
- Cluster 3 consists of those who have a positive attitude towards mobility information and ICT. Therefore we called it "flexible-open minded". Respondents who belong to this cluster are mostly young, use the internet above average and have high levels of formal education. With respect to their mode choice they are very flexible.

The most interesting finding of the typology based analysis concerns the degree to which people are ready to change their travel behavior if they get relevant mobility information. It is surprising to see that members of cluster 1 are willing to follow radio broadcasted detour recommendations – which represent a still "simple" way of information – for their way to work more than the average (46.9% compared to 39.8%) (figure 11.4). At the same time, the so-called "flexible-open minded" are mostly more ready than others to change their transport mode in the case of a breakdown of the transport system – at least if advanced mobility

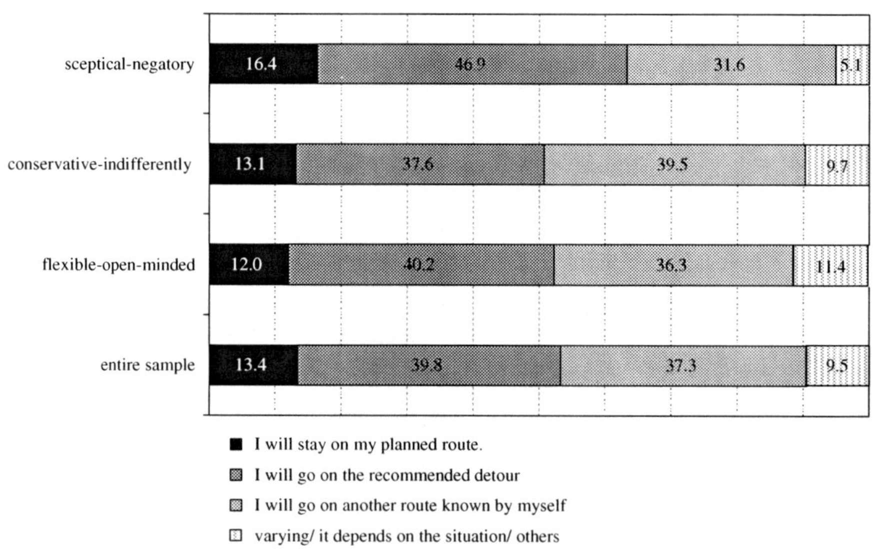

Fig. 11.4 Reaction to traffic advices on the radio depending on cluster type (n = 1510)

information is available. Given this improved information, the cluster 3 members are even prepared to switch from car to public transport use (figure 11.4). The lowest degree of active reaction to mobility information appears for the "conservative-indifferent" cluster.

11.6 Conclusion

Concerning the convertibility of travel behavior by mobility information services, the DLR study reveals that there is a considerable share of road users who are generally willing to change their travel behavior for optimization purposes. This willingness does not very much concern the shift to other modes of transport, particularly not the shift from car to public transport. But there is an important disposition to follow recommendations – and thus change initial "decisions" – with respect to the routes. This is true for both public transport and car use. It is of particular interest, of course, for the case of road traffic where re-routing could bring about a better utilization of the road network. The look at current behavior shows that the compliance of detour recommendations should already occur by the simple transmission via radio. It could be more than doubled if advanced services were available – at least for trips to work (figures 11.4 and 11.5).

The data show that a general request for mobility information exists throughout diverse social classes. Additionally the assessment of people's readiness to optimize

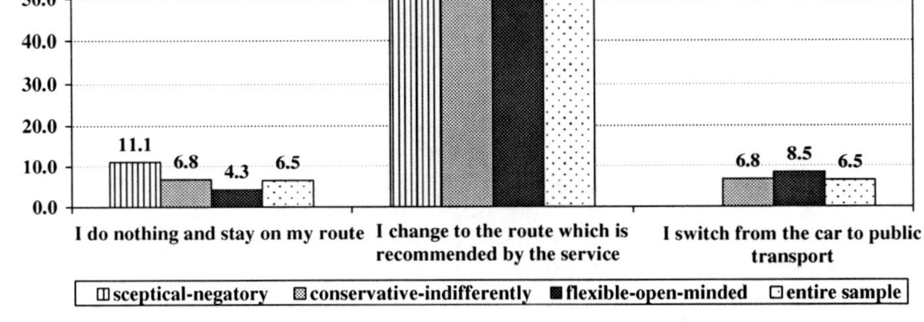

Fig. 11.5 Reaction to virtual situations occurring for trips to work using the private car by cluster type (n = 212)

their travel behavior with the help of information has shown that socially "weaker" groups are lead by the same objectives and are following more than the average travel recommendations. However, they will only be able to do that if they have access to these services. The more the accessibility of relevant information will be reduced to individual devices like navigation systems or PDAs, the more it has to be expected that some groups will loose the chance for optimization. While people with lower incomes and lower degrees of formal education apparently want to optimize their travels too, they are restricted to the use of media which offer "public" access like the radio does. If traffic management will use mobility information for a system improvement, these groups have to be addressed, too.

Notes

[1] In Europe (EU25), 36% of the population use the internet (USA: 55%), 74% possess their own mobile phone (USA: 49%). In Germany, the availability of a mobile phone among the teenage population reaches already 90% (Shell, 2002); almost 90% of young people under 20 use the internet regularly.

[2] It is noteworthy, however, that 82.3% of all respondents agree to the statement that mobility information might influence their route choice on long distance trips.

[3] Please note that the names that were given to the clusters do *NOT* characterise the respondents as persons but only their attitudes and behaviours in relation to the choice of transport modes, to mobility information and to ICT.

References

Aarts, H.; Verplanken, B.; van Knippenberg, A. (1997): "Habit and use in travel choices" In: Acta Psychologica, 96, p. 1–14.

Bargh, J. A., Gollwitzer, P. M., Lee-Chai, A. Y., Barndollar, K., & Troetschel, R. (2001): "The automated will: Nonconscious activation and pursuit of behavioral goals" Journal of Personality and Social Psychology, 81, p. 1014–1027.

Franken, V.; Lenz, B. (2004): "Nutzeranforderungen an Verkehrsinformationsdienste als Grundlage für technologische Entwicklungen", IMA 2004, Braunschweig. Germany.

Harms, S. (2004): "Besitzen oder Teilen. Sozialwissenschaftliche Analyse des Car Sharings", Zürich, Switzerland.

Hirschhausen, Ch. et al. (2001): "Ökonomische Analyse von Verkehrsinformationsdiensten", Technische Universität Berlin, Fachgebiet für Wirtschafts- und Infrastrukturpolitik, Berlin, Germany.

Kenyon, S.; Lyons, G. (2003): "The value of integrated multimodal traveler information and its potential contribution to modal change", Transportation Research F 6 (2003), p. 1–21.

Kirsch, W. (1998): "Die Handhabung von Entscheidungsproblemen", Münchner Schriften zur angewandten Führungslehre, München, Germany.

Lenz, B.; Franken, V. (2005): "Utility and demand of e-information for travel", 6th International Conference on Environmental Engineering, Vilnius, Lithuania.

Nobis, C.; Lenz, B. (2005): "Wechselwirkungen zwischen Kommunikations- und Verkehrsverhalten", Befragung des DLR-Instituts für Verkehrsforschung, Technischer Bericht (in preparation).

Schnabel, A. (2005): "Gefühlvolle Entscheidung oder entscheidende Gefühle. Emotionen als Herausforderung an Rational Choice Theorien", Kölner Zeitschrift für Soziologie und Sozialpsychologie, 57, p.278–307.

Shell 2002: 14. Shell-Jugendstudie. München, Germany.

Verplanken, B.; Aarts, B.; van Knippenberg, A.(1997): "Habit, information acquisition and the process of making travel mode choices", In: European Journal of Social Psychology, 27, p. 539–560.

12 Shared ride trip planning with geosensor networks

Silvia Nittel[1], Stephan Winter[2], Arda Nural[1], Trang Cao[2]
[1]*Department of Spatial Information Science and Engineering, University of Maine, USA*
[2]*Department of Geomatics, The University of Melbourne, Australia*

12.1 Introduction

Recently, advances in technology miniaturization with regard to short-range wireless communication networks such as Bluetooth, tiny computers in the size of a quarter, and the development of microsensors that can be flexibly attached have created a novel type of powerful small computing platform, so-called sensor networks. The computing nodes in sensor networks collaborate and communicate with each other in an ad-hoc way based on availability, short-range communication, power and computing resources and the tasks at hand to accomplish. As a consequence, these inexpensive miniature sensing and computing devices can be embedded in many natural and man-made devices. Each device can connect to an infinite network of other devices to perform tasks without human intervention such as temperature or toxidity detection.

This chapter investigates how this technology enables new types of social behavior. In particular, we investigate a specific novel application of the type of technology for inner-urban trip planning, envisioning that each actor is equipped and supported by a node in a sensor network. Imagine Hillary who just has missed her bus to a doctor's appointment today Figure 12.1. Around Hillary the traffic is floating. Now, she is glad to have subscribed to a service that mediates between her current transportation needs and vehicles going into her direction in an unpreplanned, ad-hoc manner. Hillary opens her mobile phone/hand-held device, and starts the shared ride application. The local program on the mobile phone acts as an agent (or transportation client), announcing the need for a ride from Hillary's current location to a her destination. The local agent negotiates with ad-hoc offers from other vehicles in the neighborhood, and determines and books an optimal offer. Soon after, Hillary sees a friendly car driver stopping for sharing a ride to the

H.J. Miller (ed.), Societies and Cities in the Age of Instant Access, 179–194.

Fig. 12.1 If the bus is missed riding with cars becomes an alternative—and this requires ad-hoc trip planning

train station. The ride takes her a first leg of her trip, and she will be able to catch her train to the city in time. Hillary will not come late to her appointment today.

We envision an ad-hoc shared ride system that integrates the transportation capacities of all types of (volunteering) vehicles in urban traffic in order to identify a trip for persons with an ad-hoc travel demand. The system shall assign persons, or *transportation clients*, to vehicles, or *transportation hosts*, with matching travel plans and free transportation capacities, in an ad-hoc manner. Trip planning is complex in such transportation networks. In contrast to a public transportation system, all transportation hosts act autonomously with individual routes, and they may act based on their own individual policies. They become available, and might disappear from the system. This property makes the transportation network highly dynamic and unpredictable. Therefore, any (centralized or local) trip planner in this network has knowledge that is temporally limited to the current state.

In the traditional trip planning literature, optimal trip planning can be accomplished in a centralized fashion if the routes and the schedule of the transportation hosts in the network are known. However, these strategies are no longer applicable in such highly dynamic and constantly changing environments with individual routes and strategies of automobiles and transportation clients. We pose the hypothesis that using the local communication strategies and adaptive collaboration of mobile nodes in a geosensor network allow for near-optimal trip planning in a highly dynamic transportation network. If one chooses peer-to-peer communication and limits the spreading of messages to a more local neighborhood, the trip planners have also spatially limited knowledge of the transportation supply. Still, trip planners can make reasonable decisions, but they may review and revise their trip plans periodically. With each review, they look at temporally and spatially updated information about transportation supply, since they as well as the hosts may have moved in the meantime. We expect that this approach of trip planning is fully scalable; i.e. if more transportation agents and clients enter the system, incomplete transportation network knowledge a system still can deliver near-to-optimal trips.

Such a shared ride trip planning system causes many questions. In this chapter, we focus on the efficiency and effectiveness of different communication and negotiation strategies between nodes. Hence, we will concentrate on only one wayfinding and route planning strategy. This paper extends our previous work (Winter and Nittel, 2005) by specifying a simulation environment for investigating different communication and negotiation strategies. The current research objective is: how far can we spatially limit the knowledge of a route planner without loosing too much in the quality of the traveled trip? This question is particularly important since broadcasting in mobile peer-to-peer communication networks is network bandwidth consuming and needs to be minimized.

We designed and implemented a simulation environment to understand and demonstrate the implication and correlation of the spatial limitation of information dissemination between nodes in the network and quality of the traveled trip according to the chosen cost criterion. The simulation environments enables simulating shared ride trip planning with different parameters, such as density of hosts, or diameter of neighborhoods. We expect to find a negotiation and communication strategy that adapts to a specific situation, and guarantees reasonable route qualities given this situation.

Shared-ride systems have to consider also some other implications and challenges prior to any real-world realization. They comprise, for example, *trust* and *safety, liability*, economic incentives and business models, urban mobility and access and *privacy*. When we look into trip planning we are aware of all these other aspects, but leave them for further work.

The chapter is structured as follows. Section 12.2 defines the technology of geosensor networks; Section 12.3 describes the behavior of the involved agents, the trip planning task, and the resulting need for information gathering and negotiations. Section 12.4 translates this into a simulation environment, and summarizes first results. Conclusions are presented in Section 12.5.

12.2 Geosensor networks

The continuing development of ubiquitous wireless communication technology, including miniaturization of computing platforms and the development of microscale sensors, is enabling new computer applications that would not have been possible in the past. Recent and projected advances in small, low-cost microelectronic and mechanical systems (MEMS) with limited on-board processing and short-range wireless communication capabilities are also changing the way we collect and process information about the physical world. Today, networked sensor nodes can be constructed by using open source and commercial components at the size of an inch or smaller such as the Berkeley/Intel Mica Motes.

Large collections of untethered, battery powered computing nodes with various sensing functions can be distributed over a geographic area, and measure traffic and road conditions, or environmental activity at a fine-grained temporal and spatial scale that was not possible in the past. Since such sensor nodes are tiny and the limited battery capabilities allow only for short range wireless communication, they must communicate with peer sensor nodes in their spatial proximity. Projecting the continued miniaturization, it is not expected that sensor nodes connect to a centralized computing server to upload or stream data directly; they might, however, communicate with a local 'base station', i.e. a more powerful sensor node with larger processing, storage, communication and energy capabilities. In general, information is routed via multiple communication network 'hops' to a centralized server or in-network micro-server, or the information is processed in the local geographic environment of sensor node's location.

Integrating both location-based computing and sensor network technology, we can envision sensor nodes that are aware of their geographic location, equipped with diverse sensors, mobile, and communicate with nodes in spatial proximity about information that they sensed or information or resources they need from other nodes. Sensor nodes can be mobile by either being self-propelled or by being attached to moving objects, like automobiles, USPS packages, or even humans. Sensor networks in which nodes are aware of their geographic location, and environmental phenomena are captured via on-board sensors are so-called *geosensor networks* (Nittel et al., 2004).

Efficient information routing is a significant research challenge in geosensor networks, and sensor networks at large to accomplish tasks, preserve energy and not 'cloak' the network bandwidth with too many messages. In contrast to information routing in today's communication networks, which is address-based (IP addresses), routing in sensor networks is data-centric. The goal is to distribute information only to sensor nodes that need the information or that can be a source of information. Another aspect of data-centric routing is *scalability*: if the number of nodes in a sensor network increases to thousands of nodes, a decentralized, peer-to-peer information dissemination and data collection strategy can provide efficient information distribution and eliminate the bottlenecks of a centralized database or service architecture.

By adopting the geosensor networks paradigm as the basis for an ad-hoc ride share system, each vehicle can be thought of as a mobile geosensor node, able to sense information about its own location, its own trip needs and resource availability or necessity, process this information, and communicate it to other moving objects in its neighborhood. Potentially, this decentralized collaboration can offer improved reliability and performance, since there exists no centralized service provider acting as a bottleneck to information dissemination and processing.

We distinguish between three different communication strategies between mobile nodes: *Flooding* is a strategy in which each geosensor node that has a request or receives a message about a travel request passes on the information to every other node within its communication range. The second approach is referred to as an *epidemic*, in which each node only inform *n* other agents about events. The third approach is *location-constrained*, in which information is only passed on in proximity to the travel need, and then discarded.

12.3 Shared ride trip planning

In a shared ride trip planning system using geosensor networks, we are interested in assigning clients to hosts such that the clients get quickest to their destinations. Other factors include the number of transfers for a client, his/her wait time in between rides, and required fares or a reward system for car drivers to offer rides.

12.3.1 Constraints of trip planning

In a local trip planning task we need to specify (a) the transportation demand of a client, (b) the transportation supply of the hosts, (c) the planning task, (d) the communication needs of the involved parties, including the content of messages, and (e) the communication strategies. Without limiting generality we choose the clients to be responsible for their own trip planning, and the transportation hosts to be reactive only. As constraints we consider the following: we expect to find hosts with different 'behavior' such as public transportation, taxis, or private automobiles. Some of the transportation hosts show unpredictable behavior in a sense that we do not know their routes, availability or willingness to offer rides ahead of time. Also, hosts have different capacity for accommodating transportation clients.

Overall, we can assume that we have to design a shared ride system in a continuously changing environment, e.g. cars appear or 'disappear' (become unavailable because they reached their destination), or they free up capacity or have no capacity. For such an environment, we further assume that no centralized server or planning application exists, but that each client and hosts in the system performs planning tasks solely based on local and short-term knowledge about its environment.

Reformulating our hypothesis, we will prove that mobile ad-hoc geosensor networks are an effective and efficient way of shared ride trip planning, i.e. clients can achieve optimal route planning and trip decision based on this type of information, and with the limited communication and planning horizon. We can further investigate questions of quality of routes (number of stops, or wait time), appropriate negotiation and information dissemination strategies, and optimal route selection in such a highly dynamic environment.

12.3.2 The transport demand of a client

In ad-hoc trip planning, clients have a transportation demand from their current position to a destination. We assume that clients apply a simple trip planning heuristic: they look for shared rides along the geodesic to their destination. A client is interested to reach a destination for optimal (minimal) costs. The cost function depends in general on the client's context, and may concern, for example, travel time, trip fare, number of transfers, or reputation of hosts. Without limiting generality we choose travel time in our simulation. Thus, clients that follow only the shortest path to their destination can formulate their demand in form of a sequence of street segments. Each segment can be attached with a time stamp for the anticipated earliest departure time.

12.3.3 The transportation supply of hosts

Transportation hosts travel autonomously, and independent from actual client demand. They do not announce their travel prior to their start, they have their individual travel plan (a route consisting of a sequence of street segments with time stamps), and this travel plan can have any form, including stops, being non-shortest, containing cycles, or traveling back and forth. We assume that although willing to take passengers, they are not willing to make detours for these passengers. Future positions of hosts may be influenced by traffic conditions, and hence may differ from their current plans. Furthermore, some hosts like cars have a limited passenger capacity.

12.3.4 The planning task

In our systems, clients are the planning agents. They need to know about available transportation supply to choose the optimal available rides. Here, hosts are reactive in the planning process. They only have to maintain their bookings and observe their passenger capacity, and be willing to advertise this information.

At a specific point in time, a client can at most know which hosts are currently traveling, where they are, what their current booking status is, and what their travel intentions are. A client can not know with certainty the future positions of currently traveling hosts, their future booking states, and cannot see which new hosts will enter traffic next. With other words, at any point in time a client can determine an optimal route according to the current state, but in hindsight this might not have been the overall optimal one.

Given that knowledge, clients apply a simple pattern matching technique to filter out supplied street segments that are on their route, and that have a time stamp at their own anticipated earliest departure time or later. From this candidate set, clients select the ones that start earliest (assuming that all hosts travel with about the same speed); otherwise they choose the ones with the earliest arrival times at the ends of the segments. Note, that the chosen segments can cover only a part of the client's trip, and they can have temporal gaps (waiting times at transfer points) and spatial gaps (segments for which currently no supply is known).

After planning their trip, clients need to book these trips with the hosts. Since all travel plans are reviewed and possibly revised from time to time, clients need to be able to cancel previous bookings as well.

12.3.5 The negotiation needs

Communication and negotiation is needed in ad-hoc trip planning for two purposes: (a) collecting the knowledge about the current transportation supply, and after planning a trip (b) booking or canceling specific hosts. We group these tasks requiring communication in a *negotiation cycle*. The negotiation cycle happens periodically during traveling. Since it is relatively short, we can simplify that the negotiation cycle alternates with a *traveling cycle*.

The need for a negotiation cycle means that in contrast to most other studied problems in mobile ad-hoc geosensor networks our service requires two-way communication for negotiation and assignment. Studies in the dissemination of information (Nittel et al., 2004; Wolfson and Xu, 2004) provide basic ideas, but do not handle two-way negotiations. Hence, we need radio-based communication strategies that efficiently spread messages and efficiently return answers.

The task of message broadcasting itself is a resource consuming process (communication bandwidth and drain on the battery). Since the communication bandwidth in urban communication networks is typically limited and used for many different applications at the same time, minimizing the number and range of broadcasting message has priority. For that purpose we propose additional strategies for the communication process.

The negotiation cycle is different from the communication strategy, which is defined by more technical considerations. For one and the same negotiation cycle, different communication strategies can be deployed, and result in varying performance of the negotiation cycle.

In our chosen definition, a negotiation cycle consists of the following steps: (a) a client advertises his/her requests, (b) hosts listen to requests, and make offers, (c) clients receive offers and select one or more of them, and (d) clients respond to a host with an accept of an offer, booking a ride. With (d) the negotiation cycle is finished.

Other aspects of the negotiation are an update of the agreement, and a potential cancellation (by either the client or the host).

12.3.5.1 Requests

Hosts publish their potential transportation supply only if there is demand. That means, hosts act only on requests from clients. The request from a client is specific, i.e., the request contains the full information of the route to be traveled as identified in 0. If hosts receive such a request they can evaluate the relevance of their own travel route and capacity with regard to the request, and respond only if they can contribute to the client's demand.

12.3.5.2 Offers

Hosts respond only to a request with an offer if they can contribute to the client's specified demand.

12.3.5.3 Booking and cancellations

The client performs trip planning on basis of the collected offers. The number of collected offers is already reduced to the set of potentially contributing, i.e., spatially overlapping host routes. Clients then book specific segments from selected hosts, and create a booking message for that purpose specifying the host and the segments.

Cancellations concern bookings from previous negotiation cycles. They cannot be dealt with in the manner of the other messages mentioned so far because the communication network has changed, and the host to be addressed may be outside of communication range at the time of the cancellation, which is necessarily another negotiation cycle than the original booking. Even a multi-hop link could be broken. Hence we apply a different cancellation strategy.

12.3.6 Communication strategies

Communication in mobile ad-hoc geosensor networks can be short-range radio-based (Zhao and Guibas, 2004). Messages are broadcasted, and are received only by nodes within radio range. If messages shall be sent further they have to be forwarded by nodes, thus, establishing by multi-hop communication. The receiving nodes also broadcast in a circular radio range, therefore, the original nodes will hear the message again. However, since they recognize the message's id they will ignore it, and only new nodes will process the message, and rebroadcast it to nodes in farther distance from the originating node.

We assume that the nodes communicate in synchronized, relatively short communication windows, and go into a sleep mode in the meantime. Thus, we can assume that all mobile nodes exchange messages every 5 minutes, and put themselves into a sleep mode in-between. These communication windows define the frequency of the negotiation cycles of the shared ride trip planning. In the following we discuss the considerations for communication strategies for the different steps of the negotiation protocol.

12.3.6.1 Requests

We define that the communication strategy begins with clients advertising their transportation needs, while hosts are passive. The client's request is broadcasted without an addressee but with a message and a client id. It disseminates to neighboring nodes in the network within the hardware-specific local neighborhood. Since we assume short-range communication the request is first received by hosts in close proximity (ca. one city block). However, these might be also be the most relevant hosts to offer rides in this dynamic neighborhood, and negotiation with distant hosts is less promising due to the highly dynamic nature of the system.

Nevertheless, the requests need to be forwarded to more than immediate neighbours, and we consider several communication strategies:

Flooding. a client always informs all other hosts within its communication range of all information about it travel needs; all receiving nodes forward the request to all other nodes in their communication distance, and so forth.

Epidemic. a client informs only the first *n* randomly chosen hosts it encounters within its communication neighborhood about its travel needs; receiving nodes also only forward to the first *n* randomly chosen hosts they encounter.

Spatial Proximity. a client informs hosts within its communication neighborhood and hosts within a certain threshold spatial distance *d* of its destination still rebroadcast the request.

12.3.6.2 Offers, bookings and cancellations

Since hosts respond only to a request if they can contribute to the client's specified demand, an offer is addressed to a specific client within the message header, and the client's approximate location. Since this message sending is addressed directly, fewer hops are necessary; only a directional message sending is necessary. Similarly, clients' booking messages that specify the host, and the segments to be reserved are routed back to the specified host. Booking have to be confirmed in every communication cycle, or otherwise will be cancelled by hosts and clients automatically for the next negotiation cycle. This way, client and host bookings are always kept consistent, and, in fact, no cancellation message is needed.

12.4 Agent-based simulation of trip planning

To investigate the proposed shared ride trip planning systems, we implemented a agent-based simulation environment. The simulation emulates the behavior of the transportation clients and hosts introduced in Section 0. Fundamental assumption is that all information exchange within a negotiation cycle takes place in one communication window.

For simplicity, the simulation happens in a rectangular street network. In this network all hosts are moving with the same speed, i.e. one "street" segment per time unit. After each time unit they are located at "street" intersections a new negotiation between clients and hosts takes place. Furthermore, radio range shall be limited to the four-neighborhood of each intersection (± one row, or ± one column).

12.4.1 The simulation in an example

Consider Figure 12.2. with client **C** and hosts **1–7**. To model a negotiation process, we first switch from the street network view to a communication network view.

Figure 12.3. shows how the agents are connected via the communication networks. We assume that the broadcast range of each node is that of one street segment. Due to obstruction via buildings, the connectivity in a diagonal fashion is limited. Thus, **C** can communicate directly with **2**, and **2** can communicate with **1, 5**, and **4** as well as with **3** and **7**. Both groups of nodes can communicate with node **6**. Thus, one of the shortest paths of **C** to node **6** is between **C, 2, 1** and **6** (solid line in Figure 12.3).

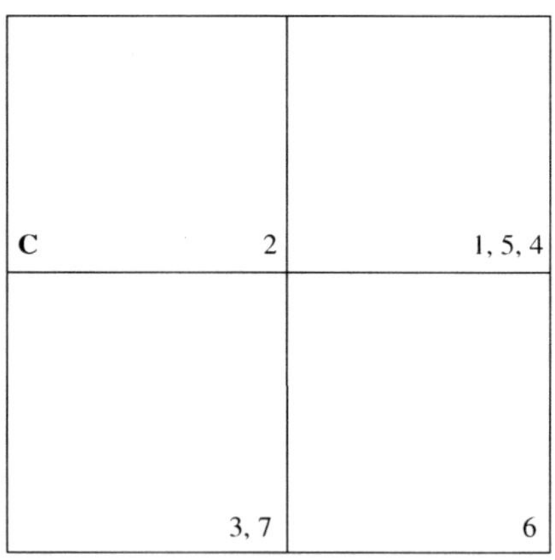

Fig. 12.2 A client C and seven hosts in a transportation network (snapshot)

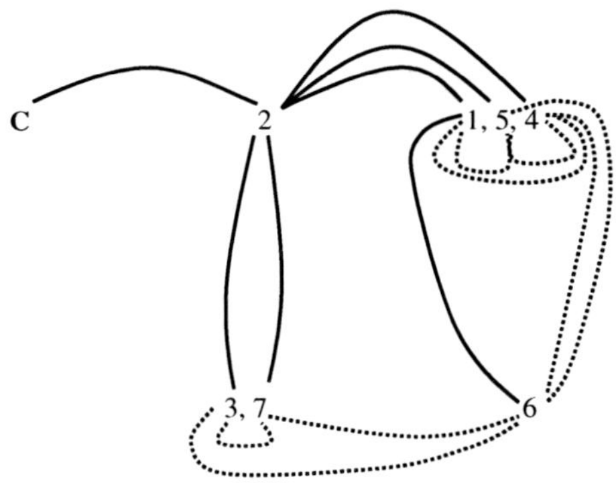

Fig. 12.3 The network of connected agents (neighborhood graph) and a shortest path tree (solid edges)

On the communication network we can demonstrate the three phases of each negotiation: sending *requests* (**r**), sending *offers* (**o**), and sending *booking messages* (**b, c**).

12.4.1.1 Requests

Let us assume that negotiations are not spatially limited. Messages will be forwarded as far as possible. With the single client, the client's request is broadcasted by:

sender	message	receiver
C	r	2
2	r	**1, 5, 4, 3, 7**, [C]
1	r	[2], [5], [4], **6**
5	r	[2], [1], [4], [6]
4	r	[2], [5], [1], [6]
3	r	[2], [7], [6]
7	r	[2], [3], [6]
6	r	[1], [5], [4], [3], [7]

In this table, the agents that receive the request for the first time (on the shortest path) are printed bold; the other agents that hear the message but who have heard the message already are printed in brackets. Only when an agent receives a request for the first time it broadcasts it. This means, in this situation each agent in the connected network broadcasts once. In other words, with the flooding strategy the number of broadcasts of a request is equal to the number of agents in the client's connected component.

Now let us consider the protocol of network 'hops' created with broadcasting. This protocol is attached to the request *r* itself. **C** creates the original request, and adds its own id. Node **2** then receives the request, and attaches its own id to the request so it will ignore future receipts of the message.

agent	received request	forwarded request
C		"r, C"
2	r, C	"r, C, 2"
1	r, C, 2	"r, C, 2, 1"
5	r, C, 2	"r, C, 2, 5"
4	r, C, 2	"r, C, 2, 4"
3	r, C, 2	"r, C, 2, 3"
7	r, C, 2	"r, C, 2, 7"
6	r, C, 2, 1	"r, C, 2, 1, 6"

With other words, each recipient knows the shortest path back to the client sending the request. Alternatively, each message can have a unique message id, based on the originating node's identifier, and a time stamp, and the listening nodes manage a local index of messages that they have broadcasted already. In this case, new routes back to the originating nodes need to be found.

12.4.1.2 Offers

Let us assume that some agents will respond to a request by making an offer. An offer, in contrast to an unaddressed request, is an addressed message to a specific recipient: the requesting client. Each offer shall travel along the shortest path (minimal number of broadcasts). For that purpose the offer contains the reversed protocol of the request as an address. Only agents on the list will forward the message.

In our example hosts **6, 3**, and **2** are going to make an offer to **C (o6, o3, o2)**. In the table below, the hosts in parenthesis are receiving a message, but are not on the address list, and hence, do not forward the offer. For example, node 6 broadcasts a message, which is received by nodes **3, 7, 1, 5**, and **4**. Only node **1** identifies itself in the shortest path, and rebroadcasts. This message is heard by node **5, 4, 2**, and **6**, but only node **2** reacts. With other words, each offer causes a number of broadcasts equivalent to the length of the shortest path branch between the offering host and requesting client. The set of broadcasts for these offers consists of:

sender	message	receiver
6	o6	(3), (7), **1**, (5), (4)
1	o6	[(5)], [(4)], **2**, [6]
2	o6	[1], [(5)], [(4)], (3), (7), **C**
3	o3	(7), (6), **2**
2	o3	(1), (5), (4), [3], [(7)], **C**
2	o2	(1), (5), (4), (3), (7), **C**

12.4.1.3
Bookings

The requesting client collects all offers, and selects the optimal one(s). This choice has to be booked with the offering client(s). In our example client **C** is going to accept an offer of host **3** (**b3**). Then the set of broadcasts consists of:

sender	message	receiver
C	b3	2
2	b3	(1), (5), (4), **3**, (7), [C]

With other words, each booking causes a number of broadcasts equivalent to the length of the shortest path branch between the client and the offering host.

Client **C** would also like to cancel a previous booking with host **7** (**c7**). Note that **C** has currently only offers from **6, 3**, and **2** in hands, and hence, does not know where **7** is. Host **7** may even be disconnected (it is not in our example). Hence, previous bookings—if not confirmed in this cycle—will time out automatically before the next negotiation cycle.

12.4.2
Specification of the simulation

The example discussed above gives reason for the following specification of an algorithm. Two measures emerged in the example as critical:

- The number of agents in the client's connected component.
- The lengths of shortest path branches (number of edges) between the client and any connected host.

- For the lengths of shortest path branches between the root (client) and any host Dijkstra's algorithm (Dijkstra, 1959) can be accomplished, computing the shortest paths between a single source and all destinations.
- After each negotiation cycle the agents travel according to their current travel plans. The client moves only when he has found a ride. After each travel phase a new negotiation cycle starts, until the client finally reaches the final destination.
- Other parameters of the simulation are the numbers of hosts, i.e., the degree of competition for rides, and the numbers of hosts, i.e., the traffic density.

12.4.3 Simulation with different communication strategies

So far the simulation applied the flooding strategy. But the same simulation can be applied with spatially limiting communication strategies, i.e. a parameter m specifying the radius of the communication neighborhood. If $m = 1$ the simulation realizes the short-range strategy, and if m is larger the simulation realizes a *mid-range strategy*. (The flooding strategy can be considered as the special case of $m = \infty$.)

For describing the behavior of the simulation, the parameter m becomes part of the request message. Each agent receiving such a request determines the number of the previous hops h (length h of the protocol), and forwards the request only as long as $h < m$. The rest of the simulation behavior (offers, bookings and cancellations) remains unchanged.

With other words, for spatially limited communication strategies the neighborhood graph is formed by the shortest path tree, cut at level m. The rest of the algorithm remains unchanged.

12.4.4 Simulation results

In our simulation, we observed the following results and trends (more detail on the simulation and results can be found in (Winter and Raubal, 2006)): we can show that the solution of ad-hoc communication and negotiation with spatially and temporally limited knowledge is *effective* and *efficient* for shared ride trip planning. While an unconstrained communication strategy is most effective (i.e. robust), this strategy is inefficient from a bandwidth standpoint. This strategy is also not feasible: the necessarily short communication windows limit the number of hops of messages. In contrast, the short-range communication strategy (one hop) is the most efficient, but least effective. Compared to the two, mid-range communication strategy proves the hypothesis: it is *effective* (e.g., for 1.56 hosts per street network node 10% longer trips than with unconstrained, but 66% shorter trips than with short-range) and it is *efficient* (e.g., for 1.56 hosts per node 5 times more messages than with short-range, but 9% of the messages with unconstrained communication—and this number is steeply decreasing with an increase of host density).

12.5 Conclusions

We have developed and specified a simulation environment for shared ride trip planning in large transportation networks. This agent-based simulation allows investigating the quality of the clients' trips depending on a communication strategy. For that purpose communication strategies are investigated under different traffic densities as well, coming up with recommendations of critical parameters m for specific traffic situations.

The sketched simulation environment can be extended in various directions, to consider additionally factors such as non-gridded street networks, multiple clients and their competition for transportation supply, or alternative trip planning strategies. The simulation environment can also be used for testing the consequences of individual behavior and preferences, such as mutual interest in client and host reputation. Some work in the direction of relaxing the wayfinding strategy and still limiting the communication effort is done in (Winter and Raubal, 2006), based on the simulation specification presented in this paper.

An interesting question in the realm of social change is the reputation of such a system, and trust-building mechanisms. Since in a distributed peer-to-peer system no centralized registry knows who traveled with whom, anonymity is of concern. For this reason, we currently are working on a safety measure that builds upon the ability to reconstruct any shared ride from the distributed, local memories of the agents that participated in the negotiation process.

The more important questions are those of social effects of such a system, which are far beyond this rather technical solution. It increases the mobility and access of citizens, in particular in disadvantaged suburbs or from disadvantaged groups (such as teens or the elderly). It is compatible with smart public transport systems, and can be combined with them. Other questions, such as economic impact, or effects on urban dynamics with such a system, come on the agenda as well.

Acknowledgements

The initial idea of a peer-to-peer shared ride service was born in a conversation with Monika Sester, Hannover, at AGILE 2002.

References

Dijkstra, E.W., 1959: A Note on Two Problems in Connexion with Graphs. Numerische Mathematik, 1: 269–271.

Nittel, S.; Duckham, M.; Kulik, L., 2004: Information Dissemination in Mobile Ad-hoc Geosensor Networks. In: Egenhofer, M.; Freksa, C.; Miller, H.J. (Eds.), Geographic Information Science. Lecture Notes in Computer Science, Vol. 3234. Springer, Berlin, pp. 206–222.

Winter, S.; Nittel, S., 2005: Shared Ride Trip Planning with Geosensor Networks. In: Brox, C.; Krüger, A.; Simonis, I. (Eds.), Geosensornetzwerke - von der Forschung zur praktischen Anwendung. IfGIprints, Vol. 23. Verlag Natur & Wissenschaft, Solingen, Germany, pp. 135–146.

Winter, S.; Nittel, S., 2006: Ad-Hoc Shared-Ride Trip Planning by Mobile Geosensor Networks. International Journal of Geographical Information Science, to appear.

Winter, S.; Raubal, M., 2006: Time Geography for Ad-Hoc Shared-Ride Trip Planning. In: Aberer, K.; Hara, T.; Joshi, A. (Eds.), 7th International Conference on Mobile Data Management (MDM'06). IEEE Computer Society, Nara, Japan, pp. 6–6.

Wolfson, O.; Xu, B., 2004: Opportunistic Dissemination of Spatio-Temporal Resource Information in Mobile Peer-to-Peer Networks, First International Workshop on P2P Data Management, Security and Trust. DEXA, Zaragoza, Spain, pp. 954–958.

Zhao, F.; Guibas, L.J., 2004: Wireless Sensor Networks. Elsevier, Amsterdam.

Mobile Information Services

13 Mobile ICT in public spaces and its impact on privacy

Karsten Weber, Ricarda Drüeke, Axel Schulz
European University Viadrina, Frankfurt (oder), Germany

13.1 Introduction

The insight that people worry about their privacy and the free flow of sensible personal related information in a wired world obviously is not new. However, answering the question what might happen if the whole environment is computerized is difficult. Mobile information and communication technologies (ICTs) will enable citizens to use information and communication services literally everywhere and anytime; simultaneously, it implies the massive production, processing and distribution of personal information. Therefore, the rising number of mobile ICT devices correlate to a corresponding growth of data created during mobile communication. Furthermore, technologies like RFID[1] could be used to control our behavior particularly in situations related to consuming. Hence, these technologies cross physical, individual and social boundaries by creating consumer profiles, which include what people have purchased as well as when and by whom these items are used. Additionally, it is very likely that in the near future mobile ICT will be embedded in a smart environment, an idea that is often called *ambient intelligence, ubiquitous* or *pervasive computing*. In such environments, ICT devices could not be recognized as separate artifacts with communication and computing capabilities; on the contrary, the ideal of that technology is to be invisible to its users (e.g. Beigl, Gellersen, Schmidt, 2001). It will provide communication and computing services in private as well as in public space without being recognized as ICTs.

Therefore, we believe that those technologies may change citizens' habits, because it will be even more difficult to control the flow of personal information than ever before. Moreover, people have to rely on and trust these new technologies and the people in charge – they simply will have no other choice. In ubiquitous computing environments, the risk to "accidentally" loose personal information or being spied will grow, consequently the notion of public sphere or space, as Habermas

H.J. Miller (ed.), Societies and Cities in the Age of Instant Access, 197–207.

and others coined it, as a place where free individuals gather and deliberately decide about public affairs, is threatened.

In this chapter, we examine the impact of mobile ICT on privacy in public spaces and argue that the advent and implementation of such technologies will increase the tendency that state authorities as well as private companies try to control public spaces. Therefore, we briefly explore some state-of-the-art mobile ICT that can be used for surveillance. The chapter addresses the extensive debate about privacy in cultural and social studies, and it briefly describes the functions of privacy for public spaces. Additionally, we introduce a study regarding the behavior of students using mobile ICT in public spaces.[2] We conclude that some behavioral changes are already observable but that further research is needed to decide whether this will alter the notions and functions of public spaces significantly.

13.2 Mobile information and communication technology

Laptops, mobile phones, personal digital assistants (PDAs), and the like are already everyday objects that are considered tools to enhance working abilities of professionals and to raise quality of life. A feature shared by most of these tools is the capability to transmit data wireless. Currently, one will find a whole bunch of mobile wireless broadcasting technologies like GPRS,[3] GSM,[4] UMTS,[5] WLAN,[6] WiMAX,[7] and so on. The areas that are covered are growing continuously; therefore, users can expect that at least in most industrialized countries they can use one of these technologies to access, for instance, the Internet. Simultaneously, the number of available applications based on such technologies is increasing as well. Once those new technologies came into existence, a new billion-dollar market with new applications and services will be created. Yet, it is note-worthy to point out that today mobile phones, as owned by almost every child, teenager, or adult living in an industrialized country, are quite established as everyday-life tools because they provide a service which was strongly demanded and sometimes even desperately needed.

But technology changes continuously – recent mobile phones are equipped with more computing capabilities, larger displays, more software like personal information managers (PIMs) or electronic games, charge-coupled device (CCD) cameras, global positioning system (GPS) functionality, and the like: "mobile phones are associated with various additional services and games, which are becoming part of everyday life." (Ahas, Mark, 2005: 547). One application of combinations of mobile phones and GPS is, for instance, a so-called *social-positioning method* (SPM). Ahas and Mark stress that "SPM will become very wide-spread in the future and will fundamentally change public life and administration" (2005: 547). It is easy to imagine that it can be used to track people and to monitor their behavior: briefly speaking, mobile ICT can be used for surveillance. Now, due to the new opportunities to communicate in public spaces we would like to ask whether new ICT will affect the notion of privacy as well as whether and how life in public places will change in an environment of omni-present control.

13.3 The impact on privacy

Although we would like to scrutinize the impact of mobile ICT on public spaces at first it will be necessary to think about privacy. In contemporary literature, one will find many definitions. For example, Gary Marx(2001: 161) emphasizes the distinction and relation of public and private. However, such a distinction does not clarify what the terms private and public mean. Moreover, unfortunately, there is no such thing like a clear definition. The notion of privacy seems to be a somehow ill-defined and fuzzy patchwork built of concepts like data protection, anonymity, or autonomy.[8]

13.3.1 Notions of privacy

Many concepts of privacy were developed and discussed more or less from a philosophical and of course a legal perspective.[9] However, there are other notions of privacy. For instance, privacy is sometimes discussed from a psychological perspective because different types and functions of psychological privacy affect individuals' behavior (e.g. Pedersen, 1997, 1999) and it is very likely that psychological aspects play a major role in the change of individual and societal attitudes towards privacy. Moreover, most currently discussed concepts of privacy strongly are linked to western thinking and culture (e.g. Newell, 1998; Kaya, Weber, 2003; Olinger, Britz, Olivier, 2005) and therefore are difficult to generalize. However, due to lack of space, we will not discuss the psychological and cultural dimensions of privacy with regard to theory any further, although in our survey we used some questions about ethnic origin.

For our purposes we will focus on informational privacy (e.g. Spinello, 2003: 143), which is already addressed in Samuel D. Warren und Louis D. Brandeis' article 'The Right to Privacy' published in 1890. Here privacy is defined as the right to be let alone: "To have privacy is to be left alone and invisible to public scrutiny" (Tunick, 2001: 520). Now, to protect privacy, either all other persons have to refrain from intruding our private realms or one is able to control the flow of information about oneself. Nevertheless, often it is inevitable that others gain access to personal related information, for instance, at the workplace, in public transport, or in restaurants. Obviously, locations differ due to the number of persons present or to their knowledge about each other. However, being in public one cannot know which and how many persons will have information about oneself. This is why we do not behave the same way in public as in private space – we adjust our public behavior to the expectations of others and to social and moral conventions of the embedding society. Because, generally speaking, public space does not provide informational privacy we try to control our behavior in order to control the flow of personal related information about ourselves.

13.3.2 Impacts on privacy

Now, ICT massively reduces one's informational privacy as Barrera and Okai (1999: 1) stress: "To be in cyberspace is to be recorded. Digital activities and objects are nothing but an ensemble of traces and records. [. . .] Those digital footprints can be, by nature, reconstituted, recreated and saved indefinitely. Where a vast number of activities in traditional space are inherently non-traceable, cyberspace actions are the traces themselves." If we leave a physical space, e.g. a bench in a park, it is almost impossible to track us. Leaving a place in real space in principle means to leave it forever; being in cyberspace however implicates to be there forever because everything one ever have done there can be tracked. Now, furnishing real space with (mobile) ICT is going to change how we can act in real space because it then will have characteristics of cyberspace – in principle, every action will be tractable indefinitely. Thus, the advantages (mobile) ICT provides to us could be overridden by the disadvantages of infringements into our privacy, as Leonhardt and Magee (1998: 52) stress: '[. . .] location services will often become repositories of potentially sensitive personal and corporate information. *Where you are* and *who you are with* are closely correlated with *what you are doing*. To leave this information unprotected for everybody to see is clearly undesirable.' Identification and authorization of users, cost and resource management, and so on will require the production of such information – mobile ICT cannot work that due to the aim of such systems to provide location- and context-aware services; in order to adequately react to our requests service and content providers continuously have to know about our actions, behavior, and lifestyle. Then, users potentially will change their behavior according to own attitudes, social norms they believe they have to meet, perceived behavioral control and moral obligations – right that was the aim of Jeremy Bentham's Panopticon (e.g. Reiman, 2004; Whitaker, 1999). Mobile ICT provides the technical means to make Panopticon as well as Big Brother a reality – it inherits surveillance by default and by design (Bennett, 2001).

13.3.3 Notions of public spaces

Debates about informational privacy already show strong relations between space and privacy: it cannot be guaranteed everywhere in real space. However, it seems that humans need shelter against public attention and social control (e.g. Cooke, 1999) and only private spaces can provide that last resort. Here, other persons are not able to learn anything about one's behavior nor are they able to criticize it. Given that other persons or society as a whole are not concerned nobody is entitled to monitor our behavior in private space; social and moral norms are suspended; relationships of persons in private space shall not be a matter of public intervention.[10] Here, 'public' denotes more or less the same like 'society' or 'community'; not a physical space is meant but a social environment in which people have to act and live. The relation of private and public or of citizen and society is constituted by moral, social, and legal norms as well as by rights and duties – technology, as indicated before, can change the range of privacy as one of these rights.

Furthermore, we have to mention another notion of the term 'public'. Historically speaking, one of the first distinctions of private and public can be found in the ancient Greek division of *oikos* and *polis*. Within the *oikos* the head of the family ruled with absolute power as master over his wife, children, and slaves. However, within the *polis* interests of citizens had to be balanced; that was done by the application of rules and laws which were made by democratic decisions – or at least, that was the Aristotelian ideal of decision-making processes in a just society. However, in his *Politea* Plato suggested a completely dissimilar society. The division of *oikos* and *polis* does not exist, private sphere is abandoned, and community shall control all aspects of personal life. Closer to our times are the Amendments of the US Bill of Rights; particularly the Fourth Amendment is important to the distinction of private and public. John Stuart Mill in turn stressed that there has to be a kind of space that cannot be compromised by society – without it, men could not develop creativity and uniqueness. Correspondingly, for Hannah Arendt private space is a refuge that men certainly need; the public covers the complete political, societal, and economical live. Finally, Jürgen Habermas defined 'public sphere' subsequently to Immanuel Kant as space in which rational persons deliberately make decisions beyond the influence and infringements of society's systems such as bureaucracy, economy, church, or state.

13.3.4 Impacts on public spaces

As often stressed, the public sphere is necessary for democracy – without it, democracy cannot work. Here, citizens shall be involved in political decision-making, control their representatives and state authorities, and even out their differing and conflicting interests. Yet, mass media already changed the characteristics of public sphere because citizens almost do not gather anymore to freely articulate their opinions; instead, mass media took that role of mediating political debates – in some sense, citizens as political actors cease to exist in media democracy, they merely are needed at the ballot booth. It is very likely that the implementation of ICT in public spaces will change our behavior even more extensively than mass media. Such technology can be used to track us and to collect data about us, our locations, and the whereabouts of actions. Concurrently, we will not know exactly who can access such information. We will loose control over the flow of personal related information and thus will not have informational privacy anymore. Certainly, it would be unreasonable to trust those who possibly have access – already now we face privacy invasions. In times like ours characterized by real threats of terrorism, imagined danger, even hysteria, fear and uncertainty state authorities do and will try to use any means to protect state's integrity. But that can entail collateral damage – a social climate of mutual suspicion, which makes it likely that citizens will refrain from performing actions like protesting against wiretapping, privacy invasions, or other restrictions of civil rights due to their fear to be sued as dangerous for society's security. Even if this scenario maybe is somehow exaggerated, the intuition lying behind is not. History shows that surveillance always changed citizens' behavior in public spaces – that is just the aim of surveillance. Now, mobile ICT will help to create an environment, in which no difference of surveillance and attention by service providers can be drawn and in which there will be no last resort for privacy. Thus, from a social or political scientist's point of view the question has to be answered whether people worry about such scenarios; as citizens we have to decide whether we would like to live in an Aristotelian or Platonian society.

13.4 Mobile phones in public spaces: A survey among students

As Pedersen (1999: 402) points out there are limits to one's behavior under surveillance of other persons or a group: it is likely that concessions with respect to own habits must be made due to preferences, aims, and desires of others. Only when people are alone they feel that they can do whatever they want to do. Subsequently, Pedersen (1997, 1999) argues that from a psychological point of view there are different types and functions of privacy people aim for. Keeping that in mind, his research seems to strengthen our hypothesis that there are some behavioral effects concerning privacy when mobile phones are used in public spaces, because there in principle and in practice it is impossibly to act unobserved – one cannot phone unheard when other people are nearby. Therefore, now we briefly describe a quantitative study among students.[11]

13.4.1 Research aims

To analyze the impact of mobile ICT on privacy we wanted to learn about privacy related behavior, attitudes, and perceptions of individuals who use mobile ICT in public spaces. Our primary aim was to examine the perceptions of individuals regarding their privacy and their expectations about privacy when using mobile phones. We focused on situations in which we expected that users could satisfy their personal needs and expectations about privacy: e.g. being alone, being without control, feeling secure. Hence, we wanted to know whether the respondents would share the situation of having privacy with others. Additionally, we sought to get information about how and in which cases individuals feel that their privacy can be 'lost', 'invaded', 'violated', and so on. As well, we wanted to evaluate the attitudes and practices with regard to the use of mobile phones. We asked, for instance, whether and how users adjust their behavior when being in public and phoning cannot be done unobserved; another question was which level of privacy user try to achieve in public while using mobile phones. A further rationale was to examine whether individuals are willing to 'sell' their privacy to get other benefits and which kind of personal related information they would provide to other persons, institutions, or companies. As preferences for privacy may vary depending on individual, cultural, and social factors we investigated the link between nationality, gender, cultural background and users' attitudes towards privacy.

13.4.2 Results

Not very surprisingly, privacy issues are quite important for the majority of the respondents: more than 50% of the respondents aim frequently or always for privacy; more than 45% of the students sometimes aim for privacy. To learn about individuals' attitudes towards privacy we asked the students about a statement concerning privacy they mainly agree with. Eight statements were presented reflecting different aspects of privacy – one of them had to be chosen. Each term addresses a particular notion of privacy that shall reflect the conception of privacy respondents have (see below). According to the answers, the most important aspect of privacy is to be let alone (23.2%) – which corresponds to Warren and Brandeis' definition of privacy, followed by the aim not to be disturbed (18.9%), and the aim to be secure (14.5%). In general, it seems that the conception of privacy that respondents have is a quite traditional one.

Presupposing that the importance of privacy might have effects on the behavior related to mobile phones, we asked in which way the aim for privacy affects the use of mobile phones in public. There is a latent ambiguity, since using a wired phone took place in private space, but now, using a mobile phone is rather a public action. Due to that change of context, we wanted to learn whether users alter their aim for privacy, their attitudes towards privacy, or their behavior in public. At first sight, the goal to have privacy and the conditions in public contradicts each other because striving for to be let alone or not to be disturbed traditionally are related to the metaphor of a room of one's own – which is private not public. Furthermore, the first four aspects of privacy ('Nobody can look at me', 'It is quiet', 'I am alone', 'I can run free') that were addressed in our questionnaire are connected with a spatial understanding of privacy that, generally speaking, cannot be realized in public. The other four statements ('I feel secure', 'Nobody disturbs me', 'I can keep my secrets', 'Nobody controls me') address aspects of privacy which are related to control theory of privacy implying control of flow of personal related information. Again, that is hard to realize in public.

To get some detailed information about behavior in public spaces, we analyzed attitudes and behavior related to mobile phone use, which is per definition rather a public action compared to the use of a wired phone at home in private space. We offered short descriptions of situations, concerning behavior in public, in which the respondents might feel disturbed by other persons. Each situation had to be judged by a range from −3 ("strongly disagree") to +3 ("strongly agree"). Thus, we learned that the most important aspects for individuals using a mobile phone are no surveillance while phoning, absence of unwanted audience, control flow of information. Respondents feel rather disturbed by other persons if they have to discuss private issues on the phone loudly. Situations in which other persons use mobile phones produce the same picture: in case that other persons phone or discuss private issues loudly on the phone the respondents feel disturbed. They want to be alone or undisturbed while phoning in public. Obviously, these aims are hard to fulfill in public particularly when using mobile ICT. However, respondents take care to defend their privacy – for instance, they go away from other persons while using a mobile phone. This proves that persons develop strategies to protect privacy in public and that awareness of privacy issues affects individual behavior. Because individuals need to feel secure in the midst of a smart environment we suggest that for the acceptance of mobile ICT it will be necessary that such aims for privacy have to be taken into account.

13.5 Final remarks

Already in 1991 Mark Weiser (1991: 94) proposed that "[t]he most profound technologies are those that disappear." His vision was that "invisible" ICT should assist people in everyday life and should work like small intelligent autonomous robots. In 1991 this vision was far away from realization but today, its advent seems to be very likely. Now, information can be collected and transmitted without knowledge, intention, and – most times – without awareness of people. On the one hand, this was exactly Weiser's vision: the door should know before hand that it should lock itself when one is leaving the apartment; the refrigerator should tell you before your weekend shopping that you run out of fresh milk. On the other hand, that opens the door for surveillance and poses questions where and how these artifacts should be used.

Our research showed that people are aware of their privacy and try to avoid to 'loose' information, or try to protect it. However, the potential invisibility of future mobile ICT could decrease such awareness. One example that raises such concerns is the project NSOPR of the United States Department of Justice. In July 2005, it launched an online "National Sex Offender Public Registry" that shall offer ordinary people a tool that lists sex offenders and shows their current state of residence on the Internet; people all around the world do have access to this website. Now, let us suppose that this service is combined with the capabilities of mobile ICT to determine the exact location of persons that are listed: obviously, those persons will loose privacy almost completely. In case of sex offenders, this maybe seems to be acceptable but we have to realize that such surveillance technology can applied to all of us.

Although citizens rejected the technology in that last case, industry currently develops countless applications of mobile ICT for an emerging billion-dollar market – the average consumer of the future will own "hundreds or thousands of mobile and embedded computing devices." (Beresford 2005: 3) Consequently, the opportunities to use and misuse mobile ICT will become manifold. Whether this will change public spaces in a direction that privacy will completely fade away and therefore will create a Platonian kind of society should be subject to further research as well as to citizens' political deliberation.

Notes

[1] Goldrum et al. (2005) provide in their rather technical article an overview of what RFID (Radio Frequency Identification) is, how it works, and how it can be used. Hunter (2005) describes current developments.

[2] The authors are currently engaged in the project "Mobile Internet Services and Privacy", which is part of a joint project of the European University Viadrina Frankfurt/Oder and the Research Institute "Innovations for High Performing Microelectronics" (IHP) funded by the German Federal Ministry of Education and Research (BMBF). The aim of the authors' project is an analysis of changes of attitudes towards privacy, particularly in situations in which individuals use mobile ICT devices.

[3] GPRS (General Packet Radio Service) is a service available to users of GSM mobile phones (cf. Ghribi, Logrippo, 2000).

[4] GSM (Global System for Mobile Communications) is the common standard for mobile phones in the world. The standard data transmission rate in GSM networks is 9.6 KBit per second (e.g. Miceli, 2003).

[5] UMTS (Universal Mobile Telecommunications System) is one of the third-generation (labelled as "3G") mobile phone technologies. UMTS networks provide a bandwidth of up to 384 KBit per second (Lehr, McKnight, 2003: 353)

[6] WLAN (Wireless Local Area Network) is getting the most important and most often used wireless technology in many business areas. Airports, railway stations, urban districts, schools, universities, and the like are equipped with WLAN. Recent WLAN standards offer bandwidths up to 108 MBit per second across distances of some 100 meters (Lehr, McKnight 2003: 355)

[7] WiMAX (Worldwide Interoperability for Microwave Access) is considered to be the alternative wireless technology for "last mile access" in rural areas, e.g. in Scandinavia. A WiMAX base station covers an area of about 50 km in radius with a bandwidth of up to 70 MBit per second (Hoymann, 2005).

[8] Again, Marx (2001: 161) stresses that "[. . .] the mental cacophony associated with the rich variety of empirical configurations seen with electronic surveillance and other forms of information technology stems from the failure to differentiate between, and note the inter-relations of various dimensions of the public and private." According to Diane Michelfelder (2001: 129) there are a couple of problems: "When philosophers with an interest in privacy move beyond discussing its common-sense meaning to a consideration of more complex issues, [. . .], they quickly find themselves in a terrain characterized by a lack of clear agreement over a variety of fundamental aspects of privacy, including its scope, definition, and value."

[9] One will find different attitudes towards the legal status of privacy. For example, Kenneth Brown (2000: 1) emphasizes that "privacy and the "the right to privacy" are loosely defined." He adds that

from an US point of view the term privacy is neither defined in the Bill of Rights nor in the amendments of the Constitution. Contrary to this skeptical view Lloyd Weinreb (2000: 1) holds that there cannot be any doubt that a right to privacy exists in US legislation. In the EU one will find legislation that shall protect personal related information.

[10] Many feminist scholars (e.g. Pateman, 1992) argue that privacy often was and is misused to hide infringements into rights particularly of women and children. Criticism is concentrated on the binary-coded disjunction of private and public sphere. Hence, the private sphere is related to women, domestic work and family – in contrary to the public sphere, which is related to political deliberation, gainful employment and civic liberties (cp. Fraser, 1989). Thus, unequal changes of men and women are perpetuated and private affairs (and maybe relationhips of violence and exploitation) are absconded from societal control. Therefore, one important aim of most feminist scholars is to create a concept of privacy which protects it for women and man without an ideologically powered attribution of certain spheres to women (e.g. Allen, 1998).

[11] We carried out our survey in winter term 2004/2005 at the European University Viadrina in Frankfurt/Oder, Germany. One of the important characteristics of this university is the large number of foreign – particularly Polish – students: about one third are Polish, more than the half are German students. We distributed the questionnaires in seminars and lectures of the three faculties of the European-University Viadrina (economics, cultural studies, and law). Hence, we got 629 completed questionnaires of about 4000 students enrolled at the university.

References

Ahas, R. and Mark, Ü. (2005), Location based services—new challenges for planning and public administration? *Futures*, **37**, 547–561.

Allen, A. (1998), Privacy, in *A Companion to Feminist Philosophy* (ed. A.M. Jaggar and I. Young), Blackwell, Malden.

Barrera, M.H. and Okai, J.M. (1999), Digital Correspondence: Recreating Privacy Paradigms. *International Journal of Communications Law and Policy*, **3**, http://www.ijclp.org/3_1999/ijclp_webdoc_4_3_1999.html, last visited 03-15-2005.

Beigl, M., Gellersen, H.-W. and Schmidt, A. (2001), Mediacups: experience with design and use of computer-augmented everyday artefacts. *Computer Networks*, **35 (4)**, 401–9.

Bennett, C. J. (2001), Cookies, web bugs, webcams and cue cats: Patterns of surveillance on the world wide web. *Ethics and Information Technology*, **3 (3)**, 197–210.

Beresford, A.R. (2005), Location privacy in ubiquitous computing. Technical Report, **No. 612**, Cambridge, online available at http://www.cl.cam.ac.uk/Research/DTG/lce-pub/public/arb33/UCAM-CL-TR-612.pdf, last visited 08-15-2005.

Brown, K. (2000), The Internet Privacy Debate. *International Journal of Communications Law and Policy*, **6**, http://www.ijclp.org/6_2001/ijclp_webdoc_11_6_2001.html, last visited 03-15-2005.

Cooke, M. (1999), A space of one's own: autonomy, privacy, liberty. *Philosophy & Social Criticism*, **25 (1)**, 23–53.

Fraser, N. (1989), *Unruly Practices: Power, Discourse, and Gender in Contemporary Social Theory*, Polity, Oxford.

Funk, J.L. (2005), The future of the mobile phone Internet: an analysis of technological trajectories and lead users in the Japanese market. *Technology in Society*, **27**, 69–83.

Ghribi, B. and Logrippo, L. (2000), Understanding GPRS: the GSM packet radio service. *Computer Networks*, **34**, 763–779.

Goodrum, P.M., McLaren, M.A. and Durfee, A. (2005), The application of active radio frequency identification technology for tool tracking on construction job sites. *Automation in Construction*, (in press).

Hoymann, C. (2005), Analysis and performance evaluation of the OFDM-based metropolitan area network IEEE 802.16. *Computer Networks* (in press).

Hunter, P. (2005), London terrorist attacks heat up identity card debate and highlight uncertainties over their efficacy. *Computer Fraud & Security*, **7**, 4–5.

Kaya, N. and Weber, M. J. (2003), Cross-cultural differences in the perception of crowding and privacy regulation: American and Turkish students. *Journal of Environmental Psychology*, **23 (3)**, 301–9.

Lehr, W. and McKnight, L. W. (2003), Wireless Internet access: 3G vs. WiFi? *Telecommunications Policy*, **27**, 351–370.

Leonhardt, U. and Magee, J. (1998), Security Considerations for a Distributed Location Service. *Journal of Network and Systems Management*, **6 (1)**, 51–70.

Marx, G. T. (2001), Murky conceptual waters: The public and the private. *Ethics and Information Technology*, **3 (3)**, 157–69.

Michelfelder, D. P. (2001), The moral value of informational privacy in cyberspace. *Ethics and Information Technology*, **3 (2)**, 129–35.

Miceli, A. (2003) *Wireless Technician's Handbook*. Artech House, Boston, second edition.

Newell, P. B. (1998), A Cross-Cultural Comparison of Privacy Definitions and Functions: A Systems Approach. *Journal of Environmental Psychology*, **18 (4)**, 357–71.

Olinger, H. N., Britz, J. J. and Olivier, M. S. (2005), Western privacy and ubuntu: influence in the forthcoming data privacy bill, in *Ethics of New Information Technology, Proceedings of CEPE 2005* (ed. P. Brey, F. Grodzinsky, L. Introna), CTIT, Enschede/The Netherlands.

Pateman, C. (1992), Feminist Critiques of the Public/Private Dichotomy, in *Justice in Political Philosophy, Volume II* (ed. W. Kymlicka), Edward Elgar Publishing, Aldershot, first published 1981.

Pedersen, D. M. (1997), Psychological Functions of Privacy. *Journal of Environmental Psychology*, **17 (2)**, 147–56.

Pedersen, D. M. (1999), Model for Types of Privacy by Privacy Functions. *Journal of Environmental Psychology*, **19 (4)**, 397–405.

Reiman, J. (2004), Driving to the Panopticon: A Philosophical Exploration of the Risks to Privacy Posed by the Information Technology of the Future, in *Privacies. Philosophical Evaluations* (ed. B. Rössler),Stanford University Press, Stanford.

Spinello, R.A. (2003), *Cyberethics. Morality and Law in Cyberspace*. Jones and Bartlett Publishers, Sudbury.

Tunick, M. (2001), Does Privacy Undermine Community? *The Journal of Value Inquiry*, **35 (4)**, 517–34.

Weinreb, L. L. (2000), The Right to Privacy. *Social Philosophy and Policy*, **17 (2)**, 25–44.

Weiser, M. (1991), The Computer for the Twenty-First Century. *Scientific American*, **265 (3)**, 94–104.

Whitaker, R. (1999), *The End of Privacy*. The New Press, New York.

14 The dimensions of locational privacy

Scott A. Bridwell
Department of Geography, University of Utah, Salt Lake City, UT

14.1 Introduction

Mutual advancements in wireless communications and geospatial technologies are creating real-time, pervasive tracking environments capable of capturing detailed information about individual activities in space and time. These advances will potentially provide benefits for individuals, researchers and commercial interests. Among the more exciting possibilities are location-based services (LBS). LBS are information services dependent upon the geographic location of users to deliver spatially adaptive content such as maps, routing instructions and friend finding (Hjelm, 2002; Schiller and Voisard, 2004; Küpper, 2005). Common applications include emergency response, personal navigation assistance, fleet management and recreation. LBS may also serve as a mechanism for collecting disaggregate activity-travel data from an unprecedented number of individuals, providing social scientists and planners detailed information regarding spatiotemporal patterns of mobility, activity participation and social interaction in urban environments (Dykes and Mountain, 2003; Miller, 2005b; Mountain, 2005).

However, the development of ubiquitous location sensing creates serious concerns for individual location privacy. Information privacy generally refers to the claim that individuals should have control over the information collected about them (Westin, 1967). Reductions in personal information privacy are generally a product of two types of innovations: innovations for observing information about people and innovations for sharing this information with others. For LBS, users are mobile and are located through the presence of location-aware devices, often wireless cellular telephones. These devices obtain detailed positional estimates for individuals and distribute this information to a broader network through established communication protocols. If positional estimates are sampled with frequent regularity, movement traces are created identifying all locations visited by individuals in space and time. Given the increasing distributed nature of information services, these observed locations may be transferred among multiple entities within the context of a single transaction.

H.J. Miller (ed.), Societies and Cities in the Age of Instant Access, 209–225.

This scenario suggests significant reductions in locational privacy, as users may have less control over what is known of their whereabouts (past, present, and future) and thus their activities and behavior. Locational disclosure—the ability to identify or infer sensitive characteristics of people from locations they occupy in space—may result from the analysis of LBS-derived tracking data and used for unethical practices such as 'Spatial Spam'. In extreme situations, power relations such as geoslavery may develop in which an authority controls the spatial behavior of another individual (Dobson and Fisher, 2003). Nonetheless, it is difficult to dismiss the benefits of LBS even in the face of these concerns. Therefore locational privacy is a vital research area concerned with the management practices of spatiotemporal information that balance the conflicting objectives of collecting detailed, individual-level data and preserving individual privacy. These efforts are complicated by the fact that privacy is an intuitive yet ambiguous concept and that researchers are still grappling with methods for reasoning about spatiotemporal phenomenon. Some of the predominant research questions include:

- What data should be collected about individuals and released to others?
- How might this data be used inappropriately, and how might this cause harm to individual participants?
- What methods may decrease the likelihood or risks of abuse?
- How do these methods impact the quality of analyses, models and applications dependent on the data?

This chapter explores the dimensions of locational privacy in an attempt to outline standards that can be used to evaluate the concept more precisely. The chapter is organized into five sections. The next section discusses general principles of locational privacy that have been defined by various organizations and regulatory efforts. Section 3 discusses the developments of safeguards used to implement these principles for aspatial and spatial data. Section 4 discusses the impacts of privacy in continuous tracking environments based on a time geographic perspective. Section 5 concludes with a summary of the chapter and suggestions for future research.

14.2 Locational privacy principles

Although often discussed, the concept of privacy is difficult to define precisely. The discussions in the literature often embrace more of a spirit or theme rather than a coherent semantic meaning. According to Westin (1967), a major component of privacy is the ability to move about anonymously, while the conflicting objective of surveillance is to be informed of an individual's location at all times. Others more explicitly suggest that locational privacy is the ability to prevent undesired entities from knowing one's past, present, and future locations (Beresford and Stajano, 2003). However, efforts at precise definitions leave a great deal of room for interpretation and ambiguity, making standardization of policies and legislation problematic. These ambiguities are given greater clarity by considering location privacy as a subset of information privacy, with specialized requirements for information hiding and disclosure limitation.

Many of the standards developed to address location privacy thus far closely resemble other information privacy standards, particularly those in place for e-commerce on the Internet. These standards vary to some degree in their source, content, and objectives but do share some common ideas (for more exhaustive reviews see Cuellar, 2002; Hjelm, 2002, pp. 309–323; Morris, 2002; Spiekermann, 2004). These can be summarized into five governing principles, provided in Figure. 14.1. According the principle of *consent*, users must be allowed to control: (i) what data can be collected; (ii) who it can be shared with; (iii) when it can be collected; (iv) how long it can be stored and (v) for what purposes. In the process, organizations must provide users *notice* to make them aware of the data collection practices. This is often facilitated through an 'opt-in' or 'opt-out', in which users consent to allow information about them to be collected and are notified of data practices on a contractual basis. Once a user consents and is notified of data collection practices, there must be proper *access* to the stored information. According to this principle, users should also be able to edit and change any stored information deemed incorrect or invasive. Each of these principles suggests that the communication between a user and an organization is critical and the semantics used to convey privacy should be clear, explicit, and consistent.

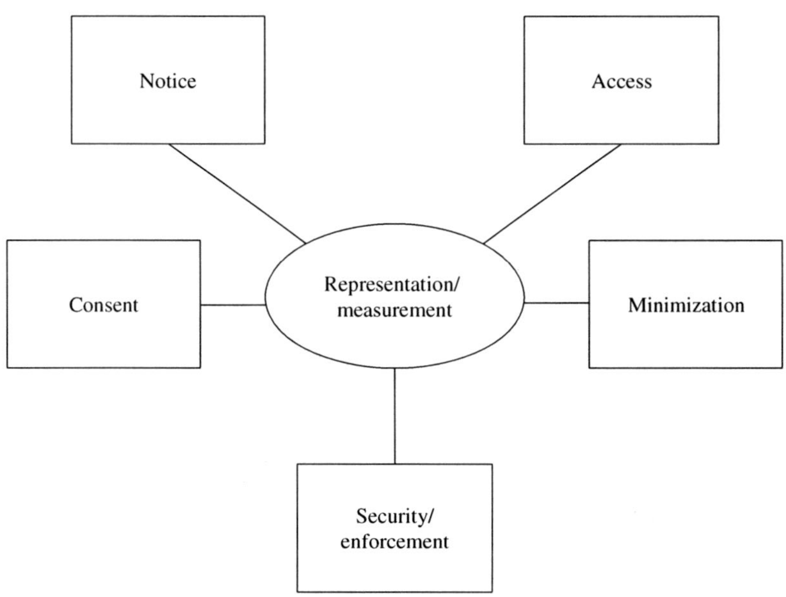

Fig. 14.1 Locational privacy principles

The principles discussed above are largely concerned with the interaction between users and data collection entities (often service providers). Additional principles address internal procedures and practices. Most standards and privacy efforts recognize the need to *securely* gather and transfer permissible data in a manner that enforces the integrity of the consent and notice provided. In terms of security, this has inspired work in intelligent information architectures particularly user appropriated software agents, encryption techniques and intervening middleware (Choy et al., 2000; Beresford and Stajano, 2003; Myles et al., 2003; Hull et al., 2004). The concept of *minimization* requires an organization to only collect data that is necessary to achieve service objectives. There must be a justification for the need for the information to be collected and used. Each of these principles is contingent upon proper representation of people based on the measurement practices for obtaining the desired information. This can be used to provide more flexible strategies for providing users with control over their information.

14.3 Privacy safeguards

The principles rooted in the tradition of existing information privacy frameworks are intended as protecting mechanisms that give individuals greater autonomy over their personal information. These principles are largely driven by widespread commercial concerns with little consideration for the benefits of collected information. In doing so they often fail to consider the impact that privacy preservation might have on the quality of analyses that rely on this information. This is in part because the collecting entities (e.g. credit card companies, web sites) are often seen as a greater threat and more likely to use collected data for unethical practices. Alternative perspectives consider the potential privacy threat relative to the utility of the collected information. This section discusses various approached to developing safeguards to provide privacy protection while collecting and sharing data with others.

14.3.1 Data masks

Data masks are disclosure limitation strategies. Data masks perform transformations on collected data in a manner that provides anonymity to participants or introduces uncertainty and error into their measured attributes while retaining the essential properties and relations required for a given analysis. Masks have proven particularly useful in fields such as epidemiology and medicine where analyses of critical health patterns are important but data are sensitive. The first masks were developed for aspatial data, beginning with Warner's (1965) technique for randomizing user responses in socioeconomic surveys.

Mask development since Warner's work has focused on the identification of techniques and procedures for inducing error and uncertainty, identifying the privacy threats within datasets, and establishing formal standards for expressing mask specifications. The development of techniques for intentionally introducing errors has included data suppression, aggregation, rounding, and bootstrapping (Bowden and Sim, 1992). Work on privacy threats has focused on the entities involved with the masking process and the types of disclosure that may occur from disseminating data to other entities. The types of entities include participants, data custodians, investigators and target participants. *Participants* are individuals who have personal information collected about them. *Data custodians* are entities that collect the information and are responsible for maintaining confidentiality. *Investigators* are third parties who wish to use the collected information. *Target participants* are individuals that may be targeted by investigators to learn more than was initially intended with the released data. The objective of the masking process is to allow the custodian to provide data to investigators while maintaining the confidentiality of the target participants and limiting potential disclosure. Disclosure occurs when unintended information is attributed to an individual, thus individuals have less control over their personal information as new data exists that they did not explicitly consent to. Particular attention is paid to limiting *identity disclosure*, the ability to identify an individual within a data set based on attribute values, or *attribute disclosure*, the ability to identify intimate characteristics of a known individual (Duncan and Lambert, 1986; Paas, 1988). Other efforts have focused on analyzing the differences in potential disclosure between masked and unmasked datasets (Duncan and Lambert, 1986) and the effects that different types of masks have on the analyses that might be conducted by investigators (Bowden and Sim, 1992).

The diversity of mask techniques and their relation to disclosure and privacy threats prompted efforts to develop standard specifications for defining mask procedures, termed *matrix masks* (Duncan and Pearson, 1991; Cox, 1994). The intention is to provide a classification of different masking techniques as well as a means of comparing different masked datasets. The masks are formalized by conceptualizing a dataset as matrix with rows indicating observed individuals and columns indicating attributes collected about these individuals. The types of functions applied to these matrices are classified as record transforming, attribute transforming, or displacing. *Record transforming masks* delete, aggregate, or insert false rows of data to obtain a desired level of uncertainty. Attribute transforming masks insert, delete, or transform attributes or columns of the matrix. *Displacing masks* introduce deterministic or stochastic noise. The primary objective of the formal matrix mask framework is to describe and compare the privacy or disclosure vulnerability of different data sets.

The disclosure threats may be further evaluated by attempting to emulate the processes by which an investigator may link the released data to an external source that may contain sensitive or identifying information (Lambert, 1993). This is generally accomplished by considering the relationships between masked and unmasked data or the relationships between the released data and a hypothetical dataset that an investigator may possess. The objective of these approaches is to obtain quantifiable measures of the probability of disclosure risk and the potential harm that may to come to participants if their identity was revealed. These measures are evaluated solely on the ability to link dataset based on semantic attributes. An increasingly prominent method for linking external sources to released data is provided by the development of geographic information systems (GIS), creating new considerations for masks and disclosure limitation.

14.3.2 Geographic masks

Locational privacy concerns are not new to geospatial technologies. The proliferation of GIS has brought pronounced changes in personal privacy as residences and work locations have become easily georeferenced and placed in context with other spatial information (Onsrud et al., 1994; Curry, 1997, 1998; Armstrong, 2002; Monmonier, 2002). Perhaps most often cited is the use of geodemographic systems (Goss, 1995), which provide marketers a spatial expression when selling products targeting groups with specific demographic profiles. Data masks provide a series of techniques for preserving confidentiality and methods for understanding the privacy threat in sensitive datasets. However, the integrative and inferential power of geographic information requires additional considerations. The development of vast geographic datasets at increasing resolutions and levels of detail in conjunction with enhanced algorithms for analyzing and visualizing this information pose additional demands on privacy preserving methodologies (Armstrong, 2002; Monmonier, 2002). Complementing technologies, such as the Internet, make these concerns more immediate as this information is easily and cheaply distributed to a broad base of people.

Geographic masking techniques are specialized masks designed to minimize location disclosure while allowing for disaggregate spatial analyses. Location disclosure is similar to the types of disclosure discussed above, but is dependent upon the ability to assign linkages between data sources based on a set of primitive spatial relationships such as topology and distance. Participants are represented as two-dimensional point features with associated attributes. The focus of these techniques is on spatial information, and aspatial attribute information is left intact (Armstrong et al., 1999). These masks provide privacy protection by 'hiding' the true location of an individual through some type of transformation that maps the participant to a new location. Geographic masking techniques include aggregation, deterministic transformations, precision, and stochastic techniques (Figure. 14.2). The inherent spatiality of the masks gives additional mask properties including the resulting *mask vector* and a *spatial uncertainty region*. The mask vector describes the relationship between the true and masked point, with the identifiable properties of distance and direction. The distribution of the mask vectors provides insights into the behavior of the masks and can be used to evaluate disclosure risks. A spatial uncertainty region may also exist which gives an indication of the

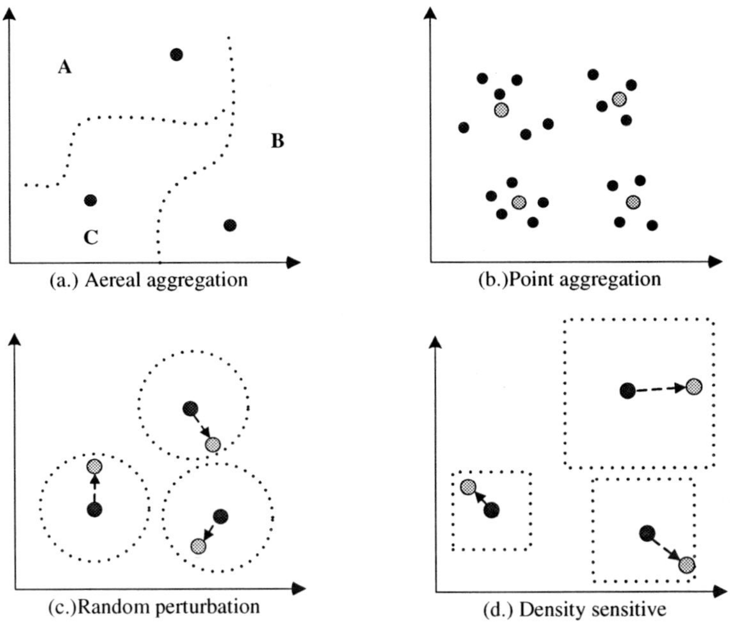

Fig. 14.2 Geographic masks (after Armstrong et al., 1999)

degree of uncertainty—defined in terms of confidence in the measured location—introduced by a given mask. The degree of uncertainty produced is theoretically inversely related to the disclosure risk.

Aggregation masks merge individual participants into a single record in the data set. Areal aggregations (Figure. 14.2a) use point-in-polygon overlay techniques to assign participants to areal units that encompass their true location. The resulting attributes will represent summaries of the aggregated records. Another possible approach is a point aggregation (Figure. 14.2b), which collapse multiple points into a single point through blurring techniques. Several types of measures can be developed to prescribe if a given aggregation scheme is adequate for preserving the privacy of the target participant. The US Census has a set of prescribed rules to differentiate among sensitive and non-sensitive regions (Federal Committee on Statistical Methodology, 2005). According to this scheme an area is sensitive if:

- *(n) threshold rule*: defines the number of respondents in a class that is less than n
- *(n, k) rule*: if a small number of participants (n) contribute to a large percentage of (k) of the total value
- *(p%) rule*: upper or lower estimates of a respondent's value are closer to the reported value than a given percent (p)
- *(pq) rule*: similar to p% but incorporates a set of prior knowledge (q)

One of the problems with aggregation masks is that they alter the structure of the data making it difficult to conduct spatial analysis techniques such as point patterns analysis and cluster detection. *Random perturbation masks* employ stochastic noise to intentionally hide the locations of users (Figure. 14.2c). Noise is introduced by generating a random number corresponding to a specified range. Variations in noise are affected by the statistical probability distribution used to generate random numbers and the algorithm used to apply the random number to the original participant location. In certain situations it may be desirable to normalize the noise according to the population density (Figure. 14.2d) to provide masks with less risk of identity disclosure and to minimize the impacts on spatial analysis (Kwan et al., 2004; Cassa et al., 2006) In these situations areas that are more densely populated will have less perturbation introduced. These methods create interesting questions about the relationship between uncertainty, anonymity and masking.

Geographic masking techniques have also been evaluated in terms of their robustness to additional tampering from 'Data Spies' and 'Map Hackers' (Armstrong, 2002). These are parties interested in compromising the masks using ancillary information. Possible inference methods include geocoding and map overlays. Beyond this, other inferences could be made through the acquisition of other resources that would compromise the mask. For instance a data set containing areas unsuitable for living could compromise a mask if masked points were overlaid and found to be in locations impossible for them to be in reality. Other possible breaches can be made through the use of 'output geographies' (Duke-Williams and Rees, 1998). In this situation spies request data for a region at multiple levels of aggregation and use differences among the data sets to infer additional information to compromise the mask. The disclosure risk may also be evaluated by considering the distance among masked locations and their 'parents' (Armstrong et al., 1999). The parent is a masked points non-masked counterpart. An analysis of the proportion of masked point's that are closer to their parents than any other location in the unmasked dataset provides an indication of the disclosure risk of the masked dataset and some of the behaviors of possible third parties.

14.3.3 Anonymity, *k*-anonymity and pseudonymity

The attributes collected about individuals can be generally classified as *key* and *sensitive* (Lambert, 1993). Key attributes correspond to attributes such as name, social security number and birth date that may act to distinguish or identify an individual from others within the dataset, this is similar to a key attribute within relational database theory. Some key attributes such as name can uniquely identify an individual. Other key attributes, such as race, income and gender act as *quasi-identifiers* that identify individuals based on a unique combination of multiple values (Sweeney, 2002). Quasi-identifiers are similar to composite keys in relational database theory. Sensitive attributes correspond to attributes such as income, medical information and credit history. Some attributes, such as geographic location, can be both key and sensitive depending on the given application and knowledge of the intentions of a third party. The objective of anonymity and pseudonymity safeguards is to explicitly disassociate key attributes from the other attributes when releasing information in an attempt to limit identity disclosures. This can be accomplished for aspatial data by removing single key fields and manipulating the member attributes of quasi-identifiers such that multiple individuals will posses the same quasi-identifier. The measure k-anonymity identifies the minimum number of individuals that share a quasi-identifier. This measure can be used with entropy-based metrics and Bayesian reasoning to model the attribute disclosure risk within the dataset. However, this process proves more difficult for spatiotemporal datasets because locations act as both key and sensitive attributes and multiple spatial events can be used to increase the probability of identity disclosures.

Spatial cloaking (Gruteser and Grunwald, 2003), adapts the concept of k-anonymity for spatial data by considering the locational representation of an object for a given duration of time as the quasi-identifier (Figure. 14.3). The method is intended for use with LBS applications in which an individual provides their location to a trusted entity or broker that then 'cloaks' the location before distributing the information to a third party. The cloak, similar to a geographic mask, transforms the representation of an individual from a point location to a region that contains the individual for a prescribed duration of time. The individual's k-anonymity is then measured as a function of the number of other individuals that are also located within this region. Using a quad-tree based method, the cloak can guarantee a desired degree of anonymity by iteratively partitioning the region anchored by the location of the individual into quadrants until the desired number k-anonymity is achieved. However, while this guarantees the degree of anonymity for a single individual, there is the potential for less anonymity when considering the cloaked regions for multiple individuals for a single temporal duration. According to the classical definition of k-anonymity, anonymity among multiple individuals is defined in terms of equality of the quasi-identifier not proximity. In order for this concept to be adaptable for geographic space, the regions defined by cloaks among all active individuals would need to create a tessellated space in which there are no overlaps among neighboring regions. Thus all individuals within a given cloaked

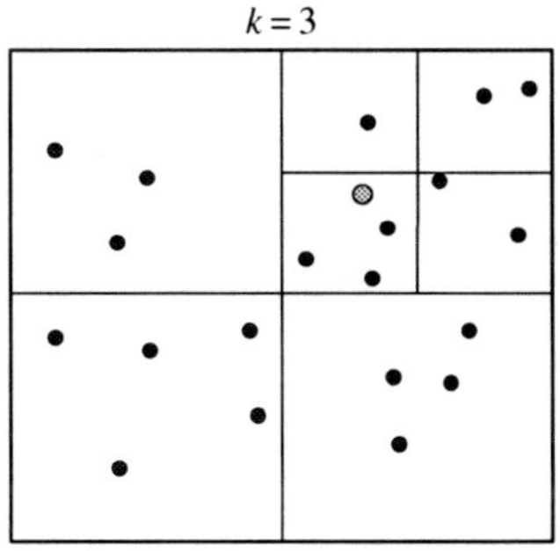

Fig. 14.3 Locational k-anonymity (after Gruteser and Grunwald, 2003)

region would be anonymous in that their spatial representation is shared. This is largely an implementation detail that can likely be rectified by a manipulating the method used to create the cloaked regions.

One of the difficulties of the majority of the methods discussed in this section is that they are predominately static and thus do not consider situations in which multiple, often dependent events are recorded for a single individual. This is essentially a scenario of tracking which will likely emerge from ubiquitous location awareness. Here there is an additional concern about how identity disclosure may occur by linking multiple events to obtain additional information about an individual. One approach to addressing tracking scenarios is the use of pseudonyms (Pfitzmann and Köhntopp, 2001). Similar to pen names used by authors, pseudonyms are fictional identities associated to observed individuals. This is based on a representation of spatiotemporal phenomenon as two entities: objects and events. Objects correspond to the individuals with persistent qualities that will not change for a prescribed temporal duration (e.g. name, birth date), while events correspond to observations (e.g. location trace, credit card transaction) that can be linked to the object. Under the pseudonymity approach, the focus is primarily on the securing anonymity in the object or individual with less consideration for the potential disclosure for the events that have transpired. Thus individuals are disassociated from the sensitive attributes at the object level, but not necessarily disassociated from the sensitive and identifying attributes at the event level.

An alternative approach is to not only dissociate key and sensitive attributes, but to also disassociate events from each other. The *mixed zone model* (Beresford and Stajano, 2003) attempts to provide anonymity for individuals at specific geographic areas by assigning individuals multiple pseudonyms that are updated when entering specific locations (Figure. 14.4). Here the tracking environment is partitioned into application and mixed zones. Locational traces are obtained for individuals when they are located in the application zone. However, upon entering a mixed zone (an ingress event), no locations are observed until they leave the zone (an egress

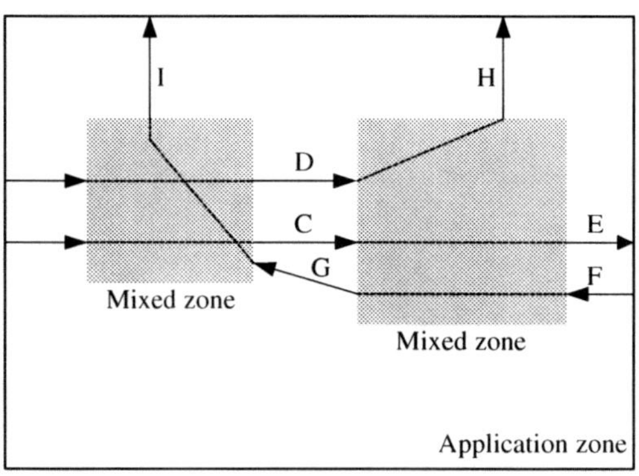

Fig. 14.4 Mixed zone model (after Beresford and Stajano, 2003)

event). A new pseudonym is given during the egress event. The anonymity for a time duration is then defined by the number of individuals within a zone at a given time. The appropriate design and configuration of the mixed zones is an important determinant for the amount of anonymity provided. Zones that are too small will only capture a small number of individuals. However, zones that are too large may compromise the intended anonymity due to increased variations in distances that objects travel in relation to their time spent in the zone. Using an information theoretic measure, Beresford and Stajano (2003) also show that knowledge of the distribution of movement transitions between zones, might also decrease the utility of this approach as some types of movements (e.g. straight) or much more likely than others (e.g. u-turns). In spite of this, the mixed zone approach is one of the few attempts to provide anonymity for individuals in densely tracked environments.

14.3.4 Negotiation and obfuscation

The privacy safeguards discussed thus far can be loosely described as disclosure limitation strategies. Within these schemes an individual provides information to a trusted entity, usually a statistical agency or LBS broker, that manipulates the representation of the information in a manner that limits the possibility of disclosure and harm. An alternative perspective is to consider a negotiation process between an individual and un-trusted entity (Duckham and Kulik, 2005a,2005b). Here the user obfuscates, or masks, their own location before providing it to an un-trusted entity that they wish to obtain a service from. Based on the degree of obfuscation and the nature of the desired service, the entity will report to the user the quality of the service that can be achieved given the information they provided. Here the user then has a choice to provide more detailed information about their location or accept the service with a reduced quality. The methods for conducting the obfuscation can be similar to the methods discussed above, however they are applied to a completely different perspective of information use and disclosure. The negotiation process described by Duckham and Kulik takes place within the context of a single transaction. It may also be desirable to model the negotiation process over multiple transactions to see how users 'learn' from their previous encounters with an un-trusted entity.

14.4 Towards a spatiotemporal model of locational disclosure

The privacy safeguards discussed in the previous section vary in terms of their definitions of privacy, anonymity and disclosure. These definitions become even more difficult to define and compare when attempting to model individuals moving and being tracked continuously. However, accepted definitions and management strategies for these concepts are essential for both supporting commercial applications and conducting scientific research on densely tracked individuals. In this section we examine some potential measures of privacy and disclosure risk based on a time geographic representation.

14.4.1 Representation of tracked objects

The time geographic or space-time perspective (Hägerstrand, 1970), provides one of the more compelling frameworks for representing tracked objects because it can capture the spatial and temporal properties of human movement, as well as the necessary conditions for these movements to occur. The *space-time path* (Figure. 14.5.) provides the preeminent conceptual structure to store observed movements of human process. These structures store individual traces of movement with respect to time. This is based on the notion that individual existence contains both temporal and spatial attributes and that this existence can be 'diagrammatically' described as a continuous trajectory in a three-dimensional absolute space. Thus rather than collapsing existence to a single location in space, a more fulfilling representation contains all locations in which an individual has occupied in space, sequenced and connected through time. The space-time path can be modeled as a set of discrete time-stamped locations (termed space-time *control points*), in conjunction with a linear interpolation function describing the location for temporal durations between measurements (Miller, 2005a). Times at which the object are

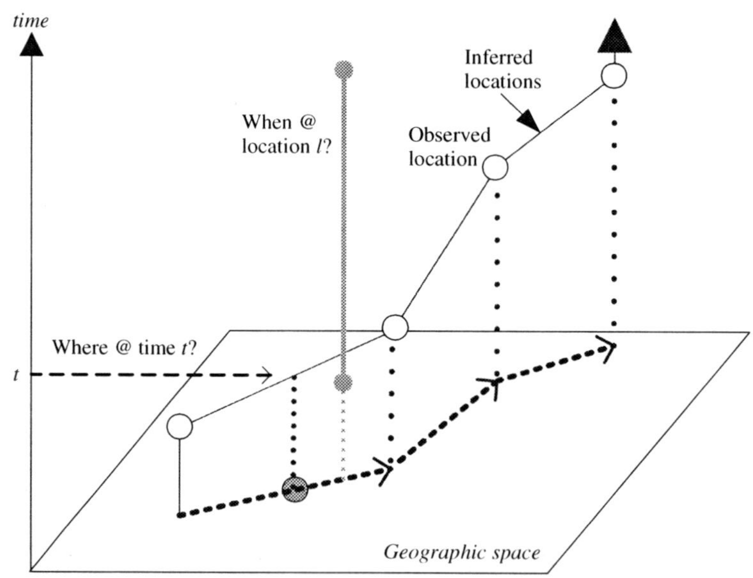

Fig. 14.5 Space-time path representation

moving are indicated by the path's slope: this indicates the speed the object is moving. When objects are stationary the path is vertical.

Because of the explicit recognition that geographic objects contain both spatial and temporal properties, the space-time path has served as one of the preeminent conceptual structures for extending GIS data models to include time. Building upon this conceptual model, more comprehensive logical data models can be created which store the locations, activities, and interactions of people (Wang and Chen, 2001). This provides support for both space-and time-dominate queries such as:

- Where is an individual at time t?
- When is an individual at location *l*?
- When will two individuals meet?
- When will an individual enter a target area?

These queries and other analytic methods provide insights into human spatial behavior but have obvious privacy consequences. It therefore presents a reasonable starting point for identifying possible measures of privacy and disclosure.

14.4.2 Representations of privacy

As discussed previously, locational privacy is essentially the ability for individuals control their personal locational information. The majority of the privacy safeguards attempt to either increase uncertainty in the measured attributes of individuals or increase the degree of anonymity by making individuals indistinguishable from each other. Given this, it is worth considering how these fundamental measures might be considered in conjunction with continuous traces. We discuss the measures in terms of their relation to limiting both identity and attribute disclosures. For the context of this discussion we conceptualize identity disclosures occurring when a single member of a set of tracked individuals can be associated with a single individual in an external data source and attribute disclosure occurring when multiple members can be associated with a single piece of information in an external source. These measures attempt to capture the potential disclosure risks based on the representations of individuals within a single data source.

14.4.2.1 Uncertainty

Uncertainty generally attempts to capture the degree of confidence in a given measurement, often expressed as a range of possible values whereby the length of the range corresponds to increases in uncertainty. When considered spatially, this is often expressed in terms of a set of possible locations in which the phenomenon or object of interest may be located. The uncertainty of the object can be considered in terms of a single moment of time or defined temporal duration. For tracked objects, such as space-time paths, the degree of uncertainty will vary naturally through time according to the behavior of the individual's movement behavior, the sampling properties of a tracking system and the spatial and temporal accuracy in the measurements (Pfoser and Jensen, 1999; Leonhardi and Rothermel, 2002). Uncertainties can be artificially introduced by adapting the previously discussed safeguards for use with space-time path representations (Bridwell, 2004).

In most applications uncertainty is seen as a problem which can potentially confound scientific analyses and decision making. In general, the greater uncertainty in a measurement the less likely one is to act upon it. Thus in the context of the privacy, the greater uncertainty in the locations of individuals may provide a basis for measuring possible identity and attribute disclosures. When considering a single moment in time, this can be expressed as the number of possible locations or the size of an area around a locational estimated for single instant. However, the discussion above highlights that the degree of uncertainty varies across time as a function of both the configuration and sampling of sensors as well object behavior. Therefore when measuring the privacy for a given period of time one must consider aggregates that summarize the uncertainty at all instants in between. Possible aggregates include: the minimum uncertainty area, the maximum uncertainty area, the average area or the percentage of time an individual is tracked with a certain degree of confidence.

14.4.2.2 Anonymity

The privacy safeguards provide three general approaches for understanding an individual's degree of locational anonymity within a dataset. When considering spatially, three general types measures can be considered: proximity, density and equality. *Proximity* measures can be used to determine the distance of the closest neighboring point, or the distance to the closest k neighboring points. Here, the distance serves as a surrogate for the degree of anonymity of an individual as shorter distances indicate a potential difficulty for an investigator to discriminate individual identities. *Density* measures consider the number of points within a given distance radius from an individual. Similar to proximity measures, the larger the number of points within the prescribed area the greater anonymity is provided. These measures can be used to inform the implementation of various safeguards, such as geographic masks, as well as various disclosure limitation tests between masked and unmasked datasets. The third measure, *equality*, strictly adapts the concepts of k-anonymity and suggest that individuals are anonymous according to shared locational representation, be they points, lines and polygons. This measure logically lends to an aggregate approach for implementing safeguards while the previous approaches lend more towards approaches that preserve the structure of individual-level data.

As the mixed zone model implies, considering anonymity when tracking individuals in space-time is much more difficult. The concept of proximal anonymity can be considered by adapting the space-time bundles from time geography. Bundles measures times in which paths are co-present in terms of space, time, direction and velocity (Pred, 1977). A bundle analysis of paths in a dataset might provide insights into the times and places in which multiple individuals cannot be identified. The density approach can be considered through segmentation functions that evaluate the density of individuals neighboring an individual for a given time or temporal duration. Multiple discrete slices would then provide an indication of how this density various throughout the course of day or extended observation time. The equality perspective may be developed by adapting Gruteser and Grunwalds's (2003) approach for integrated space-time using an octree formulation that would iteratively divide space-time into regions that would contain a prescribed number of individuals for both time and space.

14.5 Conclusions

Present trends in wireless communications and geographic information technologies suggest the emergence of ubiquitous tracking environments some time in the near future. The possibility of using LBS tracking to analyze stored human movements holds great promise for understanding the emergence of activity patterns and their relationship to broader urban problems such as traffic congestion, suburbanization and the development of social networks. In addition, persistent movement stores and data warehouses may help improve LBS development and deployment, allowing greater customization of services through the analysis of spatiotemporal trends (Smyth, 2001). However, the development of these future methods will be heavily dependent upon a platform for tracking that individuals will not feel invasive.

This chapter has addressed the dimensions of locational privacy based on the approaches developed to provide privacy protection to individuals when releasing locational information. The chapter has attempted to synthesize the various perspectives and consider how these perspective may be examined for continuous tracking situations. There is much future work to be done. Possible future avenues include adapting the safeguards for spatiotemporal representations such as those provided by time geography, testing these approaches on real-time tracking data for a large population of individuals, and developing rigorous metrics and tests for evaluating the disclosure threats present within spatiotemporal datasets.

References

Armstrong, M. P. (2002) Geographic information technologies and their potentially erosive effects on personal privacy. *Studies in the Social Sciences*, 27, 19–28.

Armstrong, M. P., Rushton, G. & Zimmermsn, D. L. (1999) Geographically masking health data to preserve confidentiality. *Statistics in Medicine*, 18, 497–525.

Beresford, A. R. & Stajano, F. (2003) Location privacy in pervasive computing. *IEEE Pervasive Computing*, 2, 46–55.

Bowdin, R. J. & SIM, A. B. (1992) The privacy bootstrap. *Journal of Business and Economic Statistics*, 10, 337–345.

Bridwell, S.A. (2004) Space-Time Masking Techniques for Privacy Protection in Location-Based Services. Paper presented at GIScience 2004, College Park, MD, October 20–23, 2004.

Cassa, C. A., Grannis, S. J., Overhage, J. M. & Mandl, K. D. (2006) A context-sensitive approach to anonymizing spatial surveillance data: Impact on outbreak protection. *Journal of the American Medical Information Association*, 13, 160–165.

Choy, M., Kwan, M.-P. & Hong, V.-L. (2000) Distributed database design for mobile geographical applications. *Journal of Database Management*, 3–17.

Cox, L. H. (1994) Matrix masking methods for disclosure limitation in microdata. *Survey Methodology*, 165–169.

Cuellar, J. R. (2002) Location information privacy. In Sarikaya, B. (Ed.) *Geographic Location in the Internet*. Norwell:MA, Kluwer.

Curry, M. R. (1997) The digital individual and the private realm. *Annals of the Association of American Geographers*, 87, 681–699.

Curry, M. R. (1998) *Digital Place: Living with Geographic Information Technologies*, London:U.K., Routledge.

Dobson, J. E. & Fisher, P. F. (2003) Geoslavery. *IEEE Technology and Society Magazine*, 22, 47–52.

Duke-Williams, O. & Rees, P. (1998) Can Census Offices publish statistics for more than one small area geography? An analysis of the differencing problem in statistical disclosure. *International Journal of Geographical Information Science*, 579–605.

Duckham, M. & Kulik, L. (2005a) A formal model of obfuscation and negotiation for location privacy. In Gellersen, H. W., Want, R. & Schmidt, A. (Eds.) *Lecture Notes in Computer Science 3693*. Springer.

Duckham, M. & Kulik, L. (2005b) Simulation of obfuscation and negotiation for location privacy. In Mark, D. M. & Cohn, A. G. (Eds.) *Lecture Notes in Computer Science 3693*. Springer.

Duncan, G. T. & Lambert, D. (1986) Disclosure-limited data dissemination. *Journal of the American Statistical Society*, 81, 10–18.

Duncan, G. T. & Pearson, R. W. (1991) Enhancing access to microdata while protecting confidentiality. *Statistical Science*, 6, 219–232.

Dykes, J. A. & Mountain, D. M. (2003) Seeking structure in records of spatio-temporal behavior: Visualization issues, efforts and application. *Computational Statistics*, 43, 581–603.

Federal Committee on Statistical Methodology (2005) Statistical Policy Working Paper 22: Report on Statistical Disclosure Limitation Methodology. Available at: http://www.fcsm.gov/working-papers/spwp22.html.

Goss, J. (1995) We know who you are and we know where you live: The instrumental rationality in geodemographic systems. *Economic Geography*, 71, 171–198.

Gruteser, M. & Grunwald, D. (2003) Anonymous usage of location-based services through spatial and tempral cloaking. *International Conference on Mobile Systems, Applications and Services (MobiSys)* San Francisco, CA.

Hagerstrand, T. (1970) What about people in regional science? *Papers of the Regional Science Association*, 24, 7–21.

Hjelm, J. (2002) *Creating Location Services for the Wireless Web*, New York:NY, John Wiley & Sons, Inc.

Hull, R., Kumar, B., Lieuwen, D., Patel-Sheider, P. F., Sahugnet, A., Varadarajan, S. & Vyas, A. (2004) Enabling context-aware and privacy-concious user data sharing. *2004 IEEE International Conference on Mobile Data Management*. Berkely:CA.

Kwan, M.-P., Casas, I. & Schmitz, B. C. (2004) Protection of geoprivacy and accuracy of spatial information: How effective are geographical masks? *Cartographica*, 39, 15–28.

Küpper, A. (2005) *Location-Based Services: Fundamentals and Operation*, Hoboken, N.J., John Wiley.

Lambert, D. (1993) Measures of disclosure risk and harm. *Journal of Official Statistics*, 9, 313–331.

Leonhardi, A. & Rothermel, K. (2002) Protocols for updating highly accurate location information. In Sarikaya, B. (Ed.) *Geographic Location in the Internet*. Norwell:MA, Kluwer.

Miller, H. J. (2005a) A measurement theory for time geography. *Geographical Analysis*, 37, 17–45.

Miller, H. J. (2005b) What about people in geographic information science? In Fisher, P. & Unwin, D. (Eds.) *Re-Presenting Geographic Information Systems*. John Wiley.

Monmonier, M. (2002) *Spying with Maps: Surveillance Technologies and the Future of Privacy*, Chicago:IL, University of Chicago Press.

Morris, J. B., JR. (2002) The elements of location tracking and privacy protection. In Sarikaya, B. (Ed.) *Geographic Location in the Internet*. Norwell:MA, Kluwer.

Mountain, D. M. (2005) Visualizing, querying and summarizing individual spatio-temporal behaviour. In Dykes, J. A., Kraak, M. J. & Maceeachren, A. M. (Eds.) *Exploring Geovisualization*. London, Elsevier.

Myles, G., Friday, A. & Davies, N. (2003) Preserving privacy in environments with location-based applications. *IEEE Pervasive Computing*, 2, 56–64.

Onsrud, H. J., Johnson, J. P. & Lopez, X. (1994) Protecting personal privacy in using Geographic Information Systems. *Photogrammetric Engineeering and Remote Sensing*, 60, 1083–1095.

Paas, G. (1988) Disclosure risk and disclosure avoidance for microdata. *Journal of Business and Economic Statistics*, 6, 487–500.

Pfitzmann, A. & Köhntopp, M. (2001) Anonymity, unobservability, and pseudonymity—a proposal for terminology. In Federrath, H. (Ed.) *Lecture Notes in Computer Science 2009*. Springer.

Pfoser, D. & Jensen, C. S. (1999) Capturing the uncertainty of moving-object representations. In Güting, R. H., Papadias, D. & Lochovsky, F. (Eds.) *Lecture Notes in Computer Science 1651*. Berlin, Springer.

Pred, A. (1977) The choreography of existence: Comments on Hägerstand's time-geography and its usefulness. *Economic Geography*, 53, 207–221.

Schiller, J. H. & Voisard, A. (Eds.) (2004) *Location-Based Services*, San Francisco, CA, Morgan Kaufmann.

Smyth, C. S. (2001) Mining mobile trajectories. In Miller, H. J. & Han, J. (Eds.) *Geographic Data Mining and Knowledge Discovery*. London, Taylor and Francis.

Spierkermann, S. (2004) General aspects of location-based services. In Schiller, J. & Voisard, A. (Eds.) *Location-Based Services*. New York:NY, Morgan Kaufman.

Sweeney, L. (2002) *k*-anonymity: A model for protecting privacy. *International Journal on Uncertainty, Fuzziness and Knowledge-Based Systems*, 10, 557–570.

Warner, S. L. (1965) Ranomized response: A survey technique for eliminating evasive answer bias. *Journal of the American Statistical Association*, 60, 63–69.

Westin, A. F. (1967) *Privacy and Freedom*, New York:NY, Atheneum.

15 Location-based services: Enabling technologies and a concierge service model

Seungmo Kang[1], Tschangho John Kim[2], Sung-Gheel Jang[2]

[1] *Department of Civil and Environmental Engineering, University of Illinois at Urbana-Champaign, Urbana, IL*

[2] *Department of Urban and Regional Planning, University of Illinois at Urbana-Champaign, Urbana, IL*

15.1 Introduction

It is widely believed that about 80 percent of public and private sectors' day-to-day decisions are related to some sort of spatial and locational consideration, leaving only a few areas that are not affected by locational considerations. The Internet puts an unprecedented amount of locational information of all kinds at a user's fingertips, information that can be used for personal production activities in a mind-boggling variety of ways (Ostensen, 2001).

Location-based services (LBS) – sometimes called location-based mobile services (LBMS) – are an emerging technology combining information technology, Geographic Information Systems (GIS), positioning technology, Intelligent Transportation System (ITS) technology and the Internet. LBS combine hardware devices, wireless communication networks, geographic information and software applications that provide location-related guidance for customers (Kang, Oh and Kim, 2006). Location-based services differ from mobile position determination systems, such as Global Positioning Systems (GPS), in that they provide much broader, application-oriented location services, which become useful in specific situations, such as the following (Kim, 2004):

- "You are about to join a ten-kilometer traffic queue, turn right on Washington Street, 1 km ahead."
- "Help, I'm having a heart attack!" or "Help, my car has broken down!"
- "I need to buy a dozen roses and a birthday cake. Where can I buy the least expensive ones while spending the minimum amount of time on my way home from the office?"

H.J. Miller (ed.), Societies and Cities in the Age of Instant Access, 227–239.

Location-based services are the new faces of the wireless Internet. Advertising and e-commerce consulting firms predict that the global LBS market will reach $8 billion by 2010, up from $981 million in 2005.[1] With this growth potential, LBS present a substantial emerging market opportunity for wireless providers. While the market for LBS seems to be rapidly emerging and some initial, yet primitive, services have been introduced in the market, there are many basic issues that have yet to be researched and developed in order to provide efficient services for users and providers.

The purpose of this chapter is to review underpinning technologies that make LBS services feasible and to introduce a new service model for expanding its area.

15.2 Enabling technologies for LBS

15.2.1 Positioning technology

Positioning technology is a fundamental technology for LBS since it enables service providers to track a subscriber's whereabouts. There are two types of positioning technologies at large: network-based and handset-based. Network-based technology includes Cell ID, Time Difference of Arrival (TDOA), Enhanced Observed Time Difference of Arrival (EOTDOA), Angle of Arrival (AOA) and Location Pattern Matching. Handset-based technology includes GPS (Global Positioning System) and A-GPS (Assisted-GPS) (Rao and Minakakis, 2003). In addition, emerging Radio Frequency Identification (RFID) technology will galvanize the field since it enables us to acquire the positions of moving objects in more accurate and responsive ways through the integration of various levels of sensor networks.

15.2.2 Geographic information systems

Since much of the utility of LBS is closely related to the manipulation of location representations and the use of location, GIS plays a pivotal role to the deployment of LBS. GIS provides LBS with efficient means to model the real world into a geographical data format either in vector or raster formats. GIS enables real world phenomena to be transformed into a digital format, either by Data Base Management Systems (DBMS) or by the DBMS functionality of GIS which manages and queries spatial data. GIS also allows LBS to display maps based on user queries and provides LBS with robust mapping capabilities including projections, spatial reference system, overlays, and others (You and Kim, 2002). Furthermore, most current GIS applications support the client-server network architecture that enables LBS to use various GIS functions and data over networks.

15.2.3
Intelligent transportation systems

Intelligent Transportation Systems (ITS) provide LBS with the methodology and technology to assist the traveler with routing, tracking, traffic and navigation information through Advanced Traveler Information System (ATIS) (see Boyce, 2002, for detailed information on traveler information system of ITS) and Car Navigation Systems (CNS). Data archiving technologies in ITS support LBS with the methodology to handle massive amounts of real-time data for LBS. Real time data collection technology in ITS and vehicle telematics become necessary tools for LBS in routing and navigation.

15.2.4
Internet technology

The Internet is a telecommunication infrastructure that makes various services, including web services, possible and available to the public. The initial motivation for development of Internet technology was supposedly *resource sharing* which is the same goal for LBS: *sharing location-based information*. The World Wide Web (WWW) has been the most innovative and pervasive Internet technology, which can facilitate different kinds of LBS technologies becoming interconnected by finding and communicating with each other. Among the various web based technologies, Extensible Markup Language (XML) based technologies have saliently influenced the development of LBS applications. Specifically, the following XML based technologies are promising for the LBS deployment: Simple Object Access Protocol (SOAP), Geography Markup Language (GML), Voice Markup Language (VoiceML), and Scaleable Vector Graphics (SVG). Since SOAP is a lightweight XML based protocol for accessing services, objects and servers in a platform-independent manner, it enables LBS to exchange any information in a decentralized and distributed environment. GML allows LBS to exchange geographical data with non-geographical data. VoiceML assists LBS technologies by conveying information to clients in a voice recognizable manner as a crucial part of route guidance services in LBS. SVG supports relatively smaller-sized graphic data without degrading the resolution, allowing LBS to be used with smaller mobile devices.

15.2.5
Wireless communication technology

Wireless devices such as cellular phones and personal digital assistants (PDAs) are potential LBS clients for receiving needed information. According to the range of wireless connectivity, there are three types of wireless networks: Wireless Personal Area Networking (WPAN, e.g., Bluetooth), Wireless Local Area Networking (WLAN, e.g., 802.11a/b/g), and Wireless Wide Area Networking

(WWAN, e.g., GSM). Among the three wireless networks, the WWAN technology enables the wireless devices to stay connected to the LBS providers over the conventional cellular networks (e.g. GSM, CDMA, etc.) while traveling. The pervasive connectedness to the service network, therefore, becomes a key element for any LBS applications. Furthermore, the WWAN speed could reach 3G kbps soon, which would enrich LBS with higher data transfer speed, richer content and wider functionality.

15.2.6 Standards

Fully operational LBS requires interoperable processing among various information services such as tracking, navigation, location transformation, rendering, yellow pages, and directory services, among others. To achieve seamless integration regardless of the computing environment, it is essential to standardize the communication mechanism of LBS related services. This type of standardization product is called an interface standard. In IT fields, this type of standard becomes more important and critical as various IT products become available for different consumers yet need to be interoperable for better efficiency in development, marketing, manufacturing, operation and management. Interface standards need to specify communication protocols, message sets and plug-in compatibility in order to facilitate interconnection between components and adjoining systems. Since each technology needs to be fully compliant with its relevant IT standards by governmental or international trade regulations, standards become a key integrator of the enabling technologies for LBS as well as a *sine qua non* technology for making LBS feasible (Figure 15.1).

In fact, there are many organizations developing standards in or related to LBS. Table 15.1 shows organizations that develop such standards. Among them,

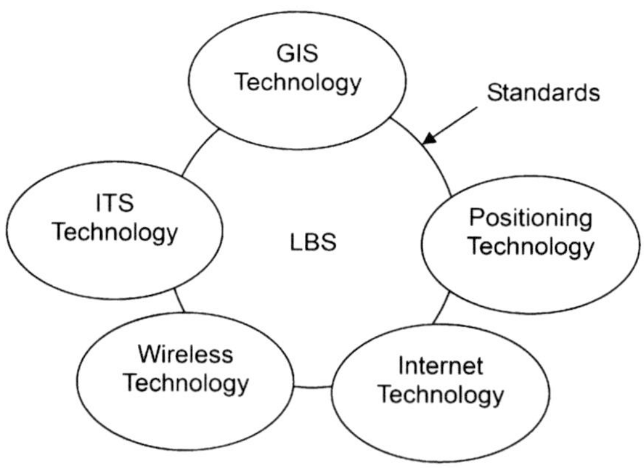

Fig. 15.1 Enabling Technologies for LBS

Table 15.1 Standards Development Organizations in LBS

Organization	Website
3rd Generation Partnership Project	http://www.3gpp.org/
CEN BT/WG 141 Intermodal and interoperable transport – Telematics	http://www.uninfo.polito.it/CEN-BT-WG141/default-English.htm
European Committee for Standardization	http://www.cenorm.be/
Federal Geographic Data Committee	http://www.fgdc.gov/
Internet Engineering Task Force	http://www.ietf.org/
ISO TC211– Geographic Information	http://www.isotc211.org/
ISO TC204– Intelligent Transport Systems	http://www.iso.ch/
Location Interoperability Forum	http://www.locationforum.org/
MAGIC Services Initiative	http://www.magicservicesforum.org/
Open Geospatial Consortium, Inc.	http://www.opengeospatial.org/
Open Location Services Initiative	http://www.openls.org/
Open Mobile Alliance	http://www.openmobilealliance.org/
SyncML	http://www.syncml.org/
Telecommunications Industry Association	http://www.tiaonline.org/

(Source: ISO/DIS 19132 Annex D).

ISO is leading in developing interface standards for LBS, which include ISO 19132: LBS reference model, ISO 19133: LBS tracking and navigation, and ISO 19134: LBS multi-modal routing and navigation. The Open Geospatial Consortium (OGC) also has developed an LBS standard for industries encompassing Navigation Services, Directory Services, Presentation Services, Location Utility Services, Gateway Services and Encodings & Protocols.

15.3 Location-based concierge service model

15.3.1 New value-added service for LBS

Growing LBS markets are demanding more integrated and value-added location based services. Also, mobile service providers continue to leverage their LBS platforms to provide more innovative and consumer-oriented services. However, current LBS offerings do not go beyond emergency 911 services, mobile yellow pages, buddy finders and car navigation services. Most are only dealing with "location information" itself without adding other types of value-added information such as price and availability information. Imagine a case in which you need to withdraw money from an ATM machine and want to find an ATM via your LBS. Mobile yellow pages might be able to provide information about the nearest machine (A) from your current position. Car navigation services could tell you the shortest path to get there. But no existing LBS model can let you know which one is the smarter option for you if there is another ATM (B) that has a fee one dollar less than (A) and is located 3 minutes further away.

The concierge service model is a way to provide a user with these types of integrated information to support smarter decisions in everyday life. Another example in a location-based service is a request to a routing and navigation service such as "finding the least cost routes using all available modes of transportation from my current position, stopping at a gas station for 10 gallons of gas, a bakery to pick up four loaves of baguette, and a flower shop for a dozen roses before arriving home," (Kim, 2004, see Figure 15.2). There could be potentially hundreds of gas stations, pharmacies and flower shops on the network. Each of these shops has different prices on similar items. Available modes may vary from automobile to bus or subway. The number of possible routes connecting each of the possible nodes in each shopping category will be more than hundreds of millions in a large-scale network. Parking availability and transit schedules should also not be ignored. In order to derive the smartest choice considering all the possibilities with given datasets and other user-based information, this model guides a user with a route and the locations of point-of-interests (POIs) to visit with minimum total cost including purchasing, stopping, and travel costs.

The concierge service model shares following characteristics with value-added services, defined by Christensen et al. (2000).

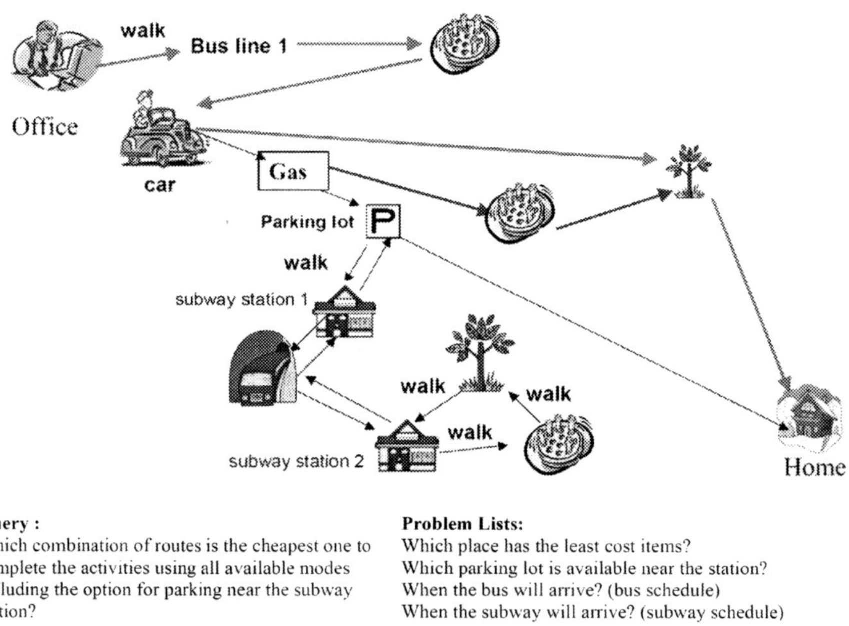

Fig. 15.2 Example of Question Lists in the Concierge Service (Kim, 2004)

1. Not a form of basic service but rather adds value to total service offering
2. Stands alone in terms of profitability and/or stimulates incremental demand for core service(s)
3. Can sometimes stand alone operationally
4. Does not cannibalize basic service unless clearly favorable
5. Can be an add-on to basic service, and as such, may be sold at a premium price
6. May provide operational and/or administrative synergy between or among other services – not merely for diversification

Location-related technologies have already advanced far enough to be implemented for this service with high accuracy as stated in previous chapters of this volume. Real-time or near real-time traffic and transit information systems including parking information are rapidly being developed by various ITS projects on a metropolitan scale. Numerous POI databases have already been constructed by private web service companies; for instance location information services of POIs are currently provided by Google Local™. Price information can be accessed by many web pages for large brick and mortar retailers as well as internet shopping sites. Aggregated pricing services also exist. For instance, Gasbuddy.com™ shows and helps users compare gas prices with just a few-hour time gap for 174 local areas in the United States.

The largest hurdle in deploying this service model on a commercial scale is the solution-search algorithm. Because the concierge service problem is a complex form of a Travelling Salesman Problem (TSP), which is an NP-hard problem (Kang, Oh and Kim, 2006), the success of this service depends on how efficiently and accurately the main algorithm gives a response to a user's query.

15.3.2 Cost components and solution algorithms

It is not easy to find the best route and POI locations having the minimum total cost even with all datasets and parameters given. To respond to a request for such a service maintaining serviceability and commerciality, a solution algorithm for the concierge service should be reasonably fast in computational time and highly accurate. Initial approaches to this problem were made by Oh (2003) and Kim (2004) and followed by Kang and Kim 2005 and Kang, Oh and Kim (2006).

In concierge service, there are three types of costs involved: 1) the purchasing costs for needed items and stopping costs including parking, 2) costs related to the time spent on the road or on transit, and 3) distance related costs such as gasoline used, wear and tear from the use of a car, and transit fares (Kim, 2004).

15.3.2.1 Purchasing and stopping cost

In the example mentioned in the previous part, there are three items to shop for, also called activities, meaning that activity one is to buy 10 gallons of gas,

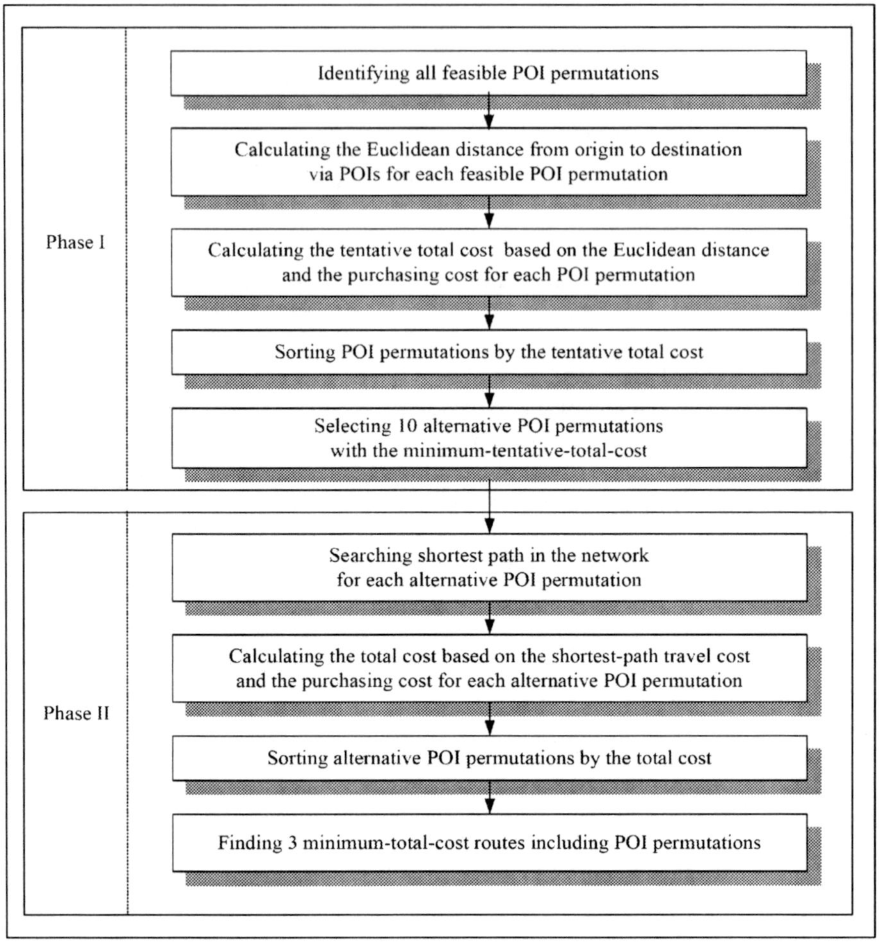

Fig. 15.3 Heuristic Search Algorithm using Euclidean Distance (Kang, Oh and Kim, 2006)

activity two is to pick up four loaves of baguette, etc. Every POIs, where activities can happen, is indexed with its location in the network and the unit price of items. The unit price for an item is multiplied by the purchasing amount in each activity. The total purchasing cost is calculated by a sum of each activity's purchasing cost at suggested POIs by the concierge service. The stopping cost consists of parking costs in the road network or discharge costs in the transit network once the user has decided to stop at a certain location. In some cases, it may include initial acceleration cost from the standstill condition or deceleration cost for stopping.

15.3.2.2
Travel cost: Time and distance related cost

Time spent on the road to visit one shop from each shopping category will depend on which mode is used, which route is followed, which shops are visited, as well as in which order the visits occur. If each traveler's unit cost of time per hour is known, the travel cost is the unit cost of time multiplied by the total time spent on the road from origin to destination of the trip. The unit cost of time can be inputted by users through the user interface system or calculated with user's other information such as income level or age and transportation mode (auto or transit) or trip purpose (commute, recreation or shopping). Also, the travel cost has to include the transfer cost when the user is supposed to change modes at a node, and the tolls in using certain highways or bridges. The distance related cost is determined by a unit cost per mile such as gasoline used, and the wear and tear from using a car, which is a function of travel speed, or transit fare.

15.3.2.3
Solution algorithms

The multimodal concierge service problem can be formulated as a minimization problem of the total cost, that is, the sum of the purchasing and stopping costs and the travel cost. The decision variables will be which route and mode the user has to use and how much of each item should be purchased at which POIs.

Unfortunately, the computational effort required to solve the concierge service problem increases exponentially with the number of variables and the size of the network. For such problems, it is often desirable to use heuristic methods to obtain approximate solutions in a reasonable time with sufficient accuracy for the purpose. Oh (2003) and Kang, Oh and Kim (2006) suggested a heuristic solution algorithm using Euclidean distance for an automobile network (Figure 15.3). They showed encouraging results in implementations in the Seoul metropolitan network of South Korea by making some candidate options using Euclidean distance and selecting the best ones with the exact cost. However, Euclidean distance methods may cause relatively large errors in some environments that are divided or disconnected by geographical objects (rivers or forest) or transportation facilities (expressways or railroads).

Kang and Kim (2005) suggested Genetic Algorithms (GAs) to solve this problem and tested it in the aggregated Chicago network (Figure 15.4). GAs are heuristic search techniques based on the mechanism of natural selection and natural genetics. They combine survival of the fittest among many string structures with a structured yet randomized information exchange, such as crossover and mutation, to form an effective search algorithm (Goldberg, 1989). The test showed very short computational time ($0.6 \sim 1.6$ sec) with relatively high accuracy (97 percent $\sim$ 99 percent).

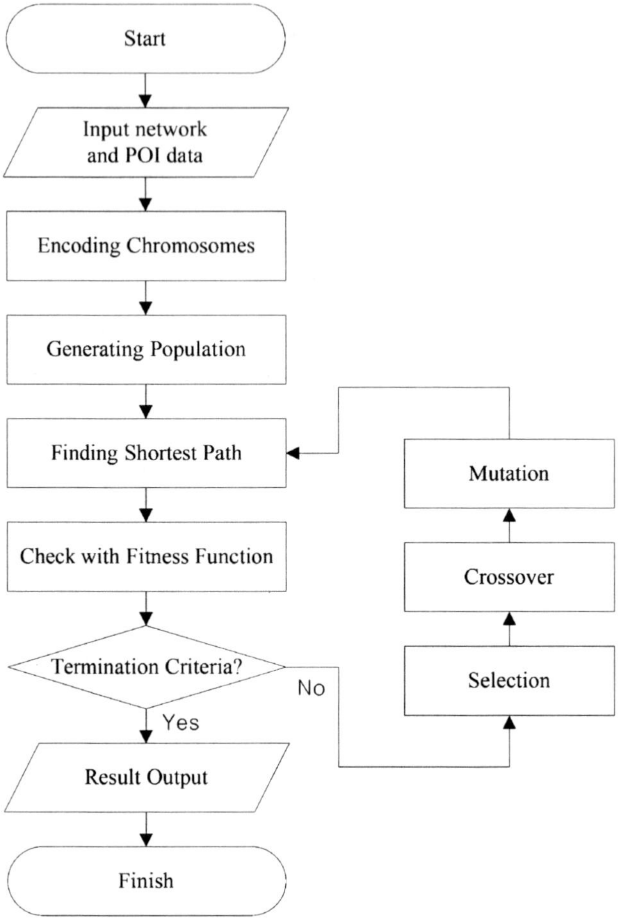

Fig. 15.4 Genetic Algorithm for Concierge Service Model (Kang and Kim, 2005)

But the aggregated test network was much smaller than real road networks at the metropolitan level, thus a necessary next step will require that the suggested algorithm be analyzed in a larger network.

15.3.3 Concierge service applications

The concierge service model can be delivered directly to end-users who need guidance to shopping points and data on the shortest paths connecting these points. Service developers can integrate this model with other information for better applications such as parking guidance, real-time traffic information, and directory services. Mobile devices, public kiosks, and car navigation systems can be utilized for these tasks through internet or wireless communication networks combined with location information technology. The following examples describe more possibilities for concierge service implementation.

15.3.3.1 Automobile navigation services

The largest market for the concierge service would be automobile navigation services. Mapping data supplier NAVTEQ has reported that the adoption rate of navigation, or the total number of vehicles offering navigation as a standard or optional feature based on the total number of available vehicles, has grown significantly in North America, reaching 37percent of vehicles sold in 2004.[2] Many shopping items and other useful information can be integrated into existing automobile navigation systems. Particularly, demand for gas price and station location information as well as parking information is very high. Gas prices change very frequently and almost every station offers slightly different prices for similar octane-rated fuels. Also, parking costs and availability information has been another primary concern for automobile drivers. Car navigation systems can be more useful if they are upgraded to include dynamic information about shopping, parking and traffic.

15.3.3.2 Mobile navigation services

In most downtown areas of metropolitan cities, many daily activities are based on walking trips. New mobile technology and services which are being developed based on LBS technologies are of great interest to concierge services because mobile devices like cell phones and PDAs can be implemented not only for in-vehicle use but also for pedestrians using these mobile devices. These systems provide facility and municipal information, e.g. weather information, shopping information, tourism information and navigation services (Eriksson, 2002) as well as walking navigation assistance. Finding the nearest Italian restaurant, the closest ATM machine with the lowest fee, or directions to the subway station indicating the subway route and the stations in which to transfer; all of those queries can be serviced by this model. Additional information can be delivered by wireless web services, such as a review for the suggested restaurant and store stock information for shopping tasks. If the service system is communicatively linked to a taxi dispatcher and an allocation system containing location data of available taxis, the user could request a dispatch directly through the mobile device and the instant service responder would be able to include the current position of the assigned taxi and expected waiting time.

15.4 Conclusion

In the past, GIS and positioning technologies have been used mostly for institutional purposes. Now, rapidly evolving wireless technology makes it possible to utilize GIS and positioning technologies for personal productivity, such as finding the best routes to a destination, finding points of interest, finding friends, finding current traffic conditions, as well as a wide variety of other convenience services for entertainment, leisure, sports, shopping, travel, local information, community interest, health, education, banking, hobbies, services, and so on. Enormous market potential can be foreseen in this field, including the market for tracking, route-finding and guiding, and notification and alert services.

The technologies that make LBS feasible include positioning technology, GIS, ITS, the Internet, wireless communication, and standard technology. The standards, especially, play a key role to integrate the enabling technologies for LBS so that the technologies become interoperable for successful LBS deployment in the foreseeable future.

The concierge service model is a new value-added service in the LBS area that is designed to provide a user with integrated information to support smarter decisions in everyday life. The new service model has to not only consider "location information" itself but also integrate it with other types of value-added information such as price, stock levels, and other availability information. Problem formulation and solution algorithms have been developed and tested in several research trials. If the various databases (price and stock information, POI information, real-time traffic data, transit service data etc.) can be constructed and integrated by wireless or mobile service providers, many of the existing hurdles for commercial application of this service model will be eliminated.

Acknowledgement

This work was supported in part by the Brain Korea 21 (BK21) project funded by the Ministry of Education, Korea and the Seoul 2010 u-Life Pilot Service Development project funded by the Seoul Metropolitan Government, Korea. Any opinions, findings, and conclusions or recommendations expressed in this publication, however, are those of the authors and do not necessarily reflect the views of the agencies that funded the project

Notes

[1] ABI Research estimates the LBS market values (from *Red Herring*, "Verizon Navigates LBS Market", January 30, 2006, http://www.redherring.com)

[2] From "NAVTEQ Reports North American Vehicle Navigation Market Growth Trend Continues," available at http://www.prnewswire.com/cgi-bin/stories.pl?ACCT=104&STORY=/www/story/06-14-2005/0003870664&EDATE=

References

Boyce, D.E.(2002) "A Memoir of the ADVANCE Project," *ITS Journal*, 7(2), 105–130.

Christensen, G., R. Duncan, P. G. Florack, (2000). *Wireless Intelligent Networking*, Artech House Publishers, Norwood, MA.

Eriksson, O. (2002) "Location based destination information for the mobile tourist." In K. W. Wöber et al. (eds.), *Information and communication technologies in tourism*, 255–264.

Goldberg, D. E. (1989) *Genetic Algorithms in Search, Optimization*, and Machine Learning, Addison-Wesley Publishing, Reading, MA,.

ISO/DIS 19132 (2006) *Geographic information – Location based services – Reference model*, International Organization for Standardization.

Kang, S., S. Oh and T. J. Kim (2006) "A Heuristic Algorithm for Solving a Multimodal Location-Based Concierge Service Problem," *Transportation Research Record: Journal of Transportation Research Board* (Accepted).

Kang, S. and T. J. Kim (2005) "Solving a Location-Based Concierge Service Problem: A Heuristic Approach using Genetic Algorithm," Presented at *Research Symposium on Societies and Cities in the Age of Instant Access*, Salt Lake City, UT, Nov. 10–12.

Kim, T. J. (1999) "Metadata for Geo-spatial Data Sharing: A Comparative Analysis," *Special Issue of the Annals of Regional Science*, 33(2), 171–182.

Kim, T. J. (2004) "Multi-modal routing and navigation cost functions for location-based services (LBS)," In D. H. Lee (ed.),*Urban and Regional Transportation Modeling: Essays in Honor of David Boyce*, Edward Elgar, Northampton, MA.

Oh, S. (2003) *A Heuristic Multimodal Location-Based Services Model: Design and Implementation*, Ph.D. Dissertation. Seoul National University, Seoul, Korea.

Ostensen, O, (2001) "The expanding agenda of Geographic information standards," *ISO Bulletin*, July 2001, available at *http://www.iso.ch/iso/en/commcentre/pdf/geographic0107.pdf.*

Rao, B. and L. Minakakis, (2003) "Evolution of Mobile Location-based Services," *Communications of the ACM*, 46(12), 61–65.

VanderMeer, J. (2001) "Location Content Drives Wireless Telecommunications," available at http://www.geoplace.com/bg/2001/0201/0201pay.asp.

You, J. and T. J. Kim, (2000) "Development and Evaluation of a Hybrid Travel Time Forecasting Model," *Transportation Research* C. 8(1), 231–256. The article was also published as a chapter in Jean-Claude Thill (ed.), *Geographic Information Systems in Transportation Research*, Elsevier Science Ltd, Oxford, UK.

16 From cyberspace to DigiPlace: Visibility in an age of information and mobility

Matthew A. Zook, Mark Graham
Department of Geography, University of Kentucky, Lexington, KY

16.1 Introduction

The information age is increasingly mobile with ever finer webs of potential connectivity overlaying the physical spaces we inhabit. While commentators have long argued that cyberspace can only be understood in reference to material places (Castells, 1996; Graham, 1998; Zook, 2000), this chapter analyzes a particularly striking example of the connections between electronic data, physicality and mobility, *i.e.*, Google's localization products. In fact, we argue that services such as GoogleLocal engender a type of hybrid space (Couclelis, 1996), which we term DigiPlace, in which digital data and physical places are continually re-combined into lived, subjective space as one negotiates through time, space and information. Particularly novel factors behind the construction and experience of DigiPlace are: the ability to access it in real time and on the move, and the impact of this electronic visibility on perceptions of physical accessibility. This chapter briefly outlines a theoretical lens through which this phenomenon can be viewed and presents a case study based on mobile access to GoogleLocal to explore changes to the way people experience and move through data, space and place.

16.2 Cyberlocal, Dataspace and DigiPlace

The real-time mixing of information, place and mobility is a relatively new phenomenon brought about by the creation of powerful database search tools (such as the Google search engine), the increasing availability of geo-coded data and the diffusion of mobile interfaces to them. Although this chapter focuses on GoogleLocal, a number of other companies such as Microsoft and Yahoo! offer similar

H.J. Miller (ed.), Societies and Cities in the Age of Instant Access, 241–254.

services, making the insights presented here extremely germane to an emerging set of information services rather than simply one specific software interface, *i.e.*, GoogleLocal.

In order to explore the theoretical implications of geo-referenced and mobile-accessed information we introduce three conceptual terms that comprise our framework of hybridized spaces: Cyberlocal, Dataspace and DigiPlace. Cyberlocal refers to digital databases (accessible via cyberspace) which are intimately linked to specific physical places. Cyberlocalities can either be formed by synchronous or asynchronous links to the physical world. Synchronous cyberlocalities are characterized by a direct feedback mechanism between physical- and cyber-space. Examples include webcams[1], live glogs[2] (first person recordings of an activity, event, or process, in which the recorder is either a participant or direct observer), or other websites with a focus on real-time information[3]. Asynchronous cyberlocalities, in contrast, consist of representations of physical places, but have no direct feedback mechanisms to account for real-time changes in those places. Examples of asynchronous cyberlocalities are local search engines (such as GoogleLocal), travel web-forums[4], and a variety of other geographically focused websites. Cyberlocalities can be entered into from any access point regardless of physical location (*e.g.*, checking a webcam focused on a highway from the desktop PC in one's office), but come into being and are intimately shaped by the properties of specific physical places (*e.g.*, the existence of a highway and rush hour traffic).

Dataspace, on the other hand, is a hybrid space constructed by the embedding of information, or code, in physical places and the resulting "automatic production of space" (Thrift and French, 2002). After Dodge and Kitchin (2004; 2005), we note three distinct types of dataspace: 'code/space', 'coded space', and 'background coded space'. Dodge and Kitchen describe code/space as dominating the production of space, "explicitly mediating sociospatial processes and experience" (2004: 198). Code/spaces such as airports, ATM machines, and subway ticket machines are all examples of how spaces can cease to function according to their designed purposes when code fails. In contrast, failures of code in 'coded space,' do not generally result in mass disruption to those spaces. Coded spaces are spaces in which code is important, yet not essential, to the ability of those spaces to serve their built purpose. Thus, "code's role is mostly one of augmentation and facilitation, rather than control and regulation" (Dodge and Kitchin, 2005: 173). Digital highway signs which update electronically based on traffic conditions, RSS feeds of stock market, weather, or forum updates onto personal computers, personal digital assistants, and cell phones, and in-car navigation systems can all be found in coded spaces. Finally, background coded space exists in situations where code can potentially mediate a solution to a problem. Once the code is activated, the space becomes either code/space or coded space. A combination of Wi-Fi, radio, and cell phone signals and the devices which can make use of them when activated are examples of background coded spaces.

While the concepts of cyberlocal and dataspace both provide useful theoretical insights, neither (in our opinion) adequately capture the situatedness and dynamism of mobile individuals balanced between the visible and the invisible, the fixed and the fluid, the space of places and the space of flows (Castells, 1996). To this end, this chapter introduces the concept of DigiPlace to analyze the merger of digital and physical space as networked individuals navigate through contemporary urban life. Wireless connections to cyberspace, have become integrated into daily routines acted out in physical places, and urbanites swim through increasingly dense (although not necessarily accessible) clouds of information (Thrift and French, 2002). In short, DigiPlace represents the real-time and dynamic mixing of specific components of the virtual and physical worlds, *i.e.*, the spatialized parts of cyberspace and coded parts of physical space.

The resulting DigiPlace, *i.e.*, geocoded and mapped cyberlocal data accessed via coded space by mobile users, provides its inhabitants with a paradoxically complex (in the amount of geo-coded information that can be queried) yet simple (generally a user will limit themselves to the top ten search results plotted on a simple map) representation of the places they inhabit. While the amount and complexity of the available data in DigiPlace is undeniably of great use, we argue that the it is the simplification and sorting of places by DigiPlace - particularly as it "fixes" or codifies users' understandings of place and place-based relations (Pickles, 2004) – that is the more compelling research topic. In theory one can access a seemingly endless stream of information in DigiPlace but in practice the algorithms used to sort and rank data presented in DigiPlace fundamentally shape one's urban experience.

Geographers and cartographers know that all maps inherently have the power to distort and influence reality (Monmonier, 1996; Pickles, 2004), and therefore are aware of shortcomings inherent to GoogleLocal and its associated DigiPlace. Less experienced users of maps, however, are more likely to accept representations at face value rather than dissect the layers of socially constructed meaning that go in to them. These include the socio-cultural-political makeup of the Internet (*i.e.*, the inherent bias of what and who is available on the Internet (Castells, 2002)), the ranking algorithms used by Google to order search results and finally the efforts by websites and businesses to manipulate those rankings. The following section examines how the DigiPlace created via a mobile phone interface to GoogleLocal can affect a user's perception of distance and visibility in this age of information and mobility.

16.3 GoogleLocal and the construction of DigiPlace

Google is one of the Internet's most accessed search engines with 62 million users per month (Nielsen/NetRatings, 2005) and is ranked by AlexaResearch (2006) as the 2nd most visited website in the world (after Yahoo!). By indexing over tens of billions of web pages and offering search results in 35 languages, the company seeks to "organize the world's information and make it universally accessible and useful" (Google, 2005). In addition to being a leading information portal to the world, Google is becoming a key means by which the general public interacts with geo-referenced information in everyday life. Google's recent development of GoogleLocal (a software interface that combines its indexed digital databases with spatial coordinates and mapping capabilities) has seemingly democratized the use of geo-coded data, making it a tool of the populace rather than the exclusive terrain of GIScience.

This accessibility of geo-coded data (*e.g.* a GoogleLocal search for a business near a particular location) lies at the heart of this example of DigiPlace. By using GoogleLocal, one can conduct a search for "Chinese restaurant" with a street address or postal code, and receive relatively accurate search results of businesses that meet this criterion. GoogleLocal utilizes data for businesses obtained from Yellow Pages directories as well as all web sites that it indexes.[5]

Figure 16.1 illustrates an example of such a search.

While a noteworthy achievement its own right, GoogleLocal achieves a undeniable saliency vis-à-vis DigiPlace through its accessibility via mobile handheld devices. GoogleLocal search results can be retrieved by a variety of mobile devices such as Personal Digital Assistants (PDAs) and wireless phones using XHTML mobile internet standards. Driving directions, phonebook lookups, business searches, and maps can all be accessed through the service (See Figure 16.2). Google also allow non-web enabled cellular phones using the more ubiquitous Short Message Service (SMS) technology to query the GoogleLocal database (See Figure 16.3).

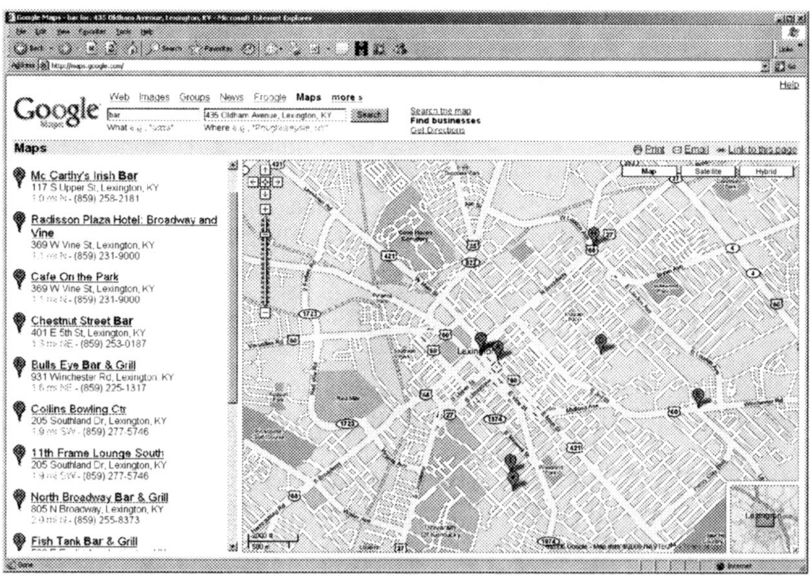

Fig. 16.1 Computer Interface to a GoogleLocal Search for a Bar near 435 Oldham Avenue, Lexington, KY

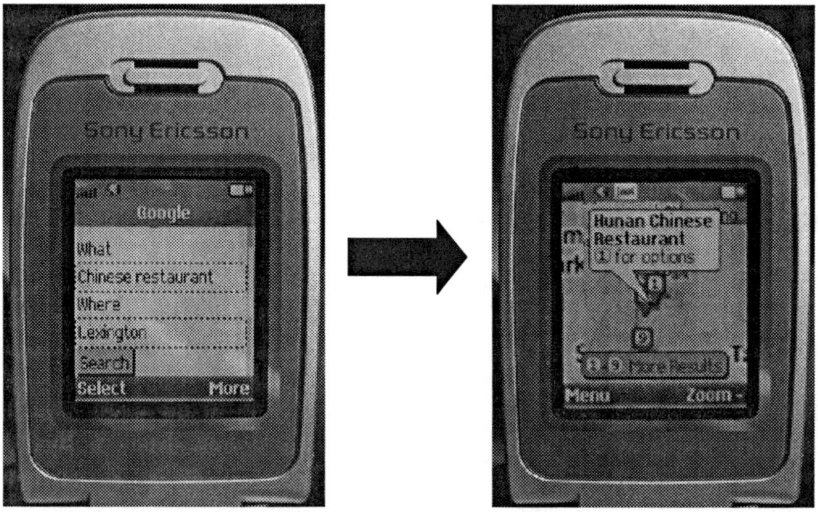

Fig. 16.2 Mobile Web Interface

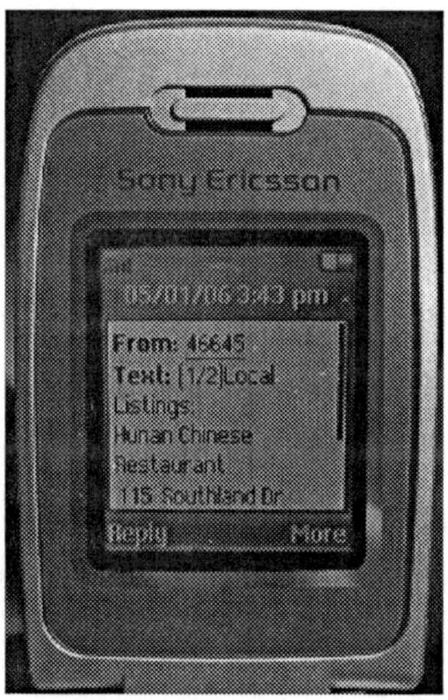

Fig. 16.3 Mobile Text Interface

A key aspect of mobile interfaces to the search engine is that only a limited number of search results are returned because of space constraints on small screens. Even if full results are available, search engine users are likely to be satisfied with the first few results that they receive (Cotlier, 2001; Cotriss, 2002; Weidlich, 2002; BBC News, 2006), making a prominent placement in Google's rankings a crucial determinant of visibility in cyberspace. One of the fundamental novelties of GoogleLocal's DigiPlace is that the importance of high rankings in cyberspace is now extended to visibility and perceptions of accessibility in physical space via the hybridized measure of distance utilized in GoogleLocal rankings.

16.3.1 Creating a hybrid measure of distance

While similar business directories have long existed on the web, GoogleLocal (and related information services) is noteworthy for the smooth integration of data stored and ranked in cyberspace with mapping technologies. GoogleLocal search results are ordered using an algorithm that takes both the distance from the geographic identifier in the search term, and the Google PageRank of the query term into account. And as Figure 16.1 illustrates, a business that is more distant in Euclidian terms (*e.g.*, McCarthy's Irish Bar labeled as "A") can be ranked higher than a physically nearby one (*e.g.*, Fish Tank Bar & Grill labeled as "I") because it is "closer" to the query location (labeled with "*" in the lower center of the figure) in the DigiPlace generated by GoogleLocal.

PageRank is a trademarked software algorithm developed by Google for ranking web pages and is central to its ability to return relevant search results. A webpage's rank is determined by the rank of all other webpages that link to it. The system "interprets a link from page A to page B as a vote, by page A, for page B. But, Google looks at more than the sheer volume of votes, or links a page receives; it also analyzes the page that casts the vote. Votes cast by pages that are themselves "important" weigh more heavily and help to make other pages important" (Google, 2005). As PageRank is combined with geographic distance, the blending of the digital and the material comes together and is actualized in new and novel ways in DigiPlace.

Although Google's PageRank system is undeniably central to the creation of DigiPlace, it is, however, an extremely opaque actor.[6] While it is understandable that Google considers the PageRank algorithm a business secret, this lack of transparency becomes problematic as Google frequently attempts to naturalize their ranking systems by referring to them as neutral algorithms which mirror structure inherent to the web. The company states that "a Google search is an easy, honest and objective way to find high-quality websites with information relevant to your search" (Google, 2006). However, such claims ignore the social construction inherent to any ranking of knowledge. And despite appeals to democracy and objectivity, Google still makes some parts of cyberspace more visible than others (and even actively make some websites invisible).[7]

Moreover, while many believe that a high PageRank equals a high-quality and popular website (an argument strongly supported by Google), a whole industry has emerged (*i.e.*, search engine optimization) to attempt to increase websites' rankings (Perkins, 2003; Van Couvering, 2004). Utilizing methods ranging from search-specific advertising (via Google's Sponsored Links) to the creation of so called "link farms" these efforts seek to highly rank one's cyber-presence in an effort to gain footing in the top ten (or even top three) results for a search, *i.e.*, some of the most "centrally" located points in cyberspace and DigiPlace. This in turn has important implications for how we experience physical places.

16.3.2 Changing notions of visibility

The implications for businesses of a hybrid distance based in part on page rank are fairly significant. Prior to widespread use of the Internet, consumers were able to locate businesses in four main ways: the visual presence of a business in the field of view of a person's daily lived geography, visual or audio media-based advertising, word-of-mouth from personal acquaintances, and phonebooks. Each of these methods of locating services affects where people spend money and also, either directly or indirectly, affects where businesses are located.

DigiPlace, as illustrated by the example of GoogleLocal, is reworking the importance of the ways in which people find businesses. A prominent street sign or high-street location are simply irrelevant to a GoogleLocal search, as only distance and PageRank are used in ordering results. Likewise, businesses without a web presence (beyond a simple listing in an electronic phonebook) now run the risk of marginalization as the use of GoogleLocal and DigiPlace expands. While businesses participating in e-commerce have long been aware of this, a combination of the hybrid distances used by GoogleLocal and the ability to access the service from almost anywhere, has made the concept of DigiPlace (and one's location in it) suddenly relevant to businesses that have hitherto been primarily concerned with offline visibility.

As a result, and as more people begin to navigate cities via the hybrid space of DigiPlace, experiences and perceptions are no longer simply based on material and spatially proximate circumstances but on the blended measures of distance and visibility provided by DigiPlace. Moreover, given the central role of ranking algorithms (*e.g.*, GoogleLocal, Yahoo!, Microsoft, etc.) in shaping and simplifying perceptions of place, private interests (*e.g.*, Google) gain a significant measure of control over visibility and fundamentally influence how one interacts with a physical place. GoogleLocal enabled DigiPlaces are thus at least partially privatized spaces, as the rankings of search results remain under the purview of Google's Pagerank. The hybrid nature of DigiPlace means that Google's private index of cyberspace has the potential to shape perceptions of even the most public of physical spaces as GoogleLocal's spatial information, in effect, becomes part of the characteristics of a place. Thus, as information based in cyberspace is accessed concurrently with movement through physical place, users simultaneously inhabit corporal (and often public) space while at the same time are immersed in virtual and privatized space.

16.4 Exploring GoogleLocal's DigiPlace

Building upon the argument that GoogleLocal's DigiPlace is introducing new measures of distance and visibility to the urban experience, this chapter explores how mobile access to GoogleLocal changes the way people experience and move through space and place. The study presented here is based on a larger database of results encompassing GoogleLocal searches for fifteen categories of urban amenities and services in ten cities. The examples included were selected not as necessarily representative samples (although they illustrate common patterns found in this research) but as exemplars of how DigiPlace contrasts with more traditional experiences of place.

16.4.1 Spatially elastic

As noted earlier, GoogleLocal creates rankings which combine PageRank with distance from a specific geographic identifier. Thus, results are ranked according to a hybrid measure of relevance which combines Euclidean distance with online popularity. The implications of this hybridization of relevance are that GoogleLocal will many times rank results which are far away from a search location higher than results that are closer. When searching for bars near one author's home in Lexington, Kentucky, the first three bars returned are a significant distance away from the search location and are located in the downtown area of the city (see Figure 16.1). Most surprising is the fact that GoogleLocal ignores all but one of nine bars located within a two block radius of the search location. This example pointedly illustrates that relevance, as defined by GoogleLocal, is not a purely geographic exercise.

Although GoogleLocal does not privilege distance as the sole deciding factor in its rankings, it is nevertheless important to understand just how its algorithms understand and formulate distance. GoogleLocal's conception of distance appears to be purely Euclidean and does not take into account street networks, traffic, congestion, speed limits, or physical barriers. For example, a search for a pharmacy in McMechen, West Virginia starkly illustrates this point (see Figures: (16.4a) a large scale map of the area, and (16.4b) a smaller scale map of the region).

GoogleLocal is able to locate nine pharmacies in relatively close proximity to McMechen. The first (and closest physically) pharmacy listed is the Shadyside pharmacy in Shadyside, Ohio. The results returned for this search are highly problematic: not because the first result is in another state, but because the first result is on the other side of the Ohio River. GoogleLocal did not consider that the closest bridge across the river is about five miles away from McMechen. This scenario becomes especially important if one considers that when accessing this information on mobile phones using Short Message Service (SMS), only the three highest ranked results are returned to the user. While similar situations (i.e. major

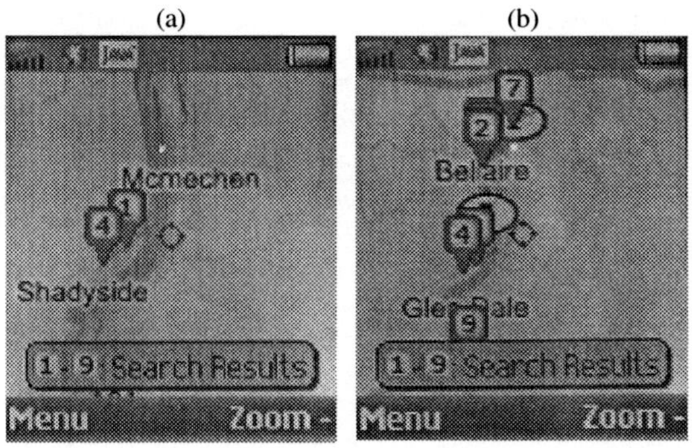

Fig. 16.4 GoogleLocal Search (via Mobile Interface) for Pharmacies in McMechen, WV

physical barriers obstructing movement) may only occur relatively rarely when searching for information, the purpose of this example is to illustrate the manner in which physical distance is factored into the GoogleLocal enabled DigiPlace. GoogleLocal's Euclidean understanding of distance, which operates outside of contextual knowledge, nevertheless affects perceptions about accessibility and mobility.

16.4.2 Individually determined

Another key component of the creation of DigiPlace is its highly individualized character, *i.e.*, DigiPlace is only experienced when a search is conducted for a specific place at a specific time. There is no such thing as an absolute or common DigiPlace as variability in search terms, location parameters and the expertise in searching (all tied to the individual inhabitant of DigiPlace) determine the parameters by which it is created. The following vignette experienced by one of the author's provides an example of an instance in which individual dimensions of hybrid spaces were brought into being.

> On a long drive home from the annual meeting of the Association of American Geographers in Chicago, a number of people in our van started to talk about stopping for a break at a well known and widespread Mexican themed fast-food restaurant. Unfortunately, on our stretch of highway, none of the Interstate-exit food signs we were passing listed that specific restaurant. To our disappointment, after driving past two successive exits, we were able to see tall illuminated signs advertising the presence of our restaurant in each town.
>
> As time passed, we started to realize that if we did not find our restaurant soon, we would be resigned to eating somewhere else. We turned on a phone and sent Google a

> text message containing both the name of the restaurant and the name of the next town as search terms. Unfortunately, we did not get the response we were looking for. The search engine did tell us, however, that the nearest restaurant was in a town ten miles from our search location (Figure 16.5a). But, this was not particularly helpful, as it gave us no specific indication as to whether the restaurant was on our route. We then used Google Local, accessed via the same phone, to perform the same search. This time the search results were displayed in map form (Figure 16.5b). We could see that we were approaching the restaurant, and could even determine which side of the Interstate it was on before we even arrived. Everyone in the van ended up eating there.

What is important about this example is that the user was able to locate, and subsequently spend money at, a business that would likely not have been found using non-digital means. In this case, the restaurant's Internet presence offset their lack of advertising presence in physical space. However, this is not to say that the restaurant has a highly visible presence in cyberspace (there were no virtual flashy billboards or large signs). The spatial knowledge about the restaurant and the user's position in relation to it, was born out of a coming together of personal positionality (i.e. the desire to eat at a specific restaurant, a specific position in time and space and the ability of an individual to access and successfully conduct a mobile GoogleLocal query) and the actual configurations of physical and virtual space at that moment. One can thus observe that any sort of comprehensive mapping of DigiPlace is impossibility, as DigiPlace encompasses the situatedness of people between virtual and physical realities, and not any sort of shared, objective, and fixed reality.

16.4.3
Temporally dynamic

Moreover, just as DigiPlace is spatially and individually dynamic it is similarly temporally dynamic. The constantly changing nature of the Internet does not allow

(a) (b)

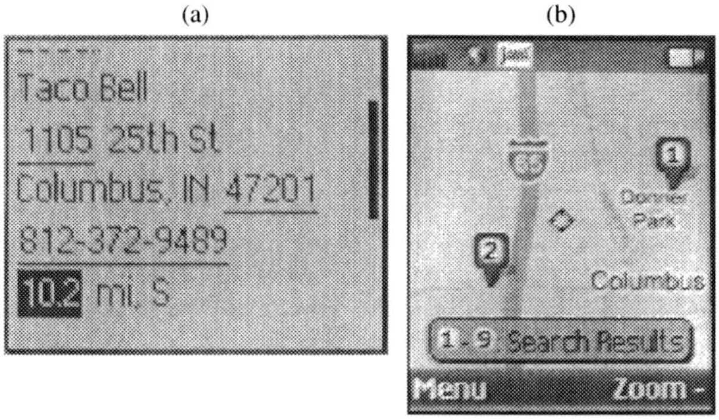

Fig. 16.5 GoogleLocal Search (via Mobile Interface) for Mexican Themed Fast Food Restaurant in Columbus, Indiana

Google's ranking systems to ever settle or stabilize. Shifting cultural, economic, and political interests and processes have effects which ripple through cyberspace. A changed PageRank score for any website results in the PageRank of every website linked to by the original site being altered. This, in turn, has subsequent effects on even more websites. Such exponential effects are constantly being calculated by the ranking algorithms and lead to a system of ranking that is far from stable. The fluidity of cyberspace combined with the dynamism inherent to physical spaces (i.e. the city is constituted in a different manner today than it was yesterday) (Harvey, 1989; Soja, 1989; Massey, 2005), results in DigiPlaces that are continually re-combined and never static.

Although GoogleLocal exhibits changing constellations of rankings for identical topic/location queries over time, the temporal dynamism of DigiPlace is most strikingly illustrated via Froogle, Google's price comparison product. Froogle allows a shopper to conduct price comparisons between in-store sales and other offerings (both online and offline) via a GoogleMobile query. Similarly to the basic Google Pagerank algorithms, Froogle aggregates price information gathered from cyberspace and ranks sites based on an algorithm that takes price, relevance to the search term, and Internet popularity (or Pagerank) into account. When combined with location information, a Froogle query creates an individualized DigiPlace for consumer products (*e.g.*, DVD players) that combines Froogle's rankings and GoogleLocal's hybrid distance. As Table 16.1 illustrate, over the course of one month, seven different retail outlets, all collected to national chains, are included in the top three search results. The order of the rankings, however, is quite volatile with stores moving up and down frequently.

This example again highlights how the cyberspace presence of a business affects its visibility in DigiPlace. A large and locally owned home entertainment store located in the downtown of Lexington does not appear in the top 100 results of this Froogle Local search. Instead well known national chains, HH Gregg, Circuit City, Kmart, Target, etc. dominate the results (See Table 16.1 and Figure 16.6). Interestingly, based on a number of Froogle searches for products such as books and consumer electronics, it was found that lower priced items were often given lower rankings than more expensive items. Furthermore, the shops selling products on Froogle and the prices of those products are by no means locked into stationary

Table 16.1 Top Three Ranked Results for the Price of a DVD Player Near 435 Oldham Avenue, Lexington, KY, April 2006

	1st Ranked		2nd Ranked		3rd Ranked	
Date	Price	Store	Price	Store	Price	Store
4/5/06	$79.99	Target-4 mi	$ 29.99	Radio Shack-4 mi	$54.86	Wal-Mart-2 mi
4/10/06	$99.99	Circuit City-2 mi	$ 79.99	Target-4 mi	$29.99	Radio Shack-4 mi
4/16/06	$29.99	Radio Shack-4 mi	$ 69.99	HH Gregg-4 mi	$44.99	Kmart-2 mi
4/24/06	$69.99	HH Gregg-4 mi	$ 119.99	Circuit City-2 mi	$34.99	Kmart-2 mi
4/28/06	$69.99	HH Gregg-4 mi	$ 119.99	Circuit City-2 mi	$74.99	Target-4 mi

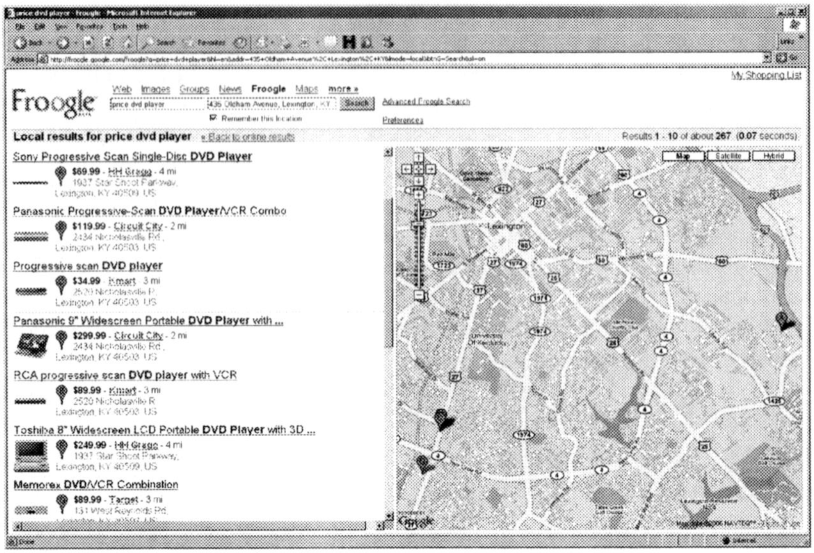

Fig. 16.6 Computer Interface to a FroogleLocal Search for the Price of a DVD Player Near 435 Oldham Avenue, Lexington, KY, April 2006

ranked positions; in fact, dramatic fluctuations in rankings for all products appeared for in the search engine results. Again, these fluctuations were often not in favor of lower priced goods. Thus, while the precise cause of these fluctuations is guarded by Google as proprietary knowledge, it is clear that a variety of factors (such as price and changing Pageranks) are constantly being recalculated and rescored.

16.5 Conclusion

Analyzing the constraints, possibilities, and potentials of space has been the work of geographers for millennia, and the new virtual and hybrid spaces emerging in the 21st century provide fertile ground for similar analyses. The encompassing cloud of mobile-accessible GoogleLocal is much more than just a floating encyclopedia; it is an interactive, dynamic space that influences how we interact with our urban environment, in short, a DigiPlace that transcends the virtual/physical binary. There remains, however, much work to be done in theorizing and empirically studying the implications of these lived hybrid spaces.

To this end, this chapter highlights a number ways in which DigiPlace created via GoogleLocal shapes users perceptions of their physical surroundings and changes the visibility of objects and business contained therein. Fundamental to this shaping is the hybrid and elastic definition of distance used in the Google's DigiPlace as measures of visibility in cyberspace are combined with physical distance. In effect, the world has made another step towards a networked society in which one is either "switched on" and participating or "switched off" and marginalized (Castells, 2002). This also extends to the individual level as the successful negotiation of DigiPlace data clouds requires both technical infrastructure (a mobile device) and knowledge (how to conduct a successful query). Thus, even as the Digital Divide in Internet access may be shrinking, new technological divides are emerging to create yet another exclusionary space.

Another important concern is the way in which the GoogleLocal DigiPlace represents the growing encroachment of private control into our lived spaces. Results filtered through Google's algorithms accord the company a large amount of control over what we see and subsequently how we interact with our environment. Although Google's algorithms are designed "to use the web's self-organizing properties to decide which things to present" (Sheff, 2004), they are not "natural", despite Google's claims to the contrary. Rather they reflect the bias and socially constructed content of the Internet and hyperlinks, and are susceptible, to one degree or another, to manipulation.

Finally, just as an understanding of the architectures of Google's algorithms is crucial for online visibility, an awareness of the structure of GoogleLocal is important for offline commerce. The increasingly intertwined nature of the virtual and the physical means that electronic visibility promises to become ever more relevant as DigiPlace expands. The extent to which Google is directly impacting consumer decision-making and consequently affecting the built environment in physical space is unknown at present, but will likely grow, as increasing numbers of people are drawn into DigiPlace as they navigate through contemporary urban life.

Notes

[1] The many webcams in and around Times Square in New York City serve as a captivating example (http://www.earthcam.com/usa/newyork/timessquare/index.php?cam=1)

[2] See for instance: photo blog of the 2004 Democratic National Convention created entirely by a cell phone (http://blogcon04.tripod.com), and a minute-by-minute match report of a soccer game (http://football.guardian.co.uk/news/matchreport/0,9752,848080,00.html)

[3] For example real-time weather (http://www.ral.ucar.edu/weather), real-time traffic (http://traffic.houstontranstar.org/layers), or a real-time map of spam (http://www.mailinator.com/mailinator/map.html)

[4] http://thorntree.lonelyplanet.com is probably the most popular Internet travel forum

[5] It is again important to note that Google is not the only company to see an opportunity in geo-data, as a competing local search engine is offered by Yahoo, while both MSN (which has also recently introduced a local search engine) and Local.com are still in early stages of development. While the discussion which follows focuses solely on the Google search engine, similar arguments could be derived from studies of alternate websites.

[6] While it is possible to email Google to ask for clarification about how their systems work, the replies are generally uninformative. For example the following text was received in response to a question about GoogleLocal categories. "Hi Mark, Thank you for your interest in GoogleLocal (beta). At this time, we don't provide this information. However, we appreciate your feedback on the usefulness of this information. Thank you for your support as we work to improve GoogleLocal. Regards, The Google Team". Received August 25, 2005.

[7] See Watts (2006) for an example of Google are actively censoring content at the request of the Chinese government.

Reference

AlexaResearch. *Global Top 500* 2006 [cited April 26]. Available from http://www.alexa.com/site/ds/top_500.

BBC News, 2006. *Search Users 'Stop at Page Three'* April 12 2006 [cited April 13 2006]. Available from http://news.bbc.co.uk/2/hi/technology/4900742.stm.

Castells, M. 1996. *The Rise of the Network Society*. Oxford: Basil Blackwell.

———. 2002. *The Internet Galaxy*. Oxford: Oxford University Press.

Cotlier, M. 2001. The Electronic Catalog: The Payoff of Paid Search Listings. *Catalog Age* 18.

Cotriss, D. 2002. Marketers Report High ROI With Paid Listings. *B to B* 87 (3):19–20.

Couclelis, H. 1996. Editorial: The Death of Distance. *Environment and Planning B: Planning and Design* 23:387–389.

Dodge, M., and R. Kitchin. 2004. Flying Through Code/space: The Real Virtuality of Air Travel. *Environment and Planning A* 36 (2):195–211.

———. 2005. Code and the Transduction of Space. *Annals of the Association of American Geographers* 95 (1):162–180.

Google, 2005. *Corporate Information* [Website]. Google 2005 [cited July 16 2005]. Available from http://www.google.com/intl/en/corporate/index.html.

———. 2006. *PageRank Explained* [Webpage]. Google 2006 [cited April 02 2006]. Available from http://www.google.com/technology/.

Graham, S. 1998. The End of Geography or the Explosion of Place? Conceptualizing Space, Place and Information Technology. *Progress in Human Geography* 22 (2):165–185.

Harvey, D. 1989. *The Condition of Postmodernity: An Enquiry Into the Origins of Cultural Change*. Oxford: Blackwell.

Massey, D. 2005. *For Space*. London: Sage.

Monmonier, M. 1996. *How to Lie With Maps*. Chicago: University of Chicago Press.

Nielsen/NetRatings, 2005. *Monthly Top 10 Parent Companies* [Website]. Nielsen/ NetRatings, 2005 [cited July 16 2005]. Available from http://www.nielsen-netratings.com/news.jsp?section=dat_to&country=us.

Perkins, A. 2005. *The Classification of Search Engine Spam* [webpage]. E-brand Management 2003 [cited August 7 2005]. Available from http://www.ebrandmanagement.com/whitepapers/spam-classification/.

Pickles, J. 2004. *A History of Spaces*. London: Routledge.

Sheff, D. 2004. Playboy Interview: Google Guys. *Playboy* 51 (9):55–60 and 142–145.

Soja, E. 1989. *Postmodern Geographies*. London: Verso.

Thrift, N., and S. French. 2002. The Automatic Production of Space. *Transactions of the Institute of British Geographers* 27:309–335.

Van Couvering, E. 2004. New Media? The Political Economy of Internet Search Engines. *Presented at the 2004 Conference of the International Association of Media & Communications Researchers (IAMCR)* Retreived from http://personal.lse.ac.uk/vancouve/research.htm.

Watts, J. *Backlash as Google Shores up Great Firewall of China* The Guardian 2006 [cited January 25]. Available from http://business.guardian.co.uk/story/0,,1694152,00.html.

Weidlich, T. 2005. *Revving Up*. Search Engine Marketing 2002 [cited 2005]. Available from http://searchenginesinfo.com/ar/marketing_revving/.

Zook, M. 2000. The Economic Geography of Commercial Internet Content Production in the United States. *Environment and Planning A* 32:411–426.

17 Paradoxical consequences of location-based services (LBS): A tetradic analysis using McLuhan's laws of media

Daniel Z. Sui
Department of Geography, Texas A&M University, College Station, TX

> Archimedes once said, 'Give me a place to stand and I will move the world.' Today he would have pointed to our electric media and said,"I will stand on your eyes, your ears, your nerves, and your brain, and the world will move in any tempo or pattern I choose."
> We are moving out of the age of the visual into the age of tactile,. . . we wear all mankind as our skin"
> Marshall McLuhan (1964)

17.1 Introduction

The accelerated development of various location-based services (LBS) (http://www.location-based-services.com) as part of the emerging ubiquitous, context aware computing, is a result of increasing integration of wireless communication and information technologies with geospatial technologies (The U.S. National Research Council, 2001, 2003). LBS today represent only the beginning of a series of technological innovations that can potentially impact society in many ways at various scales, ranging from new modes of business operations, to changes in human consumption behavior, to the growing surveillance capabilities by both government and businesses, and to the possible invasion of personal privacy (Sui, 2005). For the business community, LBS has already kicked off the so-called "mobile revolution" (Steinbock, 2005), which is engendering a drastic spatial structuring of retail and other services (Kupper, 2005). And perhaps most important of all, further penetration of LBS in people's daily lives will mark the beginning of what Bill Gates (1995) envisioned as a "documented life" or conducting business at the speed of thoughts (Gates, 1999). With more and more personal daily routines recorded at very fine spatial and temporal resolutions, massive amount of data at the individual level will be accumulated with the growing popularity of

H.J. Miller (ed.), Societies and Cities in the Age of Instant Access, 255–271.

LBS. For geographers and other social scientists, a maturing LBS industry has the potential to provide novel sources of data for researchers in a variety of disciplines and business. Yet so far, the literature on LBS has focused predominantly on its technical aspects. There still exist no comprehensive conceptual frameworks to guide our understanding of the multiple impacts of LBS on society.

Although still at a rather primitive stage, the impacts of LBS on individuals, businesses, and society at large are profound. Publications on LBS have been dominated by the computational and technical concerns (Berger, 2003; Kupper, 2005) although its potential social impacts are slowly gaining attention (Raper and Livingston, 2001; Aoyama, 2003; Sui, 2004). A key point that is conspicuously missing from the current literature is treating LBS as new media. To address this overlook, this chapter has three objectives: 1. to develop a new conceptual framework by reconceptualizing LBS as new media; 2. to conduct a preliminary analysis of the social and spatial impacts of LBS using McLuhan's laws of media; 3. to present some preliminary thoughts on how to deal with the paradoxical consequences of LBS in the age of instant access.

This chapter is organized in four sections. After a brief introduction, Marshall McLuhan's laws of media and their relevance for LBS are introduced in section two. A detailed tetradic analysis of LBS using McLuhan's laws of media is presented in section three. The chapter concludes by situating future growth of LBS in the larger context of technological innovations and their paradoxical consequences.

17.2 Laws of media: The double-helix structure of the social consequences of technological innovations

Recent developments in geospatial technologies have validated an early idea of GIS as new media (Ball, 2005) developed by Sui and Goodchild (2001, 2003). The key argument of this chapter is that LBS can (and should) also be understood as new media. Different from GIS as media in the primitive stage, LBS has pushed GIS as media to a mature stage for two reasons. First, the driving force behind LBS is wireless communication, and second, all the information accumulated in previous media can be potentially accessible via LBS. By reconceptualizing LBS as new media, we can relate the study of LBS to Marshall McLuhan's earlier work on media studies and conduct further analysis on the media and messages of LBS.

McLuhan's work was intensely debated within the academia and among the general public in the 1960s and 1970s (Gordon, 1997). Although McLuhan's influence seems to have diminished rather quickly after his untimely death in 1980, since the late 1990s there have been some consistent efforts to reassess his work in the age of the Internet (Levinson, 1999; Theall, 2001). In retrospect, McLuhan was celebrated mostly for the wrong medium back 40 years ago. As our current lives are increasingly mediated by massively connected electronic network, his contribution is more relevant than ever. As Robert Logan (quite rightly) observed "he [McLuhan] understood the Internet. He was the Internet in the sixties. The world's just finally caught up to him. He was an Internet in the sense he was in touch with the entire globe. He was wired long before the editors of Wired magazine were born (cited in Horrocks, 2000, p.1)." This is why it is even more enticing to have a peek of LBS through McLuhan's looking glass. Indeed, he was talking about LBS back in the 1960s as space, and spatial metaphors are one of the major defining features of McLuhan's thoughts (Cavell, 2002; Moss and Orra, 2004).

Space constraints do not permit discussing thoroughly McLuhan's complex ideas thoroughly. But it will suffice here to sketch a broad contour of his thoughts for analyzing the impacts of LBS. Interested readers are referred to Sui and Goodchild (2001, 2003) for further details and references.

McLuhan and many of his followers believed that there have been three dominant modes of communication throughout human history: oral/speech, writing/printing press, and the electronic medium. Each dominant medium has produced dramatically different psychic and physical impacts on both individuals and society as a whole. McLuhan succinctly summarized these impacts as the laws of media, presented through a tetradic framework (Figure 17.1) in a book co-authored with his son Eric and published posthumously (McLuhan and McLuhan, 1988). McLuhan's laws of media have four major dimensions: any innovations in the dominant mode of communication media will invariably (a) intensify/enhance certain elements of social practices in a given culture (E), while at the same time (b) making other aspects of current social practices/cultural practices obsolete (O). Furthermore, all media innovations will also (c) retrieve certain social or cultural practices long ago pushed aside (R_t), and finally (d) will undergo a reversal when extended beyond the limits of their potential (R_v). The four phases of the tetrad manifest also set the limits for the cultural impact of an artifact, by showing how a totally saturated use will produce a reversal of original intent.

McLuhan's laws of media weaves two key elements to understand his tetradic model: 1) the distinction between visual versus acoustic space and 2) the relationship between figure and ground as outlined in *Gestalt* psychology. According to

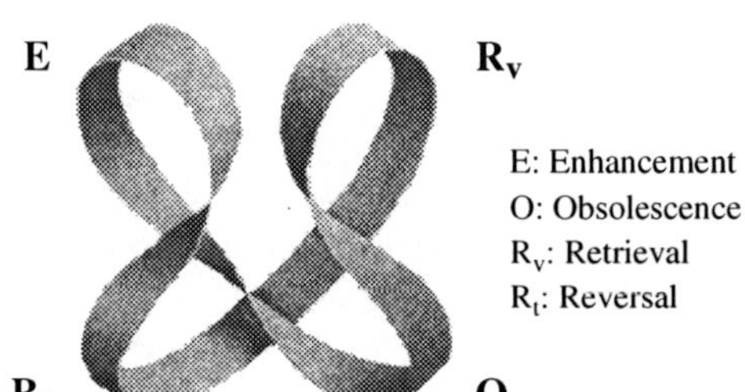

Fig. 17.1 McLuhan's law of media [modified after McLuhan and McLuhan (1988)]

McLuhan (1964), dichotomous intellectual discourses are reflections of bifurcated minds deeply rooted in the visual mode of thinking – the result of the Gutenberg (printing press) revolution. McLuhan (1989) further argue that this "visual space" mind-set leaves little room to alternatives or participation when no provision is made to accommodate two entirely different points of view. The result is usually the exclusion of alternative perspectives.

To fully comprehend the new reality mediated by the electronic medium, McLuhan and Powers (1989) called for a fundamental shift away from the values of linear thinking (visual, proportional space) to those of the multi-sensory life (the experience of acoustic space). Visual thinking places information structurally and sequentially, having separate centers with fixed boundaries. Acoustic thinking regards things as simultaneously interconnected, having centers everywhere with boundaries nowhere. Visual thinking is centered and bounded while acoustic thinking is built on holistic and organic ontology. Acoustic space has no cardinal center, just many centers floating in a cosmic system which honors only diversity. This shift demands that we engage in simultaneous understanding and integral awareness. As I have argued elsewhere (Sui, 2000), the shift from a predominantly visual to an aural metaphor is one of the major changes in intellectual discourses across the disciplines in the late 20th century. Such a shift was made possible by an array of philosophical, technological, and social changes in society.

Furthermore, it is interesting to note that McLuhan's laws of media are analogous to the double-helix structure of DNA. In the context of social impacts of technological innovations, McLuhan's laws of media weave together the dual consequences (double-helix) of new technologies – the coexisting positive and negative impacts (Figure 17.2). Just like the roles played by A, T, G, and C in the double-helix structure of DNA, different combinations of E, O, R_t, and R_v in the laws of media can be used to explain the paradoxical social consequences of technological innovations more holistically and at a deeper level.

At the phenomenological level, metaphors inspired by cell biology are often invoked to describe the swarming behavior of smart mobs in the age of instant access (Rheingold, 2002). Indeed, as the Washington Post reporter Joel Garreau (2002) so aptly described "like the bee, this evolving species (smart mobs) buzzes and swarms (p. C01)." LBS will further enhance the self-organizing capacity of groups of people, who operate like swarms of bees or flock of pigeons. Cell phones with LBS capabilities have enabled the emerging "thumb tribes" to behave in the same way as hive-based animals.

The massive literature produced so far on the social impacts of technology is full of Panglossian technophilia on the one hand (Collins and Kusch, 1998; Collins and Pinch, 1998) and Luddite technophobia sentiments on the other (Morison, 1974; Law, 1991). Different from the previous discussions on the impacts of technology on society, the double-helix structure derived from McLuhan's tetradic framework represents a synthesis of both the positive and negative consequences of changes. The four phases of the tetrad manifest different sides of the cultural life of an

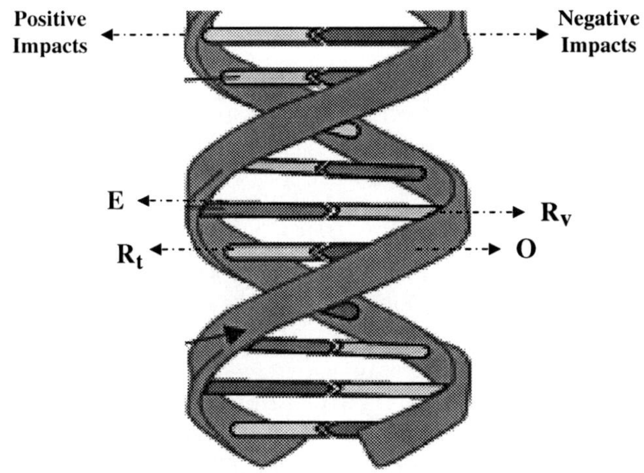

Fig. 17.2 The double-helix structure of social impacts of technological innovations

artifact. A tetradic approach enables us to reposition ourselves and move away from the monolithic linear visual image to a completely different perceptive mode - the mode of the dynamic, many-centered perspective. Instead of the simplistic Utopian or dystopian views, I believe that McLuhan's inclusive and irreducible tetradic framework provides us with a better conceptual apparatus to assess the dynamics and social impacts of the current mobile revolution led by LBS in society. As an exploratory probe, the tetrads do not rest on a theory but on a set of questions that will facilitate our simultaneous understanding and integral awareness of the possible consequences. Without perceiving all the four processes in operation we will be unable to grasp the positive and negative impacts of LBS, and ultimately fail to understand the overall effects of LBS on society. So far the fragmented literature on LBS has covered the technical, business, and social aspects of this emerging technology. McLuhan's laws of media could potentially tie all the scattered elements together, thus contributing to a more holistic perspective on the social impacts of LBS.

17.3 LBS: The media and the messages

If LBS is indeed becoming new media for communicating various forms of personal and business information within society, it will be interesting to examine the impacts of LBS from a media theoretic perspective and explore to what extent can McLuhan's laws of media be applied to probe them. In this section, LBS as media is examined first, followed by a tetradic analysis of the social impacts of LBS using McLuhan's laws of the media.

17.3.1 LBS: The media

As Boorstin (1987) pointed out earlier, the supreme law of the republic of technology is convergence – the tendency for everything to become everything else in the technological world. In the case of LBS, it is a convergence of multiple media that were developed to communicate oral/voice, printed text, visual images, and geographic/spatial information (Jagoe, 2003).

In addition to position acquisition and related geospatial information processing technologies (Hightower and Borriello, 2001; Applewhite, 2002), the backbone for the media of LBS actually relies on a variety of mobile/wireless telecommunication technologies. Mobile telecommunication networks (MTN) have evolved rapidly in Europe since the late 1990s (Schiller and Vasard, 2004). From the humble start of the first generation terrestrial systems, the second generation MTN was based on the Global System for Mobile Communication (GSM), followed by the so-called 2½ generation MTN – General Packet Radio Service (GPRS). The third generation MTN, based upon the Universal Mobile Telecommunication System (UMTS), is operational. MTN bandwidth has advanced from 9.6 Kbps to 115 Kbps, and UMTS can now support 2 Mbps (Shiode et al., 2002). Industry leaders in LBS have been working on a Wireless Application Protocol (WAP) since1997. The goal of WAP is to develop a common standard that will facilitate the convergence with the existing (fixed) Internet infrastructure so that eventually we can have one mobile Internet solution that operates using the same worldwide protocol.

These new developments in MTN have also expanded the capacity for services, from focusing primarily on transmission of voice data to various applications transmitting multimedia information. In particular, the emergence of handheld, mobile and small size wireless devices such as Personal Digital Assistants (PDAs) and mobile smart phones have been enhanced and can be connected via infrared, GSM modems or radio signals to wireless networks. With the continuing improvements in both satellite-based and land-based positioning systems, LBS will most likely continue to grow in the near future. IDC (2004) reported that by the end of 2005, almost 50% of European mobile subscribers had become consumers of various different kinds of LBS. The U.S. FCC's mandate on enhanced 911 (E911) has provided new incentives for value-added LBS (FCC, 2001). Interoperability across multiple networks, platforms and devices will become more crucial for the success of LBS, but an open industry specification has yet to be developed.

Regardless whether person- or device-oriented or whether pushed or pulled, LBS can be offered to consumers at three general levels. At the first (and also the most primitive) level LBS can provide users with raw coordinates about location and navigation activities. A GPS or a cellular radio system can supply the coordinate information (Mountain and Raper, 2001). From the raw coordinates it is possible to provide a grid reference that would enable locating individuals and provide a distance and direction for navigation. At the second level, raw coordinates are combined with information from a local GIS to indicate the user location on a street map and provide a route. At the third level, LBS could use web-based information about location of interest to provide routes to navigate to such destinations of interest. Dynamic information could be converted into personalized services by matching the changing environment and historic preferences of the user.

Different from all the previous media, LBS is in principle capable of integrating all the three major modes of human communication: oral, textual, and electronic. If we follow McLuhan's argument that media are extensions of man, LBS is definitely an extension of man in the fullest sense since it may incorporate multiple senses (ear, eye, and touch). Indeed, LBS has transformed those who have access to the technology into nomads moving within a global village. Considering the emergence of affective computing, future LBS may not just evolve to more instrumental/utilitarian applications, but enable its users to explore more intimate ways at the psychic and emotional levels.

17.3.2 Messages of LBS: A tetradic analysis

Inspired by McLuhan's laws of media, four questions can be posed to decipher LBS messages. Preliminary answers to these four questions are provided here to stimulate further discussions. My goal is to make an initial attempt to better understand this emerging technology in a more holistic way, and suggest areas for future study. The following discussions are not meant to be definitive answers to these questions.

17.3.2.1 The message of LBS (E) - What does LBS enhance?

It has been long believed throughout human economic history that time is money (Faden, 1977). The development of LBS is both challenging and complementing this long-held motto. Future innovations in LBS technologies will enhance the value of locational information related to human activities (Gruber, 2005). The first important consequence of LBS is that they will lead to the commoditization of location and the emergence of mobile commerce (m-commerce) (Dholakia and Dholakia, 2004). Indeed, LBS has converted the entire world into a digital market place in the literal sense and has become an integral part of the new wave of businesses that run at the speed of thought using a digital nervous system as Gates (1999) had predicted. Goodchild (2002) stressed that location is already a fundamental component in the algorithms used to charge users for cellular services. Furthermore, carriers might pass locational information to organizations as part of location-based subscription services purchased by the customer, who would in return receive information related to the current location, such as the state of congestion on nearby roads. As a result, customers might purchase services when LBS alert them that certain locational conditions exist, such as the locations of pet stores or children recreation.

Both industry and governments have been making efforts to facilitate the further commoditization of location and promotion of m-commerce. A number of LBS providers are working together to develop standards, and promote them through organizations like the Location Interoperability Forum and the Open GIS Consortium. The US Wireless Communication and Public Safety Act of 1999 defined the conditions under which network operators can release locations to emergency services, and future legislative frameworks will likely regulate other uses of locational information. In the UK the Regulation of Investigatory Powers Act of 2001 can require recovery of locational information for intelligence purposes. All these new legal mandates could potentially stimulate further growth of LBS and its diffusion into additional services as the business community awakens to the value of locational information.

At the individual level, LBS will enhance the social use of location (Raper, 2001; Shiode et al., 2002). The pricing structure of cellular phones has encouraged very heavy use by younger cohorts (those under 30) and mostly for socializing purposes. Evidently, the short message service (SMS) is being intensively used for exchanging information about location, and facilitating meeting/finding friends. With the embedding of various mobile social software (MOSOSO), LBS can be expected to further promote communities without propinquity on the one hand, and encourage *flocking* or *swarming* behavior among social groups on the other. LBS will also further enhance the kind of virtual life style Mitchell (1999, 2003) described so vividly.

17.3.2.2 The message of LBS (O): What does LBS make obsolete?

The second proposition of McLuhan's laws of media argues that new media and technologies do more than simply extending or enhancing our social practices in many significant ways. At the same time new media also make obsolete (or "amputate") various existing social practices and human faculties. By obsolescence, McLuhan did not mean the total disappearance or outdating of certain social practices, but rather that some of them will no longer be dominant.

As increasing numbers of virtual communities are formed through instantaneous telecommunication networks, LBS will continue to fundamentally alter conventional communities and social relationships. LBS will further contribute to the erosion of social capital (Putnam, 1996) by accelerating centrifugal tendencies and the push toward atomization and solipsism. As LBS becomes more readily available in the years to come, we will witness further affirmation of individualism, rootlessness among more and more people, and a declining/decreasing sense of place. LBS has not only implications about "where" we can have access to certain services, but also "when," which has resulted in what Ling (2004) called the softening of schedule as more activities are being finalized through last-minute micro-coordination.

LBS will continue to worn-out the role of distance in society and simultaneously contribute to the emergence of what Gates (1995) called the friction-free capitalism. In retrospect, the pronouncement of the death of distance in the late 1990s (Cairncross, 1997) proved to be a little bit premature, even though the importance of distance in influencing the behavior of individuals and social practices has been greatly undermined by revolutions in telecommunications. As so many scholars have pointed out the death of distance is the result of a long historical process of time-space compression and distanciation, in which space has continuously been annihilated by time due to sharp declines in the cost of long distance communication and transportation. We are increasingly conducting our lives in a shrinking world – which McLuhan (1964) has so perceptively called "the global village (p. xii)."

For the business community, LBS also entail substantial amount of creative destruction. LBS is transforming industries forcing old market leaders out of their dominant positions. The old world with closed and vertically integrated systems is giving way to a layered and open architecture based upon new protocols and standards, as demanded by LBS. Even though the old style market leaders won't vanish over night, it is the datacom industry that will ultimately win the market battle. The world has not only been shrinking into a village, but also becoming flat during the past five years (Friedman, 2005).

LBS also poses new challenges to conventional business practices and models. In ways similar to e-commerce's challenge to brick and mortar, LBS and M-commerce will surely make certain practices of e-commerce outdated. The new thum-economy, led by M-commerce, is characterized by the emergence of location-based services delivered by a variety of hand-held terminals, such as smart mobile phones and palmtop devices. LBS is capable of accumulating massive amount of user profiles, creating new challenges to the industry. Instead of basic e-mail spam or web browsing hijacking, businesses in the age of LBS need to develop personalized one-to-one advertising with much improved spatial and temporal precision. Businesses are nowadays capable of offering diverse services to satisfy consumer needs at specific moments and locations. This naturally requires businesses to rethink space allocations in the context of LBS, as is proved by the growing emphasis on micro-marketing and multi-site and multi-service stores. In recent years, the quickest route to retailing disaster was trying to be all things to all people, an approach that was a reflection of the initial impacts of LBS. Indeed, multiplicity might be the order of the day for business in the near future.

In the new digital economy, LBS will blur the traditional boundary between production and consumption. The growth of LBS is a potent testimony of Rifkin's (2000) observation that life is an all-paid experience in the new economy. Many of the conventional business models are become obsolete. For example, new trends in the car sharing business such as FlexCar (http://www.flexcar.com) poses serious challenge for conventional car ownership and car rental business. New businesses such as FlexCar are made possible partially due to the maturity of LBS. However, the death of distance does not mean the end of geography. On the contrary, it will mark a new beginning for geography (Kolko, 2000; Malecki and Gorman, 2001) as the laws of media tells us that innovations also simultaneous retrieve older practices that have long been pushed aside.

17.3.2.3 The message of LBS (R_t): What does LBS retrieve?

The backbone of LBS relies on the real-time exchange of spatial and temporal information at the individual level. LBS entails *locational profiling*, which include the tracking of the user and of characteristics associated with the tracking – such as speed of movement, positional accuracy, and relevance of information to the user. Once a detailed locational profile is collected, it is possible to analyze tracks, to obtain estimates of speed, to make inferences about activities based on the speed and geometric nature of the track, or to transform them into density surfaces, etc. All these tasks are precisely what Torsten Hägerstrand (1970) tried to accomplish back in the 1960s and 197s0 through his time geography. Obviously, the growth of LBS has retrieved "time geography" as a research practice during the past decade and this trend will continue in the near future.

Hägerstrand (1970) is widely acknowledged to be an early pioneer of modern interest in tracking people through space and time, and the processes and constraints that govern their tracks. LBS can provide abundant *tracking* data on the daily movements of people, and is already widely used in ecology to track the behavior of animals (Goodchild, 2002). In a more generic sense, tracking data normally consists of a sequence of tuples $< x, y, t >$ ordered by t, indicating the location (x, y) of the moving object at intervals of time denoted by t. In mountainous areas, altitudinal information z may be also required. Although early attempts were made to implement Hägerstrand's framework using GIS were not successful, recent development of LBS has fully retrieved time-geography (Dykes and Mountain, 2003) Brimicombe and Li, 2006). Inquiries along this line have led Miller (2004) to develop a people-based GIS, as opposed to a location-based GIS.

Another resurgence that was enabled by LBS is the retrieval of classical locational theories. The growth of LBS has made researchers revisit many premises in the classical location theories as advocated by von Thünen, Weber, Lösch, and Christaller. Such theories deal with locational choices made by individuals or groups, and the economic and social bases for such choices in traditional economies related to agriculture, industry, and retail services. I noticed that most cellular phone service providers have chosen hexagon as the basic shape to design the spatial configuration of their network of cellular transmission towers, which coincides with the optimal shape of service areas as predicted in Walter Christaller's central place theory. So LBS has not only retrieved Hägerstrand but also Christaller. Now the resulting questions are: to what extent can location theory provide theoretical insights for choosing locations for servers, LBS devices, clients, and other components of computing networks (Goodchild, 2001)? Will new sets of theories, fundamentally different from the classic theories, be needed for LBS? These topics are obviously locate over the research frontiers to be targeted by researchers in the next couple years.

The rediscovery of time geography by LBS and the prospects of developing a people-based GIS (Miller, 2004) can potentially lead to a much improved analysis and monitoring of people's daily routines, thus helping to ease problems such as those related with traffic congestion and access to services, and improve people's quality of life. But it is precisely the possibility of a fully documented life through locational/geographic profiling (Tossmo, 2000) that is causing grave concerns.

17.3.2.4 The message of LBS (R_v): What does LBS reverse into?

The last proposition of McLuhan's laws of media states that when a medium is pushed beyond its limit, it will be reversed into the opposite of what it was originally designed for. In McLuhan's words, "we become what we behold; we first make the tools, then the tools will make us" (McLuhan, 1964). To McLuhan and many of his followers, media are not simply "making-aware" agents but also, perhaps more importantly, "making-happen" agents, through which LBS contributes to the transduction of space (Dodge and Kitchin, 2005). With growing concerns about the intrusive nature of LBS, we are already beginning to see earlier signs of LBS' reversal effects as privacy itself is rapidly evolving from a right to highly priced commodity (Davies, 1997).

Urry (2003) argued that constant mobility is one of the defining characteristics of our age. Yet what is different from the mobility of the early days of modernity is that now peoples' movements can be constantly tracked and monitored in the age of LBS (Phillips, 2003; Armstrong and Tuggles, 2005). Concerns over the possible invasion of personal privacy with the use of geographic information are becoming even more acute in the context of LBS. With personal profiling capabilities increasingly embedded, LBS could easily be reversed into an electronic version of Jeremy Bentham's *panopticon* (Simon, 2005), a place where all of us are being watched without knowing it. Although the societal benefits of using LBS to conduct electronic surveillance are enormous, like in emergency situations (Phillips, 2005), ordinary citizens should be concerned with its pernicious potential of an Orwellian Big Brother watching us wherever we go. With implementation of the federal mandatory E-911 initiative, both people and places are also increasingly "legible" (Curry, 2004). Accept it or not, various forms of *panopticon* (CCTV in malls, supermarkets, and other public spaces, cameras at traffic lights, etc.), although imperfect (Hannah, 1997), are already part of our daily lives (Gandy, 1993). LBS, if not well guarded, could make each one of us living in the perfect *panopticon*, which could have profound implications on social relations – citizens vs. state, employer vs. employees, children vs. parents, husbands vs. wives, etc. How ironic it will be if LBS, a technology originally developed to liberate us from the confines of location, can potentially also convert us all into "prisoners" of geography.

In fact, the growth of LBS, especially through the development of human tracking devices, introduces a new potential for real-time control that goes far beyond privacy and surveillance per se (Dobson, 2006). As a result, Dobson and Fisher (2003) even warned that society must address a new form of slavery characterized by location control. Geoslavery now looms as a real, immediate, and global threat. Also according to Dobson and Fisher, geoslavery refers to a practice in which one entity (the master) coercively or surreptitiously monitors and exerts control over the physical location of another individual (the slave). The inherent danger, they further argue, is that there is "the potential for a master to routinely control the time, location, speed, and direction for each and every movement of the slave or, indeed, of many slaves simultaneously (p.48)". This new form of slavery has raised the fundamental question of whether locational privacy is a fundamental human right that should be universally protected (Monmonier, 2002; Fisher and Dobson, 2003).

New services offered by companies such as TeenArriveAlive (www.TeenArriveAlive.com), World Tracker, or Tracking the World have indeed become electronic versions of the *Panopticon* and they could be easily used to geo-slave the vulnerable population of society. Other possible reversal effects of LBS may include negative environmental effects due to the increasing levels of material and energy consumption caused by ubiquity (Sui and Rejeski, 2002), personal health problems due to the excessive use of radio frequencies (RF) linked to LBS, pollution (Fishbein, 2003), and the potential for growing crime on identity theft (Slosarik, 2002).

To fix this type of problems, we obviously need more secure technologies to better protect our privacy (Gruteser and Grunwald, 2003; Myles et al., 2003; Beresford, 2005; Marias et al., 2006). We should also develop a new set of ethical rules on how to and when to "forget," and even "delete" the massive electronic information accumulated in the age of pervasive computing (Dodge and Kitchin, 2005). Perhaps, the most disturbing part for the reversal effects of LBS is that as of today, our national or international legal frameworks are lagging behind the rather rapid development of LBS, especially in the area of rights and responsibilities for the use of human tracking systems (Reidenberg, 2003; Herbert, 2006). Some general ideas drawn from the laws of robotics could provide a helpful start to the debate on such legal and ethical issues (Sui, 2004).

17.4 Summary and conclusions: Can we avoid the tragedy of the information commons?

Maturing LBS, as part of the mobile revolution, will have profound impacts for individuals, communities, and societies in the years to come. By reconceptualizing LBS as media, this chapter aims to develop a more synthetic, robust framework to better understand the multiple, complex implications of LBS. The analysis of LBS as media, conducted using McLuhan's tetradic framework, enabled us to link messages of LBS that have been scattered in the literature so far. Instead of a dichotomous characterization in an either/or mode, the development of LBS has been showing a complex set of processes of simultaneous enhancement, obsolescence, retrieval, and reversal at the individual, organizational and societal levels.

Reconceptualizing LBS as media enabled us to transcend common instrumental and utilitarian interpretations of LBS by both developers and users, and develop a more holistic approach to the non-linear relationships between LBS and society. The most clear and present danger is not related to LBS or any rapidly evolving technologies *per se*, but to our blissful ignorance of the full implications of what new technologies can potentially do TO us, because we tend to focus primarily on what they can do FOR us. The only sensible way of dealing with this inherent technological ambivalence is always to treat the LBS media as a means to higher social ends. In a larger context, the paradoxical impacts of LBS are a reflection of the paradoxes of our technology-mediated age (Handy, 1994) or what we generally consider as "progress" (Tuan, 1989). In many interesting ways, Costanza's (1999) four visions of the humanity's ecological future – Star Trek, Mad Max, Ecotopia, and Big Government – resonate with McLuhan's laws of the media in its tetradic exegesis. Also learning from the lessons of the tragedies of the commons in the environmental domain (Hardin, 1968), we need collective actions if we want to avoid the tragedy of the information commons (Onsrud, 1998).

The rapid development of various LBS led by coalescing innovations in mobile computing, wireless communications, GIS, mobile Internet and other technologies has simultaneously enhanced, obsolesced, retrieved and reversed the roles of location, space, and distance played in human affairs. It is partial, even misleading, to characterize its effects simply as either "always-on Panopticon" or "cooperation amplifier" (Rheingold, 2002, p. 1983). The take-home message of this chapter is that LBS is simultaneously creating multiple paradoxical impacts at the individual, organizational/business, and societal levels. These impacts raise some challenging ethical questions. Becoming fully cognizant of the unintended consequences of human actions is the first step towards developing geospatial ethics for the applications of LBS. We all want to savor the fruits of these technologies without succumbing to its most dangerous temptations. Whether we can achieve that delicate balance in the age of instant access is far from clear at present. Similar to the dilemmas brought by the dizzying bio-tech innovations (President's Council on Bioethics, 2003), how to capitalize the enormous benefits of LBS and at the same time deter (or at least diminish) the emergence of a dystopia of Orwellian/Frankenstein proportions remains a challenge for our society in the age of instant access.

Acknowledgement

I'd like to thank Jose Gavinha for commenting on earlier versions of this chapter. Research assistance by Shujuan Li, Zengwang Xu, and Yong Wang are gratefully acknowledged. Thanks are also due to Peter Fisher and Jerry Dobson for sending me their papers on geoslavery. An earlier version of this chapter (Sui, 2005b) was presented during Geoinformatics'04 in Sweden.

References

Aoyama, Y. (2003) "Sociospatial dimensions of technology adoption: Recent M-commerce and E-commerce developments," *Environment and Planning A*, 35(7), 1201–1221.

Applewhite, A. (2002) "What Knows Where You Are?" Pervasive Computing (Oct.–Dec.): 4–8.

Armstrong, M.P and Ruggles, A. (2005) "Geographic information technologies and personal privacy," *Cartographica*, 40(4), 63–73.

Ball, M. (2005) "Google Verifies that GIS Is Media," *GeoWorld*, December Issue, *http://www.geoplace.com/uploads/FeatureArticle/0512edit.asp*

Beresford, A.R. (2005) *Location privacy in ubiquitous computing*. Technical Report (UCAM-CL-TR-612), University of Cambridge.

Boorstin, D.J., 1987: *Hidden History*. New York: Harper & Row.

Brimicombe, A. and Li, Y. (2006) "Mobile space-time envelopes for location-based services," *Transactions* in GIS, 10, 5–23.

Cavell, R. (2002) *McLuhan in Space: A cultural geography*. Ottawa, University of Toronto Press.

Collins, H. and M. Kusch (1998) *The Shape of Actions: What humans and machines can do*. Cambridge, MA.: The MIT Press.

Collins, H. and T. Pinch (1998) *The Golem at large: What you should know about technology*. New Cork: Cambridge University Press.

Costanza, R. (1999) "Four visions of the century ahead: Will it be Star Trek, Ecotopia, Big Government or Mad Max?" *Futurist*, 33(2), 19–27.

Curry, M., Phillips, D.J., and Regan, P. (2004) "Emergency Response Systems and the Creeping Legibility of People and Places." *The Information Society, 20*(5), 1–13.

Davies, S. (1997) "Re-engineering the right to privacy: How privacy has been transformed from a right to a commodity," in P. Agre, & M. Rotenberg (eds.), *Technology and Privacy: The new landscape*. Cambridge, MA.: MIT Press, 143–166.

Dholakia, R.R. and N. Dholakia (2004) "Mobility and markets: Emerging outlines of m-commerce," *Journal of Business Research, 57*, 1391–1396.

Dobson, J. E. 2006. "Geoslavery," *Encyclopedia of Human Geography* (Thousand Oaks, CA: Sage Publications).

Dobson, J. E., and P. F. Fisher (2003) "Geoslavery," *IEEE Technology and Society Magazine*, 22(1), 47–52.

Dodge, M. and R. Kitchin (2005) "Code and the transduction of space," *Annals of the Association of American Geographers*, 95(1), 162–180.

Dodge, M. and R. Kitchin (2005) "*The ethics of forgetting in an age of pervasive computing*" Unpublished manuscript, available on-line at: http://www.geog.ucl.ac.uk/casa/martin/martin.html (accessed on Oct. 15, 2005).

Dykes, J. A. and D. M. Mountain (2003) "Seeking structure in records of spatio-temporal behaviour: visualization issues, efforts and applications," *Computational Statistics & Data Analysis, 43*, 581–603.

Faden, A.M. (1977) *Economics of Space and Time: The Measure-theoretic foundations of social science*. Iowa State University Press: Ames, IO.

Federal Communications Commission (FCC) (2001) "FCC Wireless 911 Requirements," http://www.fcc.gov/e911/as of January 2002 (Accessed Feb. 2, 2006).

Fishbein, B.K. (2003) "*Waste in the Wireless World: The challenges of cell phones*," http://www.informinc.org/cellphone.htm (accessed on March 1, 2004).

Fisher, P. F, and J. E. Dobson (2003) "Who knows where you are, and who should, in the era of mobile geography?" *Geography*, 88(4), 331–337.

Foucault, M. (1977) *Discipline and Punish: The birth of the prison*. New York: Vintage Books.

Gandy, O. (1995) "It's discrimination, stupid," in J. Brook and I. A. Boal (eds.), *Resisting the Virtual Life: The culture and politics of information*, San Francisco: City Lights, 35–47.

Gandy, O.H. Jr. (1993) *The Panoptic Sort: A political economy of personal information*. Boulder, Co.: Westview Press.

Garreau, J. (2002) Cell biology (a report of smart mobs). Washington Post (Wednesday, July 31, section C), *http://www.washingtonpost.com/ac2/wp-dyn?pagename=article&contentId=A23395-2002Jul30&notFound=true* (accessed on Aug. 25, 2005).

Gates, B. (1995) *The Road Ahead*. New York: Viking.

Gates, B. (1999) *Business @ the Speed of Thought: Using a digital nervous system*. New York, NY.: Warner Books.

Goodchild, M.F. (2002) *Final report on a Specialist Meeting on Location-based Services*, http://www.csiss.org/events/meetings/location-based/goodchild_lbs.htm (accessed on January 3, 2003).

Goodchild, M.F., 2001: Towards a Location Theory of Distributed Computing and E-commerce. In: *Worlds of E-Commerce: Economic, Geographical and Social Dimensions*. T.R. Leinbach and S.D. Brunn(Eds). Wiley, New York. 67–86.

Gordon, W.R. (1997) *McLuhan for Beginners*. New York: Writers and Readers Publishing, Inc.

Goss, J. (1995) "We know who you are and we know where you live: The instrumental rationality of geodemographic systems," *Economic Geography*, 71(2), 171–198.

Gruber, H. (2005) *The Economics of Mobile Telecommunications*. New York: Cambridge University Press.

Gruteser, M. and D. Grunwald (2003) "Anonymous usage of location-based services through spatial and temporal cloaking," in *Proceedings of the ACM/USENIX International Conference on Mobile Systems, Applications, and Services (MobiSys)*, 121–133.

Hägerstrand, T. (1970) What about people in regional science? *Papers of the Regional Science Association* 24, 7–21.

Handy, C. (1994) *The Age of Paradox*. Boston, MA.: Harvard Business School Press.

Hannah, M. (1997) "Imperfect Panopticism: Envisioning the construction of normal lives," in G. Benko and U. Strohmayer (eds.), *Space and Social Theory: Interpreting modernity and postermodernity*. Malden, MA.: Blackwell, 344–359.

Hardin, G. (1968) The Tragedy of the commons," *Science*, 62, 1243–1248.

Herbert, W. (2006) "No direction home: Will the law keep pace with human tracking technology to protect individual privacy and stop Geoslavery?" *ExpressO Preprint Series*. Working Paper 1006, http://law.bepress.com/expresso/eps/1006/

Horrocks, C. (2000) *Marshall McLuhan and Virtuality*. Totem Books, New York.

IDC, 2004: U.S. *Cellular Location-Based Services 2004–2008 Forecast and Analysis: Mapping a Mobile Future*. IDC.

Jagoe, A. (2003) *Mobile Location Services: The definitive guide*. Upper Saddle River, NJ.: Prentice Hall.

Karimil, H.A. (2004) *Telegeoinformatics: Location-based computing and services*. Taylor & Francis, London.

Kupper, A. (2005) *Location-based Services: Fundamentals and operation*. New York: John Wiley & Sons.

Law, J. (1991) "A sociology of monsters: Essays on power, technology and domination," *Sociological Review Monography* 38. London: Routledge.

Levinson, P. (1999) *Digital McLuhan: A guide to the information millennium*. New York and London: Routledge.

Ling, R. (2004) The Mobile Connection: The cell phone's impact on society. San Francisco, CA.: Morgan Kaufmann.

Malecki, E.J. and Gorman, S.P. (2001) "Maybe the death of distance, but not the end of geography: The Internet as a network," in S.D. Brunn and T.R. Leinbach (eds.), *The Worlds of Electronic Commerce*. John Wiley, New York: 87–105.

Marias, G.F., L. Kazatzopoulos, C. Delakouridis and P. Georgiadis (2006) "Applying privacy on the dissemination of location information," *Telematics and Informatics*, 23, 211–225.

McLuhan, M. (1964) *Understanding Media: The extension of man*. Sphere Books, London.

McLuhan, M. and McLuhan, E. (1988) *Laws of Media: The new science*. University of Toronto Press, Toronto.

McLuhan, M. and Powers, B.R. (1989) *The Global Village: Transformations in world life and media in the 21st century*. New York: Oxford University Press.

Miller, H. J. (2005) "What about people in geographic information science?" in P. Fisher and D. Unwin (eds.), *Re-Presenting Geographic Information Systems*. John Wiley, London: 215–242.

Mitchell, W.J., 2003: *Me++: The Cyborg Self and the Networked City*. Cambridge, MA.: MIT Press.

Mitchell, W.J., 1999: *E-topia: "Urban life, Jim–but not as we know it."* Cambridge, MA.: MIT Press.

Morison, E.E. (1974) *From know-how to nowhere: The development of American technology*. Oxford: Blackwell.

Moss, J. and L. M. Morra (2004) *At the Speed of Light There Is only Illumination: A reappraisal of Marshall McLuhan*. Ottawa: University of Ottawa Press.

Myles, G., A. Friday, and N. Davies (2003) "Preserving privacy in environments with location-based applications," *IEEE Pervasive Computing*, 2(1):56–64. *http://dx.doi.org/10.1109/MPRV.2003.1186726.*

Onsrud, H.J. (1998) "The tradgedy of the information commons," in *Policy Issues in Modern Cartography*, Berlin, Elsevier Science, 141–158.

Phillips, D.J. (2005.) "Texas 9-1-1: Emergency Telecommunications, Deregulation, and the Genesis of Surveillance Infrastructure." *Telecommunication Policy*, 29, 843–856.

Phillips, D.J. 2003. "Beyond Privacy: Confronting Locational Surveillance in Wireless Communication." *Communication Law and Policy 8*(1): 1–23.

President's Council on Bioethics (2003) *Beyond Therapy: Biotechnology and the pursuit of happiness*. New York & Washington, D.C.: DANA Press.

Putnam, R.D. (1996) "The strange disappearance of civic America," *The American Prospect*, 24, 34–46.

Reidenberg, J. (2003) "Privacy wrongs in search of remedies," *Hastings Law Journal*, 54, 877–898.

Rheingold, H. (2002) *Smart Mobs: The next social revolution*. Cambridge, MA: Perseus Publishing.

Rifkin, J. (2000) *The Age of Access: The new culture of hypercapitalism, where all of life is a paid-for experience*. New York: J.P. Tarcher/Putnam.

Rossmo, D.K. (2000) *Geographic Profiling*, Boca Raton, FL, CRC Press.

Schiller, J. and Vasard, A. (eds.) (2004) *Location-Based Services*. Morgan Kaufmann, San Francisco, CA.

Shiode, N., Li, C., Batty, M., Longley, P. and Maguire, D. (2002) The Impact and Penetration of Location-Based Services. http://www.casa.ucl.ac.uk/working_papers/Paper%2050.pdf.

Simon, B. (2005) "The return of Panopticism: Supervision, subjection, and the new surveillance," *Surveillance & Society*, 3(1), 1–20.

Slosarik, K. (2002) "Identity theft: An overview of the problem," *Justice Professional*, 15, 329–343.

Steinbock, D. (2005) *The Mobile Revolution: The making of mobile services worldwide*. London: Kogan Page.

Sui, D.Z. (2004) "Are laws of robots applicable to location-based services (LBS)?" *GeoWorld* (formerly *GIS World*), September Issue, 22–24.

Sui, D.Z. (2005a) "Will ubicomp make GIS disappear?" *Computers, Environment, and Urban Systems*, 29(1), 361–367.

Sui, D.Z. (2005b) "The media and messages of location-based services: Death of distance or the revenge of geography?" *Journal of Geographic Information Science*, 10(2), 166–174.

Sui, D.Z. and D.J. Rejeski (2002) "Environmental impacts of the emerging digital Economy: The E-for-Environment E-Commerce?" *Environmental Management*, 29(2), 155–163.

Sui, D.Z. and M.F. Goodchild (2001) "Are GIS becoming new media?" *International Journal of Geographical Information Science*, 15(5), 387–390.

Sui, D.Z. and M.F. Goodchild (2003) "A tetradic analysis of GIS and society using McLuhan's laws of media," *Canadian Geographers*, 47(1), 5–17.

Tenner, E. (1996) *Why Things Bite Back: Technology and the revenge of unintended consequences*. New York, NY.: Knopf.

Tenner, E. (2004) *Our Own Devices: The past and future of body technology*. New York: Alfred A. Knopf.

The U.S. National Research Council (2001) *Embedded, Everywhere: A research agenda for networked systems of embedded computers*. Washington, DC: National Academy Press.

The U.S. National Research Council (2003) *IT Roadmap to a Geospatial Future*. Washington, DC: National Academy Press.

Theall, D.F. (2001) *The Virtual Marshall McLuhan*. Montreal: McGIll_Queen's University Press.

Tuan, Y.F. (1989) *Morality and Imagination: Paradoxes of Progress*. Madison, WI.: The University of Wisconsin Press.

Urry, J. (2000) Sociology beyond Societies: Mobilities for the twenty-first century. New York: Routledge.

Social and Economic Networks

18 The evolving social geography of blogs

Sucharita Gopal
Department of Geography and Center for Cognitive and Neural Systems, Boston University

18.1 Background: The blogging phenomena

A blog is a frequently updated webpage, typically consisting of brief entries presented and archived in reverse chronological order. Blog posts are primarily textual, but many include photos, and other multimedia content. Most blogs are interlinked in that they provide links to each other as well as other sites on the Internet. Many are interactive since they act as communication channels between the author and readers of the blog. They are the fastest form of online communication resulting in the rapid formation of new virtual communities. Blogs have gained enormous popularity during and after the 9/11 event. As of July 2005, Technocrati reports that there are 14.2 Million Blogs. A new blog is created every second and 80,000 blogs are created daily (Source: Technocrati, 2005).

The focus of this chapter is understanding how blogs are leading to the formation of new cyber societies, with a specific objective of uncovering the spatio-temporal structure of two representative blogospheres. It takes a spatial science perspective to examine the growth and evolution of blogs using Hägerstrand's time geographic framework (Hägerstrand, 1970). Two types of blogs are analyzed. In the first type of blogs, communities coalesce around special interests such as politics, law, technology, gadgets, religion, etc., (for example, Instapundit, Boing Boing). The second type of blog is influenced by geography and is set up for local action as in political campaigns (for example, Moveon.org's Operation Democracy). Influential leaders, message content, external links, time, actors (bloggers) and other variables of the diffusion process are analyzed in both contexts. The results are visualized using UCINET 5 and Pajek. This research will ultimately help us explore how new communication channels are affecting the social geography of the Internet and how virtual communities are linked to local or regional geography.

H.J. Miller (ed.), Societies and Cities in the Age of Instant Access, 275–293.

18.1.1 The genesis of blogs

Weblog or Blog is as old as the Web itself (see Blood, 2002a). The precursor to weblogs can be traced back to Mosaic, the first widely available web browser, which maintained a "What's New" page from June 1993 to June 1996 displaying links (Blood, 2002a). Many web enthusiasts and web designers adopted a similar format to record their digital wanderings through cybespace, links with commentary with the new "stuff" being displayed on the top. These sites acted as *filters* of interesting Internet content, that proved useful in the absence of online search engines. According to Wikipedia (Wikipedia.org), the term "weblog" was coined by Barger in 1997. The more popular term "blog" was coined by Merholz in 1999 meaning "to edit one's weblog or a post to one's weblog". By 1998 several of these sites added a list of links to similar sites and referred to themselves as weblogs. By the end of 1998 the longest list of weblogs contained only 23 sites. Blogs were the turf of the web nerds since they were difficult to create and maintain and required an ability to code or use HTML.

Blog use and popularity spread when Pitas, the first free build-your-own web tool, was launched in July 1999; in August of the same year, Pyra released Blogger and other web-based tools that extended the means of making weblogs to those whose only technical knowledge was to operate a web browser. By the end of 1999, weblogs were too numerous to count. The genre of weblogs continued to evolve as many of the new blogs went beyond filtering other content on the Internet to creating their own content. Blogging gained wider acceptance and spread to other countries and languages. Many blogs specialized on subject matter, location, author's opinions and perspectives, and so on. The foundation of each weblog reflected the personality and motivation of its author—each site is unique as its content is determined each day by the decisions of the author regarding what to write and what to link to. By 2002, there were perhaps 500,000 blogs. Blogs became more widespread in the United States, particularly since 9/11. Recent Pew Internet and American Life research (Pewtrust.com) estimated that 11 percent of American Net users have read blogs that is created by about 2–7 percent or roughly 2.4 and 8.4 million bloggers. Net users with a college degree are the most likely bloggers.

Blogging has become so ubiquitous that many companies are implementing them. AOL implemented a version in 2003. In the same year, Google bought Blogger (from Pyra Labs) to claim its share in the blogosphere. In 2004, Nokia announced that it was adding LifeBlog to its phones. Blogs have demonstrated their usefulness in assessing social and political trends, aiding in world-wide crisis communication and in garnering political support and funding (Efron, 2004).

18.1.2 Blog types

The style of weblogs ranges between two extremes: filters and blogs (Blood 2002a, Blood 2000b). Filters are the classic form of weblogs which focus on links to interesting places on the Internet as discovered and presented by a dedicated web surfer. Blogs, on the other hand, refer to a form of Internet publishing that can range from individual on-line diaries or personal journalism to organized campaigns run by political groups, media and corporations (Efron, 2004). However, most definitions of blogs incorporate one or more of the following characterizations: blogs are webpages containing dated entries in reverse chronological order. Often blogs include a list of external links (*attribution*) to news articles, documents (Herring, Bonus, Scheidt, & Wright, 2003) as well as links to other blogs (*blogroll*) and to sites integrated into each posting. Inter-textuality is achieved through hypertext linking between blogs. In blogs, old content remains accessible through archived entries while RSS or XML feed promote ease of syndication.

Krishnamurthy (2002) classified blogs into four types, along two dimensions: personal vs. topical, and individual vs/community. Major blog sites oriented to politics (Instapundit or DailyKos) and technical matter (Gadgets) are "heavy hitters" as they receive thousands of hits a day. Such sites are topical maintained by an individual or community. Herring et al., estimate that a majority of blogs (over 70%) are written by ordinary people on largely personal themes and have much smaller readership. Howard Dean, one of the democratic candidates during the Primaries, used blogs to disseminate political opinion. Both Left and Right winged bloggers were responsible for pushing the Trent Lott story into the news leading to his eventual resignation as Senate Majority leader in 2004. Citizen journalism run by bloggers is changing the way news gets reported. For example, bloggers' accounts of Katrina's devastation became part of the mainstream media broadcasts. We will examine the content and relevance of geographical diaries of the hurricane maintained by mainstream media as well as citizenry impacted by it.

18.1.3 Tracking blogs

There are a number of automated weblog analysis that can track blogs and provide indicators of their popularity and influence. These can be grouped roughly into three categories: 1) services that track and allow subscription to blogs and RSS3 feeds; 2) services that rank blogs according to popularity, usually deter- 3 Really Simple Syndication (also Rich Site Summary or RDF Site Summary) is a format for syndicating news and the content of news-like sites (http://www.xml.com/pub/a/2002/12/18/diveinto- xml.html).

18.1.4 Blogstypes

Blogs are not created equal: Web services such as Technorati provides a basic measure of blog readership via the number of incoming links from other blogs and, in so doing, make an inference about the relative status of a blog to other more or less widely linked blogs. Technorati updates its list of the top 100 blogs. Table 18.1 shows the top 10 blogs in July 2005 list that range from political

Table 18.1 The Top 10 Blog Sites

Top 10 Sites	Links
Boing Boing: A Directory of Wonderful Things	16,913
Daily Kos: State of the Nation	11,245
Drew Curtis' FARK.com	10,623
Gizmodo: The Gadgets Weblog	10,509
Engadget.com	10,268
Instapundit.com	9,927
PostSecret	7,847
Talking Points Memo: by Joshua Micah Marshall	7,796
Davenetics* Politics Media Musings	7,346
Dooce	7,327

(Source: http://www.technorati.com/ pop/blogs/)

to technical topics. Truth Laid Bear Ecosystem (Bear, 2005) provides an index of registered blogs using an evolutionary animal metaphor. Blogs are ranked by incoming links from other blogs registered in the Ecosystem. Blogs are ranked hierarchically from top – *predator* InstaPundit described as a *Higher Being* down through tiers including *Playful Primates, Flappy Birds, Lowly Insects* all the way to *Insignificant Microbes* that subsist without a single incoming link to their name.

18.1.5 Utility of blogging and other communication data in social sciences

The new form of connectivity 24/7, including emails, blogging, mobile and other communication, is providing unprecedented data on interaction and behavior among people that enables the social scientist to model individual and group behavior. The resulting datasets are several orders of magnitude larger than data from survey or other traditional data gathering techniques. In addition, they are amenable to both qualitative (Wellman, 2001) and quantitative analysis. For example, a project at MIT Media Lab called Reality Mining (http://reality.media.mit.edu/) is collecting machine-sensed environmental data pertaining to human social behavior (Eagle, 2005, Eagle and Pentland, 2005). The MIT researchers are using datasets assembled from mobile phones to investigate conversation context, proximity sensing, and spatio-temporal location of large communities of individuals. In the future, many personal blogs include geotags or geographical coordinates because of the increasing ease of integrating geospatial reference (http://www.blogmapper.com) as well video or images (http://www.flickr.com/photos/tags/geotagged/).

18.2 Cultural geography of the blogosphere

Cybergeography is the study on the spatial structuring of computer networks (including blogs) that are connected to the Internet. It encompasses a broad spectrum of geographical phenomena, such as the study of the physical infrastructure, the information traffic and the demography of virtual communities (Castells, 2000; Dodge, 2004). Some recent research describes the spatial distribution of blogs.

18.2.1 What is the spatial distribution of blogosphere?

Blogging at this point in time is a global phenomenon. A popular blogging site called Livejournal.com has more than 1.5 Million registered bloggers. This site provides demographic, spatial and link statistics for its bloggers. Figure 18.1 shows the global distribution of bloggers based on data from Livejournal.com. The United States, Canada and the UK are the three top blogging nations followed by other countries in the European Union and Brazil in Latin America. Africa ranks low in the blogosphere distribution.

18.2.2 Does blogging differ by zip code?

Are there blog hot spots? Is there an underlying spatial process that leads to these concentrations? To address this issue, blogs have to be linked to an actual geographical location (zip codes or latitude and longitude). A service called Blogmapper using open formats lets a user link a location to his/her blog. The

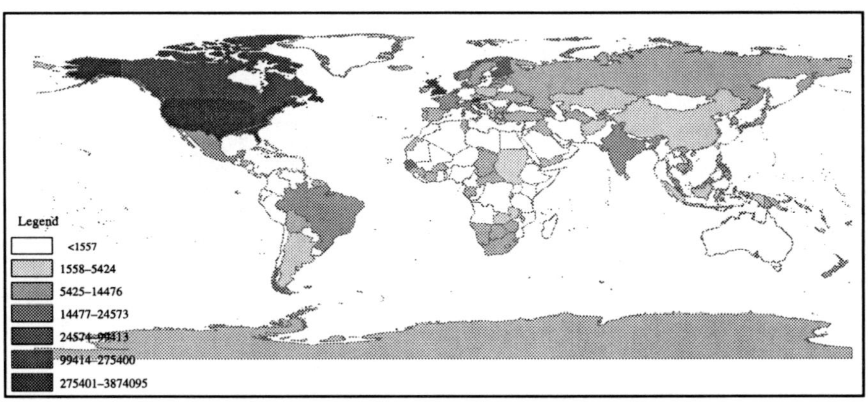

Fig. 18.1 Global Distribution of Bloggers (based on Livejournal.com data)

technique works by first transforming the RSS feed for the blog into RDF. Then a RDFMapper mapping engine is employed to produce the final interactive map. (see www.blogmapper.com for a demonstration and how to link blogs to location). Using a similar technique, Lin and Halavais (2004, 2005) describe the spatial distribution of American bloggers registered with two popular weblog hosting services called *Livejournal* and *Diaryland.* Their map Figure 18.2 reveals that the spatial distribution of bloggers correlates positively with the distribution of the U.S. population, with most bloggers heavily concentrated in large cities in coastal areas and their surrounding suburbs including New York, Boston, Los Angeles, Chicago, San Francisco, Washington, D.C., and Philadelphia. In contrast, the Midwest has a dearth of blogger concentrations. The majority of blog-clusters were found in 26 three-digit zip code units. Traditional high technology clusters such as the Bay area and Boston as well as new technology clusters such as Austin, Houston, Atlanta, Orange County, Calif., the region east of Phoenix (Mesa, Chandler and Tempe), Las Vegas, and Portland have larger blogger group. Suburbs and regions surrounding big cities such as Detroit, Washington, D.C., San Francisco, Boston, Phoenix, Los Angeles, Dallas, and Seattle have larger blogger groups compared with their city centers. The average household incomes in most of the 26 three-digit zip code units was around $59,000, which is higher than the national average and in six of those 26 areas, average household income was more than $100,000. Some concentrations of bloggers were found in cities where household incomes are relatively low. But thse cities were either college towns like Berkeley, Calif.; Bloomington, Ind.; Madison, Wis., and Tallahassee, Fla., or cities like Austin, Texas; Orlando, Fla.; Atlanta and Worcester, Mass., in which 18–34-year-olds make up more than 25 percent of the population.

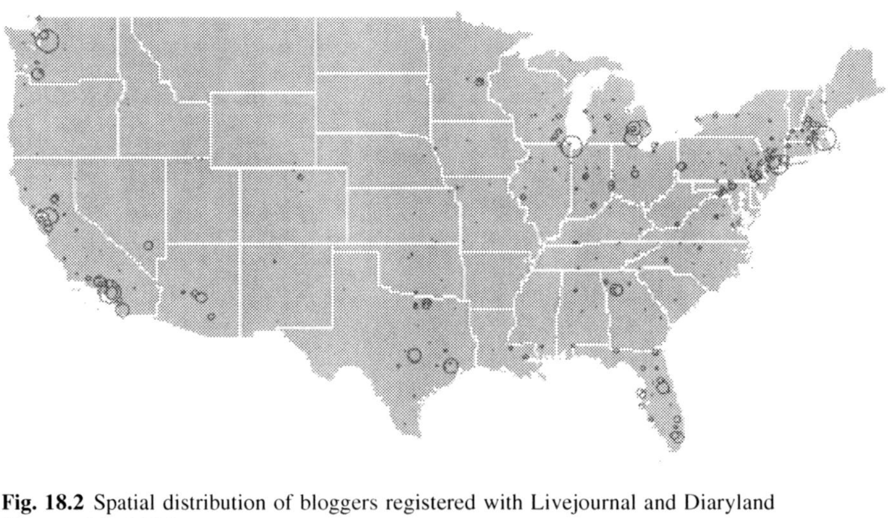

Fig. 18.2 Spatial distribution of bloggers registered with Livejournal and Diaryland (Source: Lin, 2004)

18.2.3
Can blogs be related to demographics?

Kumar et al. (2004) analyzed the location demographic characterstics (age) and interests of bloggers using the Livejournal database. The largest group of bloggers are between 16–24 years of age. Kumar et al., segment the blogger population into 300 densely connected interest clusters. The interests (and friendships) of the bloggers are correlated with age. Their analysis of interconnections between bloggers reveals that location (in other words geography) explains 55% of the links. Interests alone explain 45% of the interconnections while only 30% of friendships are between bloggers of different ages, from different locations, and with no expressed shared interests.

Blogging provides a new type of data relating to individual behavior. This type of data combined with available demographics (for example, claritas.com and its geodemographic segments) may provide new perspectives on individual and group behavior that has relevance to marketing (Businessweek, 2005) as well as politics, sociology and communication.

18.2.4
Can virtual interactions lead to real world interactions?

Castells (2000) notes that, "Cities do not disappear in the virtual networks. But they are transformed by the interface between electronic communication and physical interaction, by the combination of networks and places". Members of virtual community networks have ample opportunities to meet and know each other outside their online experience. A grass-roots effort based in New York City has established a web site with the specific purpose to organize local gatherings about anything, anywhere. Since 2002, Meetup has been the forum for over 100,000 clubs with 2 Million members. In Spring 2005, there were 2,400 Meetup meetings a week (http://press.meetup.com/). For example, Seattle Weblogger Meetup Group meets monthly and has 328 members in and around the city including bloggers covering urban culture (Seattlest.com), politics (SoundPolitics.com and Horsesass.org) or techno-babble (chris.pirillo.com). Blogs form a new type of media that can join the interaction in the real world with face-to-face meetings in the physical realm of real space thus impacting the social geography of the area.

18.3
Datasets and methodology used in the present research

Representing and analyzing the blogosphere involves both qualitative and quantitative research techniques. The former draws heavily from sociology and other social sciences while the latter draws from social network analysis (SNA) and graph theory. SNA has developed a set of methods for mapping and analyzing relations among people, communities and organizations. The seminal work of Milgram (1967) was one of the first on SNA. It estimates that every person in the world is only six "edges" away from every other. This work set the stage for investigations into social networks and algorithmic aspects of social networks. SNA is grounded in graph and systems theory and has provided a convenient tool for studying networks in physical and social sciences, including on the web (Berkowitz, 1982; Adamic, 1999; Albert and Barabási, 2002). Most recently, SNA methods have begun to be applied to blogs. Guervos, et al., (2005) mapped the blogs of 200 members, belonging to Blogalia weblog hosting site, using the metaphor of a neural network to identify community features. Kumar, et al. (2003) observe and model temporally-concentrated bursts of connectivity within blog communities over time, concluding that the 'blogspace' has been expanding rapidly since the end of 2001, "not just in metrics of scale, but also in metrics of community structure and connectedness". Adar, et al. (2004) and Gruhl, et al. (2004) identify blogs that initiate "information epidemics" and visualize the paths specific infections take through blogspace. Marlow (2004) used SNA to identify "authoritative" blog authors. These authors represent opinion leadership and authority in the popular press and turn out to be members of the A-list, who are most linked to other bloggers. On the other hand, the graph theory approach has allowed investigators to analyze the topologies of the Internet (Newman, 2001a; Albert and Barabási, 2002) and produced data that has allowed researchers to better understand distributed denial of service attacks, challenges and changes in routing systems and has afforded insight into dynamic networks (Barabási, 2002; Newman, 2001a, 2001b).

Although most work in SNA assumes a static topology, dynamic networks incorporating time, provide greater insights into dynamics and evolution of the blogospheres. The present research adopts Hägerstand's time geography framework to model blogging as a process in space and time. Hägerstand's framework (Hägerstrand, 1970; Kellerman, 1989; Miller, 1991) views space and time as inseparable, a useful construct in modeling the blogosphere. The Space-Time-Cube visualization that allows the display of Space-Time-Paths (STP) can be used to model the trajectories (Kwan, 1998) or footprints of the bloggers. Bloggers are more likely to meet fellow bloggers in their blogroll than others outside their space-time paths. Activity bundles representing places or locations where people meet is similar to a phenomenon called "bursts" (Kumar et al., 2004; Kleinberg, 2002) in the blogosphere, a period of intense activity.

Two types of blogs are analyzed. In the first type of blogs, communities coalesce around special interests such as politics, law, health, technology, gadgets, religion, etc., (such as Instapundit, Boing Boing). For this research, blogs related to AIDS, patients and their networks, are identified and a subsample of 146 blogs (active in August 2005) was selected for detailed analysis. We describe the topology of this network including its structure and leaders. The second type of blog is influenced by geography and is set up for local action as in political or community campaigns. Preliminary analysis of blogging in the wake of Katrina's devastation in the Gulf Coast is analyzed to describe how local and national blogs by media, citizen journalists, and individuals are being interlinked in new and interesting ways. UCINET 5 (Borgatti et al., 2002) and Pajek (Batagelj and Mrvar, 1999) software packages are used for analysis and visualization.

18.4 Representing and analyzing the blogosphere

Social network graphs attempt to represent the strength of social ties between parties. The structure of a typical blog is centered on local community interaction between a small numbers of bloggers. Members of the blog usually list one another's blogs in a blogroll and might read, link to, and respond to content of on other community members. Blogosphere networks can be visualized in terms of nodes (bloggers) and the link information derived from the "blogroll." Our AIDS related dataset is visualized using UCINET program. A single directed binary network shows the flow of information between blog authors (their blog identities) and their readers. Figure 18.4 shows the simplified network with nodes and links.

18.4.1 Opinion leaders and influencers

Network theorists have introduced a large number of *centrality* indices to describe the varying importance of the vertices in a network according to one criterion or another (Wasserman and Faust, 1994) such as *degree, closeness, and betweeness* (Freeman, 1977, 1979). Degree measures the number of ties to others, (i.e. row or column sums of adjacency matrix). Closeness measures the number of ties to others (i.e. row or column sums of adjacency matrix). Betweenness measures the number of geodesic paths that pass through a node (i.e. the number of "times" that any node needs a given node to reach any node by the shortest path).These indices are of great value in analyzing the blogosphere, especially the role of various authors and their influence. Not all blogs (nodes) have equal links or ties in the sample AIDS dataset. Bloggers who receive and send links are prominent or *central* figures enjoying prestige. Bloggers who have unusually high *out-degree* links are authors who are able to exchange information with many others and are often said to be *influential* in SNA. In Figure 18.3, there are at least seven individuals who are influential. The average centrality measure is around 3 degrees for this dataset. Figure 18.4 shows a simplified representation of the same dataset. Figure 18.5 shows an ego network of a blogger called *metro* receiving more than 10 links, playing a central figure in this dataset. A majority of blogs in this dataset have at least one in-link while 10 bloggers have more than 5 links each. Ego networks help in the visualization of the role that a blogger plays in the social structure of the blogosphere.

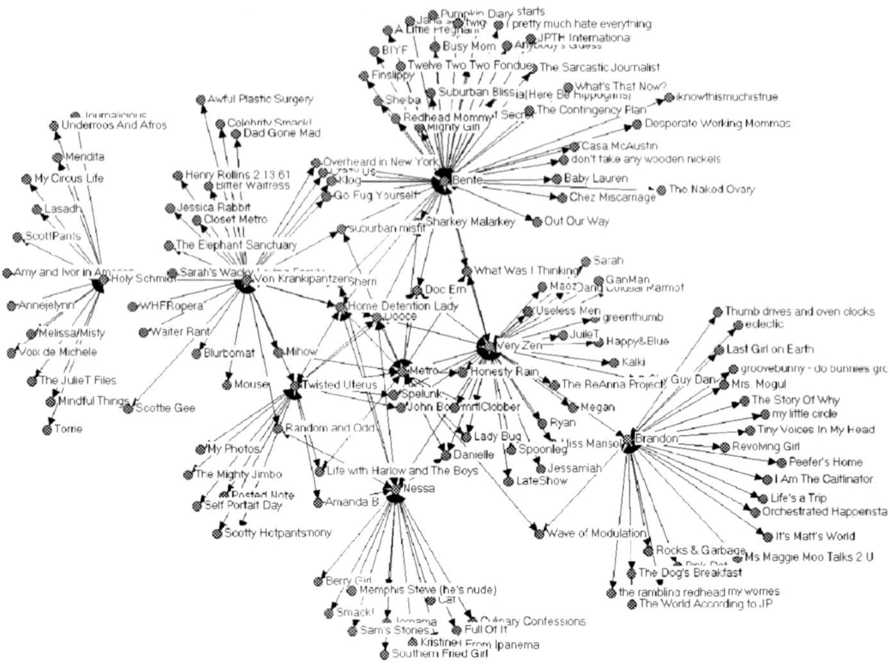

Fig. 18.3 Network Representation of the AIDS Dataset

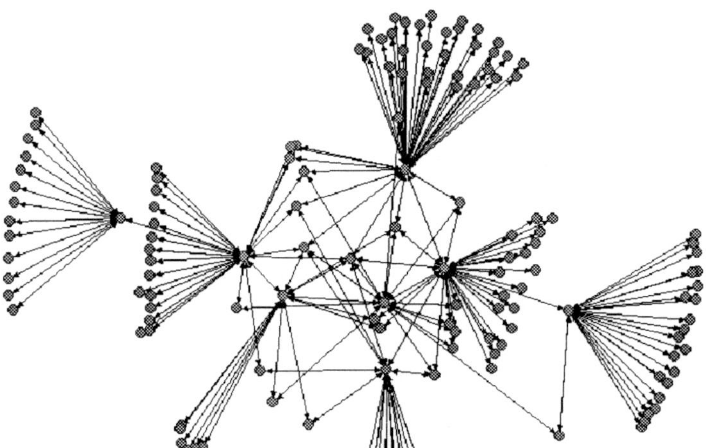

Fig. 18.4 Simplified Network Representation of the AIDS Dataset

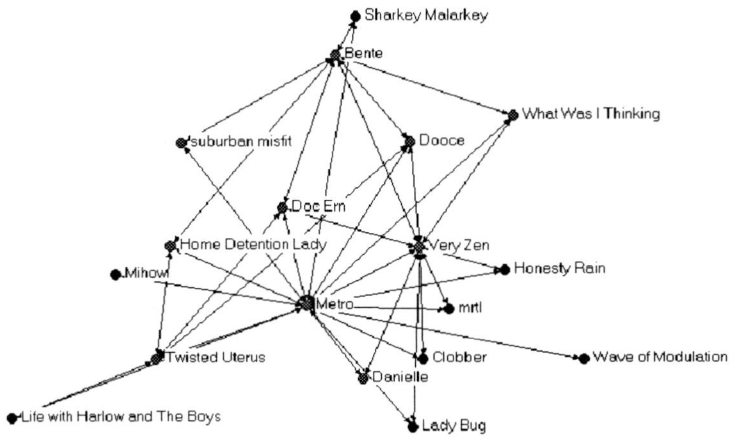

Fig. 18.5 Ego Network Representation in the AIDS Dataset

18.4.2 Community extraction

Recent research in networks has emphasized community structure, the grouping of vertices into groups, where each group is characterized by higher density of edges within groups than between them (Newman, 2004). Two methods are used to extract network structure.

Figure 18.6 shows the community structure in the AIDS dataset, extracted using hierarchical clustering technique, that develops a measure of similarity between pairs of vertices, based on the given network structure. There are 6 clusters or groups. Each group has its own influential central blogger. But the hierarchical method may not yield the best community structure if a vertex is connected to the rest of the network by a single edge. Then the issue is the extent to which it belongs to any community on either side of that edge (Girvan and Newman, 2002). Figure 18.7 shows the community structure derived using Girvan and Newman algorithm (Girvan and Newman, 2002). Their algorithm is based on edge betweenness that is defined to be the number of geodesic (i.e. shortest) paths between vertex pairs that run along the edge in question, summed over all vertex pairs. Their algorithm has three distinct features – it is a divisive method, in which edges are progressively removed from a network (in contrast to agglomerative hierarchical clustering method), edges to be eliminated are chosen by computing betweenness scores and betweenness scores are recomputed following the elimination of each edge. Figure 18.6 shows the AIDS dataset partitioned using Girvan and Neumann algorithm resulting in the characterization of 2 clusters (red and blue). The blue cluster has an important gatekeeper – a blogger called *Brandon* who in turn linked to 21 other bloggers (shown in blue) in this network.

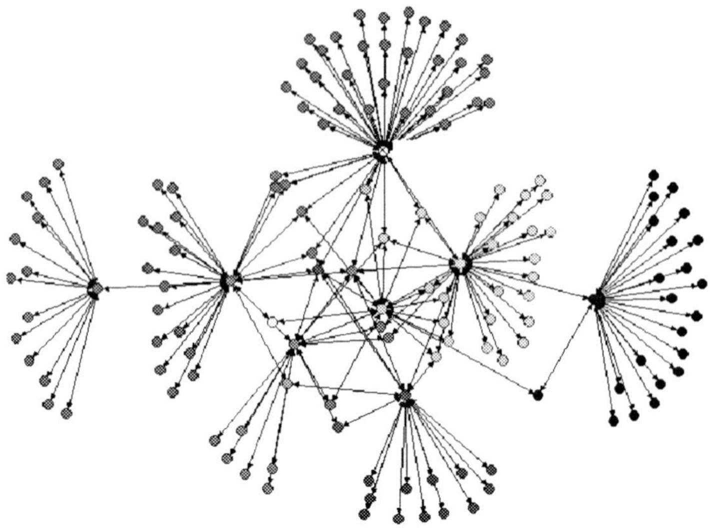

Fig. 18.6 Community Structure of the AIDS database using Factions Method

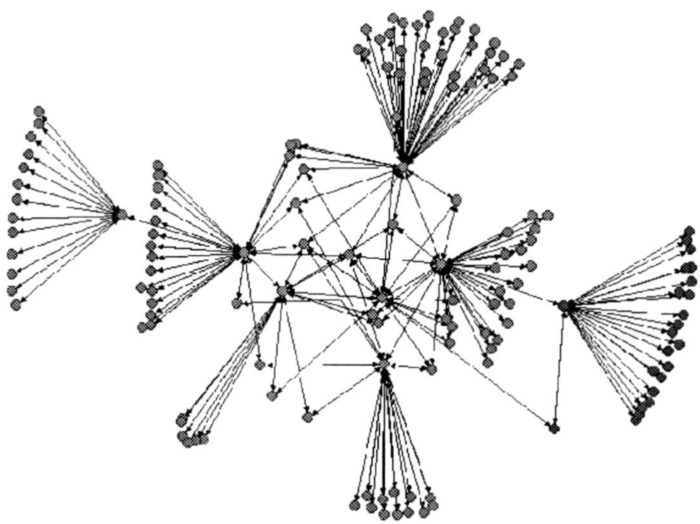

Fig. 18.7 Community Structure of the AIDS database using Girvan and Newman Algorithm

18.4.3 Further characterization of networks

The addresses were matched to geographical locations as described in section 2. We find that about 40% of bloggers in this dataset are from the West coast, and the remaining distributed in other states. The dominance of the West Coast may perhaps reflect the general trends in blogging across the country shown in Figure 18.2.

A preliminary content analysis using text retrieval algorithms (to be explored in a subsequent paper) show that the content of a vast majority of blogs consists of personal stories, experiences with the disease, health, family, friends, hobbies, as well as politics. Future research will focus on deriving community structure based on a detailed content analysis.

18.4.4 Blogging in response to natural disaster

One important aspect of blogging is the periodic brief burst in activity as bloggers respond to interesting topics. Hence burst is characterized by periods of high and low as topics rise to prominence, and then recedes. This type of "blog bursts" occurred after the Katrina disaster in New Orleans in the last weekend of August, 2005. For example, after the recent bombings in London people posted photos and eyewitness accounts online.

We conducted a content analysis examining both the content as well as locational information of bloggers. The blog statistics is shown in Table 18.2 (derived from Technocratic). By September 2, the number of blogs mentioning Katrina reached 19694 (compared with 1149 on august 25, 2005).

Table 18.2 shows a content analysis of 500 active blog sites compiled from August 30, 2005. We classified the blogs related to Katrina using text retrieval algorithms. The largest percentage of blogs related to Katrina were from media channels (such as CNN, MSNBC, and NPR) and from citizen journalists (see operationflashlite.com).

Each class had around 30 percent share of the total blogs in our sample. Influential bloggers (see Table 18.1) discussed Katrina from political, logistics and cultural perspective (Instapundit.com) and were also spearheading contributions to charities such as the Red Cross through their blogsites. Statistics on the amount of

Table 18.2 Timeline of Blogging Following Katrina Disaster

Date	Frequency of Blogs
August 20, 2005	143
August 23, 2005	207
August 25, 2005	1149
August 27, 2005	1365
August 29, 2005	8670
August 30, 2005	9025
September 2, 2005	19694

money raised through blogging is not yet available. Urban or metro blogs maintained by citizens of New Orleans showed temporal snapshots of the hurricane and the ensuing damage (for example, neworleans.metblogscom). There were some pure hurricane tracking sites such as wunderground.com that gave a continuous weather update (including storm surges, wind speeds, etc.). Personal stories were somewhat minimal given that most residents of New Orleans lost their electricity and had no Internet access. Figure 18.8 shows blog maintained by a news organization that tend to maintain a specific news format – for example, MSNBC (http://www.msnbc.msn.com/id/3217961/).

Contrast that with a citizen blog shown in Figure 18.9. Both styles of blogs served different populations and different needs at the time of the crises.

Although most work in networks assumes a static topology, we can incorporate time to gain insight into the functionality of the group itself. In the case of Katrina, blogs in the site called http://neworleans.metblogs.com/archives/2005/09/another_useful.phtml were retrieved. There are three major bloggers who posted

Sept. 1, 2005 | Updated 7:30 p.m. ET

Before and after Katrina: Now that the storm has passed, Earth-imaging satellites are getting a better fix on the damage caused by Hurricane Katrina. The QuickBird satellite, operated by Colorado-based DigitalGlobe, got a clear shot of New Orleans on Wednesday and posted before-and-after views on its Web site.

QuickBird's "after" view, captured from a 280-mile-high (450-kilometer-high), sun-synchronous polar orbit, shows dark floodwaters over highways and even the downtown golf course, as well as the water surrounding the Louisiana Superdome. We've created an interactive viewer that labels the landmarks and lets you switch quickly between the before and after views.

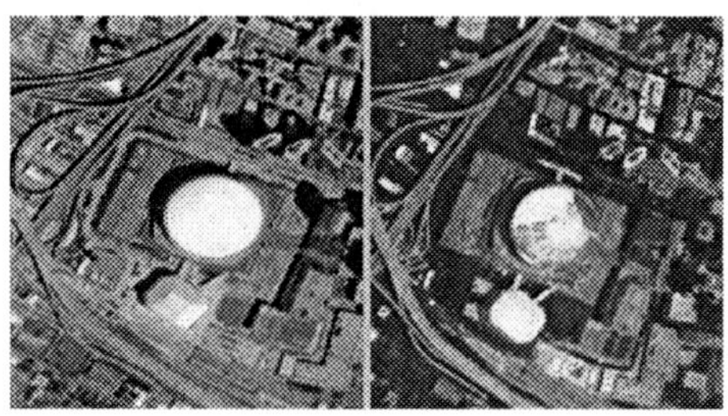

Digital Globe

Before-and-after pictures of the Louisiana Superdome, taken by DigitalGlobe's QuickBird satellite, highlight the flooding around the site as well as damage to the dome itself. The "before" picture was taken on March 9, 2004. The "after" picture was taken Wednesday.

In addition to the Big Easy pictures, DigitalGlobe is offering before-and-after views of Biloxi, Miss., which was also hard-hit by the storm. (Source: http://www.msnbc.msn.com/id/3217961/)

Fig. 18.8 A Blog Maintained by MSNBC Offering Frequent Updates

Katrina Web Log (http://www.nola.com/weblogs/nola/)

N.O. Inner-city teacher mourns

5:40 p.m.

Name: Diana Boylston

Home: (504) 914-5302 or (713) 586-2444 Room 826

Email: boylstonclark@cox.net

Subject: My Hurricane Story -- New Orleans Public School teacher mourns possible loss of students

Story: New Orleans Inner City School Teacher mourns possible loss of students
(Putting a face on a hurricane victim)

I had to leave my students behind.

Much of inner city New Orleans is filled with indigent or low-income families with no transportation. These people didn't stay in the city, for the most part, because they were "attached" to their homes. Most have little to attach to and no money or means to leave. Instead, many either rent one side of a shotgun double house or "stay" in one of the city's five huge housing projects. And that's where I had to leave my students: on the second floor, in their neighbor's apartment in the Lafitte Housing Project.

Dwight and Dwan, twin brothers who just turned 17 years old, first became my students at one of the lowest performing middle schools two years ago. Their individual stories are sad before Hurricane Katrina and maybe too intensely painful for the average parent or reader. But their reality was to call Children Services themselves last Friday when they came home to find their circumstances unlivable, once again. That day, they asked if they could live with me, but it wasn't possible Friday. We agreed to meet on Sunday and plan their future. Hurricane Katrina made all that impossible when I evacuated Saturday night.

We spoke several times trying to coordinate how to drop-off food for the storm but officials issued a curfew by 7p.m.on Saturday night making that impossible. I asked their neighbor to take them to the Superdome, but she said it was a bad experience two years earlier when they evacuated for a tropical storm and that they trusted God.

We spoke at 4:00 a.m. and the storm hit Monday morning at 5:00 a.m. We spoke a few hours later and I haven't been able to reach them since. From a hotel room in Houston, I sit tortured in from of the TV hoping to see a shot of their building or a face. The news just reported that the Orleans Parish School System would be closed for the next two to three months. What I want to know is... will my students be alive.

Diana Boylston
boylstonclark@cox.net
(504) 914-5302 (cell)
(713) 586-2444 (Room 826 till Sunday a.m.)

Fig. 18.9 Katrina Devastation Reported by Citizen Journalists

information on conditions of the approach and onslaught of Katrina between August 26–28. Citizens mostly in the region around New Orleans as well some outside responded via comments and feedback lines shown in Figure 18.10. Even this small network seeded by four authors spawned burst of blog activity. The structure of this network in Time period 1 has less than 12 links while in Time period 3 has over 100 links. We have traced the evolution of this network in static time shots (Timer period 1,2 and 3). A complete analysis shows how this network (of metro blogs) is enmeshed with media, and other blog networks (storm related)

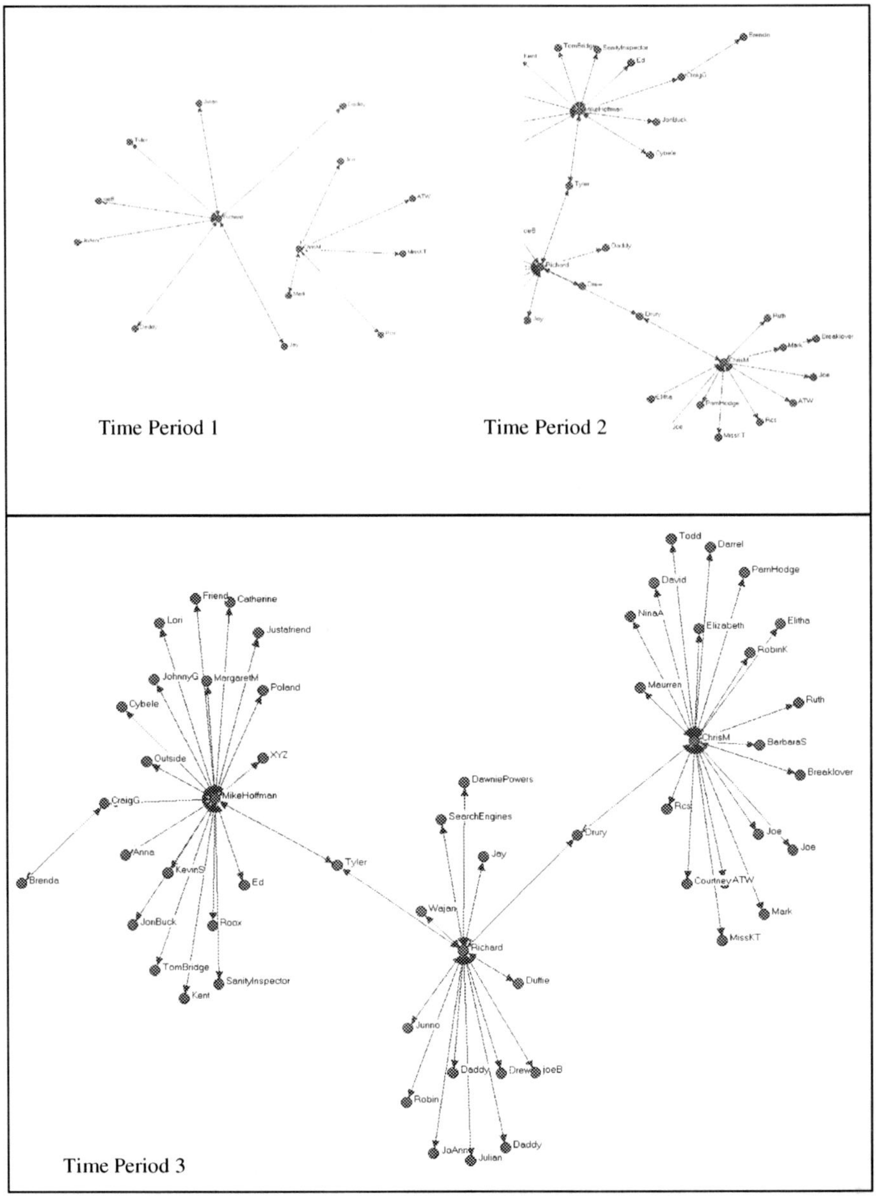

Fig. 18.10 Spatio-Temporal Evolution of New Orleans Blogosphere

via links. Hence the structure of this evolving blogosphere consists of several layers of networks that seamlessly integrated to provide viable communication channels following the natural disaster.

Table 18.3 A Classification of Blogs Relating to Katrina (August 2005)

Category of Blog	Example Blog	Percentage
Media	www.cnn.com/2005	30
Citizen journalists	www.operationflashlight.com/	30
Personal Stories	polimom.blogspot.com/	15
Legal help	www.ernietheattorney.net/	3
Neighborhood updates	www.nola.com/forums/vieuxcarre/	10
Fund raising	Instapundit.com	8
Urban blogs	neworleans.metblogs.com/	2
Hurricane tracking	www.wunderground.com	2

18.5 Discussion and conclusions

This paper has provided some background on the emerging world of blogosphere that is rapidly changing the social geography of the Internet.. We have used some fundamental concepts in spatial sciences, social network analysis, Hägerstand's space-time framework and network theory to analyze two datasets. The first dataset relating to AIDS examines the network structure including community and social dynamics. The second dataset relating to Katrina serves to illustrate the space-time evolution of blogs. Blogs have deep implications in the current paradigm of social sciences. First, they provide new and dynamic sources of data that will aid the investigators in social sciences to model dynamic individual and group human behavior. Second, blogs along with other mobile technologies are changing the way individuals are communicating in virtual world that can promote and shape real world communication. Third, blogs are social tools that are making long-range interactions between individuals in far and near locations increasingly easy. Future research in this area will examine longitudinal datasets to analyze the complex integration of various networks in the blogosphere both from statistical and social science perspectives.

References

Adamic, L. A. (1999). The small world Web. In *ECDL'99, Proceedings of the Third European Conference on Research and Advanced Technology for Digital Libraries* (pp. 443–452). Berlin: Springer.

Adar, E., Zhang, L., Adamic, L. A., & Lukose, R. M. (2004). Implicit structure and the dynamics of Blogspace. Workshop on the Weblogging Ecosystem, *13th International World Wide Web Conference*. http://www.hpl.hp.com/research/idl/papers/blogs/blogspace-draft.pdf (2005). *Proceedings of the Thirty-Eighth Hawai'i International Conference on System Sciences (HICSS-38)*. Los Alamitos: IEEE Press. 11.

Albert, R. & Barabasi, A-L. (2002). Statistical Mechanics of Complex Networks. *Reviews of Modern Physics* 74, 47 (2002).

Barabasi, A-L. (2002). Linked: How Everything Is Connected to Everything Else and What It Means.

Bear, N. Z. (Ed). (2004). The truth laid Bear ecosystem. *The truth laid bear*. Retrieved January 18, 2004, from http://www.truthlaidbear.com/ecosystem.php.

Berkowitz, S. D. (1982). *An introduction to structural analysis: The network approach to social research*. Toronto: Butterworth.

Batagelj, V. and Mrvar, A. (1999). Pajek: Program for large networks. http://vlado.fmf.uni-lj.si/pub/networks/doc/pajek.pdf.

Blogger Website. http://www.blogger.com/about.pyra.

Blood, R. (2002a). Weblogs: A history and perspective. In J. Rodzvilla (Ed.), *We've got blog: How weblogs are changing our culture* (pp. 7–16). Cambridge, MA: Perseus Publishing.

Blood, R. (2002b). *The Weblog handbook: Practical advice on creating and maintaining your blog*. Cambridge, MA: Perseus Publishing.

Borgatti, S.P., Everett, M.G. & Freeman, L.C. (2002). Ucinet for Windows: Software for Social Network Analysis. Harvard, MA: Analytic Technologies.

Businessweek (2005). *Blogs will change your business*. Businessweek, May 5, 2005.

Castells, M. (2000). *The Information Age: Economy, Society, and Culture* (three volumes). Oxford: Blackwell, 2nd edition, 2000.

Cavanaugh, T. (2002). Let slip the blogs of war. In J. Rodzvilla (Ed.), *We've got blog: How weblogs are changing our culture* (pp. 188–197). Cambridge, MA: Perseus.

Cheswick, B. (2004). The Internet mapping project. Retrieved July 29, 2004, from http://research.lumeta.com/ches/map/.

Dodge, M. (2004). Topology maps of elements of cyberspace. *An atlas of cyberspace*. Retrieved June 27, 2004, from http://www.cybergeography.org/atlas/topology.html.

Eagle, N. (2005), *Machine Perception and Learning of Complex Social Systems*, Ph.D. Thesis, Program in Media Arts and Sciences, Massachusetts Institute of Technology, June.

Eagle, N. and Pentland, A.(2005). Reality Mining: Sensing Complex Social Systems, *Personal and Ubiquitous Computing*, to appear: September 2005.

Efron, M. (2004). The Liberal Media and Right-Wing Conspiracies: Using Cocitation information to Estimate the Political Orientation of Web documents. *Proceedings of the* ***2004*** *ACM Conference on Information and Knowledge Management (CIKM).*, 390–398.

Freeman, L. C. (1977). Centrality in Social Networks: Conceptual Clarification, *Social Networks* 1: 215–239.

Freeman, L. C. (1979). A Set of Measures of Centrality Based Upon Betweenness, *Sociometry* 40: 35–41..

Girvan, M. & Newman, M.E.J. (2002). Community Structure in Social and Biological Networks *Proc. Natl. Acad. Sci. USA* 99, 8271–8276.

Gruhl, D., Liben-Nowell, D., Guha, R. & Tomkins, A. (2004). Information diffusion through blogspace, ACM SIGKDD Explorations Newsletter, v.6 n.2, p. 43–52, December 2004.

Hägerstrand, T. (1970). What about people in regional science? *Papers of the Regional Science Association*, 24, 1–12.

Herring, S. C., Scheidt, L. A., Bonus, S. & Wright, E. (2004). Bridging the gap: A genre analysis of weblogs. *Proceedings of the Thirty-seventh Hawaii International Conference on System Sciences (HICSS-37)*. Los Alamitos: IEEE Press.

Kellerman, A. (1989). *Time, space and society: Geographical-societal perspectives*. Boston: Kluwer Academic Publishers.

Kleinberg, J. (1999). Authoritative sources in a hyperlinked environment. *Journal of ACM (JASM), 46*(5), 604–632.

Kleinberg, J. (2002). Bursty and hierarchical structure in streams. In *Proc. 8th ACM SIGKDD Intl. Conf. on Knowledge Discovery and Data Mining*, 2002.

Krishnamurthy, S. (2002). The multidimensionality of blog conversations: The virtual enactment of September 11. Paper presented at Internet Research 3.0, Maastricht, The Netherlands. October.

Kumar, R., Novak, J., Raghavan, P., and Tomkins, A. (2003). On the bursty evolution of blogspace, WWW 2003, ACM.

Kumar, R., Novak, J., Raghavan, P., and Tomkins, A. (2004). Structure and evolution of blogspace, *Communications of the ACM*, 47 (12).

Kwan, Mei-Po. (1998). Space-time and integral measures of individual accessibility: A comparative analysis using a point-based network." *Geographical Analysis*, 30:3, 191–216.

Lin, J., & Halavais, A. (2004). Mapping the blogosphere in America. Thirteenth International World Wide Web Conference, New York City, May.

Lin, J. & Halavais, A(2005) Geographical distribution of Weblgos in the US. Paper submitted to *ICA general conference*, New York City, May.

Lumeta. (2004). Internet mapping project. *Lumeta.com.* Retrieved January 18, 2004, from http://www.lumeta.com/mapping.html.

Marlow, C. (2004). Audience, structure and authority in the weblog community. Paper presented at the International Communication Association Conference, May 27-June 1, New Orleans, LA.

Merelo-Guervos, J-J., Prieto, B., Rateb, F., & Tricas, F. (in press). Mapping weblog communities. Submitted to *Computer Networks.*

Miller, Harvey J. (1991). Modelling accessibility using space-time prism concepts within geographical information systems. *International Journal of Geographical Information Systems*, 5:3, 287–302.

Milgram, S. (1967). The small world problem. *Psychology Today, 2*, 60–67.

Newman, M.E.J. (2001a). The Structure of Scientific Collaboration Networks, *Proc. Natl. Acad. Sci. USA* 98, 404–409.

Newman, M.E.J. (2001b). The Structure and Function of Complex Networks, *SIAM Review* **45**, 167–256 (2003).

Pew Charitable Trust. http://www. Pewinternet.org.

Technocrati.com (2005). Top 100 blog sites. Retrieved August 27, 2005 from http://www.technorati.com/pop/blogs/.

Wasserman, S., & Faust, K. (1994). *Social network analysis: Methods and applications.* Cambridge; New York: Cambridge University Press.

Wellman, B. (2001). Computer networks as social networks. *Science Magazine, 293*, 2031–2034.

Wikipedia. (2005, August 30). Bl. *Wikipedia.* Retrieved June 27, 2005, from http://en.wikipedia.org/wiki/Blog#Blogging_begins.

19 Cell phones and places: The use of mobile technologies in Brazil

Adriana de Souza e Silva
Science, Technology and Society Program, North Carolina State University, Raleigh, NC

19.1 Introduction

Mobile phones are increasingly pervasive technologies in contemporary society. Worldwide cell phones have largely surpassed the number of PCs,[1] and appear to be surpassing the popularity of TV sets (Rice & Katz, 2003, p. 598). However, it is not possible to define a monolithic global cell phone culture. The mobile interface is used significantly differently in distinct parts of the world, depending on cultural, social, and economic local specificities. Even within specific countries and regions, there are substantial differences in the use of technology as a result of many factors, including age, culture, socioeconomic distribution, and instruction level. The use we make of technology does not depend solely on the technology per se, but is intrinsically connected to how the technology is culturally embedded in social practices.

This paper is an early exploration of the reasons for the exponential growth and social uses of mobile technologies in Brazil, a developing country, with wide-ranging social and economic diversity. This study aims at contextualizing three research questions to the Brazilian situation: (1) How do low income communities appropriate technology in unusual ways based on price policies and technology availability? (2) To what degree does the increasingly pervasiveness of cell phones transform them into social collective technologies? (3) How the initial idea of the digital divide should be redefined when cell phones replace not only landlines, but also personal computers?

With the goal of answering the above-mentioned questions, this article is structured in two main parts. The first two questions are addressed with a comparison between low income[2] and high income population[3] use of cell phones in Brazil. It looks at how cell phone usage is shaped by the availability of technology, focusing on two factors that promoted cell phone growth in the country: poor landline infrastructure and the emergence of pre-paid phones. It also gives an overview of the mobile services available in the country for the

H.J. Miller (ed.), Societies and Cities in the Age of Instant Access, 295–310.

high income populations, which are similar to those services offered in developed countries, such as the United States, Finland and Sweden. The second part re-contextualizes Manuel Castells (1999) concept of the dual city. It explores how cell phones as pervasive information and communication technologies (ICTs) in today's metropolises might actually help to bridge the digital divide, by allowing those who could not afford a personal computer access to the Internet. Throughout the paper, the study also briefly maps social use of cell phones in Brazil to those of other developing regions, such as South America and Africa.

The relevance of studying the relationships between Brazilian society and mobile technologies are drawn from several facts. First, Brazil experienced one of the fastest cell phone growth rates in the world in 2005 in terms of absolute numbers (31.4%, representing a total of 20.6 million new mobile phones) (Teleco, 2006d).[4] Second, the country is the biggest cell phone market in Latin America, with 86.2 million cell phones as of the end of 2005 (46.58% penetration rate) (Teleco, 2006b). However, these numbers lack meaning without understanding how the social-economic diversity in the country entails different cell phone usage.

Broadly, this paper contributes to the ongoing studies on social uses of mobile technologies by analyzing (1) how cell phones can be used as collective media, in opposition to most scholarly works to date that have studied cell phones as private communication technologies, suggesting that mobile phones withdraw users from the physical space in which they are (e.g. Plant, 2001; Gergen, 2002; Puro, 2002), and (2) how technology usage varies depending on socioeconomic status, and geographic position of its users.

19.2 Signs of the economic divide: Mobile phone usage and availability among low-income population

Brazil is well known for being one of the countries in the world with the greatest socioeconomic differences.[5] A report from the Brazilian Institute of Applied Economic Research (IPEA, 2005, p. 52) shows that 10% of the population owns 46% of the country's overall income, while 50% hold only 13.3% of this amount. Moreover, only 1% of population (around 1.7 million people) detains the same amount of 50% of the population: roughly 13% of the country's overall income. These economic differences influence the diverse use of communication technologies across the country, specifically of mobile telephones (Figure 19.1).

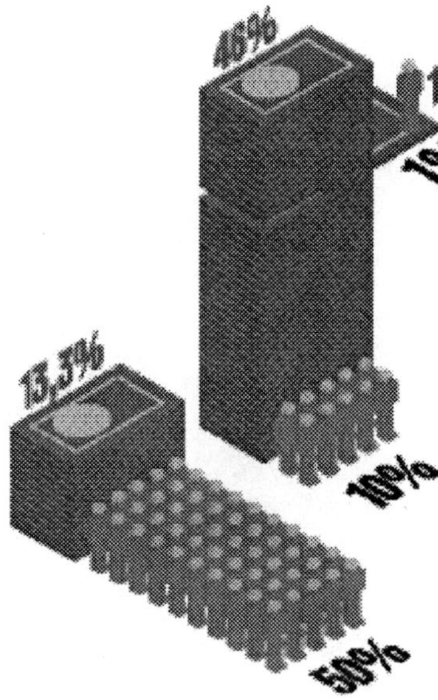

Fig. 19.1 Comparison between the overall income of the 50% poorest people and the 10% richest people in the country
(Source: IPEA, 2005, p. 52).

19.2.1
The role of technology and economic policies in cell phone growth

19.2.1.1
Poor landline infrastructure

One of the main reasons for cell phone growth in Brazil, as well as in other countries in Latin America, is the poor landline infrastructure. The same is true for many regions in Africa, where mobile phones could penetrate where fixed phone cables could not reach (The Economist, 2005). Recent data from the Brazilian National Telecommunications Agency (Anatel) shows that there was almost no growth in fixed telephones during 2004 (Teleco, 2006a). While there were almost 20 million new cell phones, the number of newly installed landlines was roughly 400,000. In September 2003 Anatel announced that the number of cell phones surpassed the number of fixed landlines (Anatel, 2003).[6]

Looking at a cell phone penetration rate map across regions in Brazil (Figure 19.2), it is clear where the lack of landlines influenced the growth of mobile phones. As it will be demonstrated later in this article, it is no surprise that the South and Southwest regions of Brazil,[7] as the country's richest economic regions, already have high cell phone penetration rate. However, it is interesting

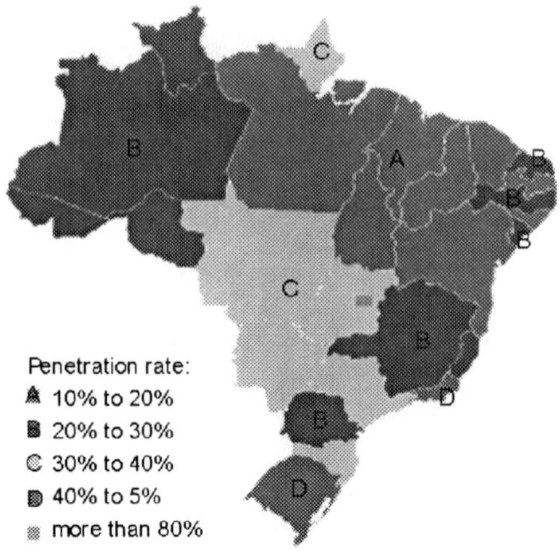

Fig. 19.2 Cell phone density in Brazil in 2004, focusing on the states of Mato Grosso (MT), Goiás (GO), and Mato Grosso do Sul (MS), which correspond to region C on the map (source: Teleco, 2004).

to observe the density of cell phones in the Central-West region of Brazil (Mato Grosso, Goiás, and Mato Grosso do Sul states) as among the highest in the nation (30–40%). Similarly, in 2004 cell phone ownership grew more in the center of Brazil, where the landline infrastructure is still precarious[8] (Teleco, 2005b).

At this point it is worth noting that mapping data by state creates a problem of regionalization – it homogenizes data within the boundaries of a single state, occluding the intra-state differences due to social class, urbanity, ethnicity, and education. As has been mentioned, there is no way to define a general Brazilian cell phone culture, since economic differences within the country are huge, and the use of technology varies with different socioeconomic statuses. The same applies to each federation state, none of which are homogeneous units. Nevertheless, it is possible to map in Brazil regions with no previous ICT infrastructure, such as the Center-West, and how this lack of infrastructure influences the development of other types of ICT, such as cell phones. As is discussed below, these regions can be represented by whole states, or by social enclaves within big cities, like the favelas (Brazilian slums).

According to journalist Andrés Velázquez (2002), "the deficient infrastructure and the difficulties imposed by geography make the cell phone the only option in many regions [of Latin America]" (¶ 4). In Brazil, until approximately 1999, fixed phones were scarce and used to take years to be installed. In 1998, the Brazilian Telecommunications Agency (Telebrás) was privatized, increasing the number of fixed phones and decreasing the installation waiting time and price. Nowadays, one has to wait approximately 10 days and pay an installation fee of roughly 100 reais[9] (45 USD) to have a fixed phone installed. However, maintaining a landline is still expensive for a large amount of the population. Subsequently, 40% of the population that earns up to 10 minimum salaries a month (1620 USD) do not have telephones at all (Teleco, 2005d). Moreover, among the population that earns up to one minimum salary[10] a month (162 USD), that is, 50% of the country's population, only 18.8% have a fixed phone at home.

Having a fixed phone means paying a monthly subscription of 42 reais a month (around 19 USD), which represents 13% of the Brazilian minimum salary, plus additional minutes spoken on the phone. Therefore, even in regions where landlines are more available, such as the South and Southeast cell phones represent less expensive options for communication. As of 2004, according to the Brazilian Institute of Geography and Statistics (IBGE/PNAD), the percentage of homes with only cell phones is already larger than the percentage of homes with only fixed lines among the population that earns up to 10 minimum salaries a month (Teleco, 2005d).

19.2.1.2
The rise of pre-paid phones

Mobile phones were commercially released in Brazil in 1990. At that time, Brazilian state-run operator, Telebrás, "demanded from subscribers a guarantee deposit of 20,000 USD just to enable a phone line" (Siqueira, 2001). Moreover, each new phone line could be acquired for an average price of 2,300 USD. Surprisingly, even under these conditions the operator managed to sell two thousand subscriptions only in Rio de Janeiro, the first city in Brazil to have a cellular network. With Telebrás privatization in 1998 the number of available lines (fixed and mobile) increased and prices decreased. However, the "killer-application" for mobile phones in Brazil was the release of pre-paid phones in the same year.

With pre-paid phones, it is possible to buy a phone card and use it until it is empty. However, even when the card is empty, the user can still receive calls – although he/she is not able to place a call. Consequently many users, with not enough money to pay for their calls, use mobile phones as a type of "free phone": a device to receive calls, and to be available for others, even if they are not able to call anybody. As a result, cell phones are generally used to receive calls, but if one needs to make a call, the cheaper option is to look for a public phone.

Because pre-paid phones eliminated the need to pay a monthly subscription and users can therefore control their costs, this type of phones became widely available in the country. As of December 2005, pre-paid phones represent almost 81% of the total number of mobile phones in Brazil (Teleco, 2006d). However, although the prepaid cellular phone is still a cheaper option for the low-income population when compared to landlines, this does not mean that mobile phones are inexpensive for most of the people. In a country where more than 50% of the population earns up to 162 USD a month (a minimum salary), buying a cell phone requires a significant amount of their budget. As an example, Nokia's and Motorola's websites,[11] the main cell phone producers in the country (Teleco, 2006b), do not offer a single device for less than 126 USD, which is barely the earnings of a whole month for the majority of the population. This price policy influences social uses of the mobile technology.

19.2.2 Social usages: On some consequences of technology development and price policies

It is possible to define 3 main social consequences of the high price of cell phone devices and calls: (1) cell phones are mostly used for voice communication and for "emergency calls", as opposed to the dominant use of text communication and Internet in places such as Norway, Finland, and Japan (Ling, 2004; Puro, 2002; Ito, Okabe, & Matsuda, 2005), (2) a general concern with cell phone theft, and especially with the cloning of cell phones, and (3) cell phone sharing in low income communities.

Having the state-of-the art cell phone device is also induced by cell phone companies, when advertising, for example, that a phone which originally costs 1229 reais (550 USD) can be purchased in 10 payments with no interests.[12] As a consequence, some users who earn up to 500 reais a month (225 USD) purchase phones that cost in average 400 USD (personal communication, October 18, 2005). However, most of these phones have pre-paid services and users keep cell phones for emergencies, since having a long conversation on the mobile would quickly use their monthly pre-paid cards. Using other cell phone features such as Internet connection and Short Message Service (SMS) also represent extra-costs and are generally outside the scope of the average cell phone user.

With the high price of the device and the service, it is no surprise that concerns about cell phone theft have been frequent on the media (Diário Popular, 2005; Imirante, 2006), and services like cell phone insurance became popular. Global System for Mobile Communication (GSM)[13] technology, now dominant in the whole world, has in fact motivated cell phone theft, according to a law project from the Brazilian government from 2005.[14] GSM phones work with SIM cards which can be removed from the device and replaced with another one, therefore disconnecting the phone and the information which it carries. Also in 2005 some operators announced the intention to create a new integrated system against cell phone theft, which would block the robbed cell phone, not allowing the same device to be re-activated in another state with a different operator (Correio do Estado, 2005).

One of the main purposes of cell phone theft is to use it as a model for a cloned one. Cloned cell phones use the same number of an already existing subscriber. The new device (generally stolen) is re-programmed so that it works as the same line of the original subscriber and call costs also go into the main subscriber's account. According to Anatel (Ucel, 2005) between January and August 2005 there were 7,380 new complaints about cloned phones. However, these numbers do not show the reality of this type of fraud, since many users do not realize they are being victimized.

Cell phone theft and cloned phones are a direct consequence of the socioeconomic inequalities in Brazil. These inequalities become evident when comparing different regions of the country (e.g. Center-West and North vs. South and Southeast), as previously demonstrated, but they are also apparent within states, especially in big metropolises such as Rio de Janeiro and São Paulo. In urban spaces, this inequality is visible through the growth of favelasslums. Favelas, especially in Rio de Janeiro, have the peculiar characteristic of being constructed on hills embedded in the wealthiest areas of the city. For example, Rocinha, the biggest favela in Latin America, is rooted in the rich neighborhood of São Conrado.

Although situated within the same urban area as the wealthier neighborhoods, favelas generally lack basic infrastructure, such as electricity, water, gas, and phone lines. Therefore, the development of informal connections is common, through which citizens acquire the services for free using clandestine wiring. Something similar happens with the telephone. Private landlines have never been widespread in regions of low income population. However, even pay phones are rare. Cell phones, in this context, filled in for the lack of private landlines and public phones through an informal appropriation of the technology. Frequently these cloned phones – used for free – are not private phones (as generally cell phones are regarded), but become collective phones used by the community. These types of clandestine connections for telecommunications in the favelas are called "diretão", which means roughly a straight call (which the user does not pay for). To this extent, we can affirm that mobile phones in Brazil are also used as social and collective technologies.

19.3 High-income population: Access to the same services and same behavior patterns as in the "developed" world?

A very different situation can be found not very far from the favelas, but within a different population. For the majority of the high income population, the cell phone generally represents the third or fourth phone and the second or third computer. The IBGE/PNAD (2004) survey mentions that 90% of the population that earns more than 20 minimum salaries a month (3240 USD) has both fixed *and* cell phones (Teleco, 2005d). Cell phone usage within the high income populations living in Rio de Janeiro and São Paulo, for example, can be easily compared to usages in other developed countries, such as the United States, Japan, and Finland.

Four examples will make this point: (1) the use of cell phones as remote control; (2) the emergence of third generation (3G)[15] cellular phones, (3) the use of Wireless Application Protocol (WAP)[16] technology to access entertainment and news via the mobile Internet, and (4) location-based mobile games.

Recently Mitsubishi, Nokia and the Brazilian company Compera announced the commercial release of services already available in countries like Japan and Finland: the use of the cell phone as a remote control (Ditolvo, 2005). The service will allow users to send commands via SMS in order to turn on and off the security alarm, lights, and air conditioning in their homes from wherever place in the world. In order to subscribe to it, users need to pay a fee of 15,000 reais (6700 USD).

Moreover, one of the two cell phone models that can be used for the service (Nokia 6681) costs 2200 reais (1000 USD) and developers admit that the target audience are the 7 million richest people in the country, mostly residents in Rio de Janeiro and São Paulo.

Vivo, the larger operator in the country, is widely announcing the commercial release of third generation (3G) cell phones. The service was initially expected for April 2005, however the company is waiting for radio-spectrum to be released by the National Telecommunications Agency (Muniz, 2005). Vivo plans to offer the same 3G services available in any other country, such as high speed Internet connection, music download, live-TV and video streaming. However, despite the massive advertisement campaign, the user who wants to enjoy high speed Internet connection in their cell phone needs to by one out of two devices enable to work with the service. One is the Motorola E815 and the other is Samsung Evolution, both costing from 1500 to 2000 reais (670–900 USD).

Another sector that is being developed by journalistic companies in Brazil, such as Abril, Selig, Terra Mobile, Folha de São Paulo, and Estado de São Paulo, is the production of news and entertainment services, such as music and videos for mobile phones (Ferreira, 2005, p. 126). Among popular services are soccer news, the download of soccer short videos, chat services, and traffic information, most of them accessed via WAP, the mobile Internet (Ferreira, 2005, p. 131). However, although some predict that this type of services will become popular in the country from 2007 on, current users still represent only 2% of cell phone subscribers (Ferreira, 2005, p. 135).

Finally, some cell phones already include location awareness,[17] opening the possibilities for the development of location-based services and games. Location-based games are popular in countries such as Sweden and Japan. Alien Revolt[18] is the first type of these games commercially launched in Rio de Janeiro, Brazil, in May 2005 by the company M1nd Corporation and the operator Oi. The game uses Java-enabled cell phones with location-awareness to transform the city into a battlefield. The game's goal involves virtually "shooting" other players with the cell phone who are within a specific radius in the city space. Moreover, players are able to see and fight with virtual alien creatures nearby represented on the radar of their cell phone screen. However, the game has not become popular yet. According to one of the game developers (personal communication, June 2006), the game had a maximum of 300 subscribers – in a city with 13 million cell phone users (Teleco, 2005c). Still not all of them would connect to the game at the same time. In a big city such as Rio de Janeiro, the reduced number of players in a multiplayer game becomes a problem, since there are large uninhabited game areas. The solution found by the game developers was the creation of virtual alien creatures, with which users could interact even if there were no other players logged in.

When asked about possible reasons for the low number of subscribers, one of the game programmers (personal communication, December 2005) pointed two: price and technology. As of January 2006, Alien Revolt only works on five types of cell phones from Nokia. Both cell phones and the prices for using the GPRS[19]

Internet connection while playing the game are relatively expensive for the average user. Moreover, the GPRS bandwidth is too limited to play the game in the Java interface. Acknowledging the limitations of the game, Alien Revolt is now being developed in a simpler SMS version which will be available in more phone models and will be less expensive for the user.

All these examples show that although high-end services are available, or at least in developmental phase, they still target a very small portion of the population, providing evidence that even within the high-income population, cell phones are still mostly used for voice communication. Moreover, it is important to keep in mind that when we look at usage of these high-end services, we are talking about less than 1% of the population.

19.4 Are cell phones redefining the digital divide? Considerations about the use of mobile technologies among low income population

A general belief is that mobile phones, not personal computers, are the technologies that will help to bridge the digital divide in developing countries in the future (LaFraniere, 2005; Markoff, 2006). The concept of the digital divide has been originally conceived in the context of the United States and applied to Internet access via desktop computers. The "divide" referred to the "haves" and "have-nots": with and without access to information (Rice & Katz, 2003; Katz & Aspden, 1998; Cooper & Kimmelman, 2001). While countries like the United States and the United Kingdom have more than half of their population connected to the Internet via desktop PCs, in Brazil this rate is only about 10%, and the whole continent of Africa does not reach 2% of Internet users via PCs (ITU, 2005).

Manuel Castells (1999) created the concept of the dual city to think about this unequal distribution and access to ICT in big metropolises. The dual city is defined as "an urban system socially and spatially polarized between high value-making groups and functions on the one hand and devalued social groups and downgraded spaces on the other hand" (Castells, 1999, p. 27). For Castells (1999), the power of information technologies enhances and deepens features present in the social structure and in power relationships of the dual city. The case becomes apparent when taking into consideration urban structures like the previously mentioned favelas in Rio de Janeiro, where two populations occupying adjacent urban areas are deeply disconnected due to the restricted access to communication infrastructure from the low income population. Within this context, the favelas become almost distinct segregated neighborhoods, having an autonomous existence inside the major city.

Castells (1999, p. 30) also refers to the space of flows as the dominant spatial logic of the network society, characterized by the simultaneous concentration and decentralization of people and activities connected by ICT. The space of flows, as the material organization of time-sharing social practices in the dual city, "links up *valued* spaces at the same time that it separates and isolates *devalued* spaces in the inner city, (. . .) where low-income communities (. . .) remain trapped" (Castells, 1999, p. 31). Within this context, ICT have generally being analyzed as increasing the digital divide.

Nevertheless, cell phone growth rates worldwide, especially in developing countries indicate that it will be mobile technologies, not the fixed desktop PC, which will bring access to the Internet and to information to people who currently do not have it. Cell phones are much more inexpensive than laptops and handheld computers, and therefore affordable for a larger number of users. Furthermore, while PCs are accessed only from specific places (except for laptops), mobile phones can be used anywhere (where there is a signal).

Even when there is no Internet access involved, just by connecting previously disconnected regions in the developing world cell phones are already noted for accelerating the economy and fostering small businesses (LaFraniere, 2005). However, increasingly cell phones have more computational capabilities, including Internet access, a fact that makes this relationship even stronger, because mobile phones are easier to carry. We have seen that services such as 3G, location awareness and WAP are indeed available in Brazil, but they do not reach most of the population because they are still relatively expensive. It is a matter of time, however, to be able to embed these services as regular functions of the cell phone also in developing countries, transforming them not only in the main telephone access point, but also in the primary Internet connection.

Although affirming that ICTs currently increase the digital divide, Castells (1999, p. 36) proposed six developmental policy initiatives that might help the interaction between information technology and urban social reform. Today it is clear that much of this gap is already being fulfilled by mobile technologies, not by personal computers.

At this point, it is valuable to re-contextualize some of Castells' proposals originally conceived to think about the integration between computers and low income communities to a mobile technologies framework. I will focus at three of the six initiatives. The first one concerns the "necessary spurring of entrepreneurialism and small business among low-income communities' residents" (Castells, 1999, p. 36). This tendency is already underway in Brazil, Africa and many other countries in Latin America. For example, in Africa, in many places which do not have electricity and fixed telephone service, small businesses and private salespersons use cell phones as ways to improve their sales (LaFranieri, 2005). Also similar to Brazil, frequently cell phones are used as public pay phones in regions where no other service is available. Gilbert Nkuli, deputy managing director of Congo operations for Vodacom Group, one of Africa's biggest mobile operators, tells that in Congo "One man uses it [a cell phone] as a public pay phone. Those who want to climb to his platform and use his phone pay him for the privilege" (LaFranieri, 2005).

The second initiative "refers to the expansion of telework not from home but from community telecenters" (Castells, 1999, p. 36). We have seen that in many favelas (slums) in Rio de Janeiro, people use mobile phones as public phones that serve the community. Because many do not have telephones at home, the "public" cell phone becomes the primary connection technology to the community and might well be used by community members for teleworking.

Another initiative is related to emphasizing "the potential of information technology for improving the educational chances of poor populations" (Castells, 1999, p. 37). A project with this goal is underway at the Massachusetts Institute of Technology (MIT), led by Nicholas Negroponte. Their 100-dollar laptop project[20] aims at developing a fully functional portable computer costing less than one hundred USD. Such computer will be sold to governments in developing nations such as Brazil, India, China, and Argentina, with the goal of helping learning and integration of the new generation in the information society. The laptops will come with built-in wireless data connection and, according to Negroponte (2006), will benefit from peer-to-peer network and possible low-cost connections to the backbone of the Internet. Nicholas Negroponte (2006) emphasizes the importance of mobility when justifying the construction of the 100-dollar portable laptop, instead of a desktop computer, to help educating children in developing countries. Portable computers allow children to bring them home, also engaging the family in school work and access to information.

As a response to Negroponte's project, which employs the Linux operational system, Microsoft announced what they consider a less expensive alternative to a laptop: transform the cellular phone into a computer by connecting it to a TV and a keyboard (Markoff, 2006). This is possible because cell phones nowadays already have computational capabilities.[21] Proposing the cell phone as the technology to bridge the divide, Microsoft is relying on the high cell phone growth rates in the world, foreseeing that mobile phones will in fact be already pervasive technologies. In Brazil, while most cell phone devices still do not have computational capabilities, Philips is negotiating with mobile phone producers for the release of a device for less than 20 USD, directly targeting the low income population (O Globo, 2005).

Considering Mark Weiser's (1996) theory of ubiquitous computing, which divided the history of computers in three main waves – mainframes, personal computers, and ubiquitous computing – it is worth noting that many countries in sub-Saharan Africa, as well as in Latin America, simply skipped the personal computer wave, jumping straight into the mobile (ubiquitous) technologies era. In 2004 (ITU, 2005) the African continent, for example, had close to 100 million total telephone subscribers, 76 million of which were mobile subscribers.

19.5 Conclusions

The numbers related to cell phones distribution in Brazil cannot be taken for granted without a deeper analysis. For a superficial observer, they seem impressive: 80 million cell phone subscribers and the fifth market in the world. However, a recent report from the Brazilian Atlas of Telecommunication (Ordoñez, 2006) attested that 43% of Brazilian counties still do not have mobile telephone service. It is therefore irrelevant to look at raw numbers without understanding how cell phones users are geographically distributed in the country, as well how mobile phones are differently used by the population according to their socioeconomic status. Interestingly, it seems that it is exactly when people cannot afford available services and devices that unusual appropriation of technology occurs.

Two general lessons can be taken from this preliminary analysis of the use of mobile phones in Brazil. The first one takes into consideration how low income populations appropriate technologies in unusual ways, depending on pricing policies and technology availability. It is widely accepted that technologies both construct and are constructed by historical, social, and cultural contexts (Ito, Okabe, & Matsuda, 2005; Wellman, 1999; Bijker & Law, 1992; Callon, 1986; Hine, 2000; Suchman, 1987). However, studying how populations with limited access to technology find ways of re-purposing its usage, especially in using it as a means of promoting sociability and communication among the community, gives us additional insights on how interfaces are culturally defined. Some examples mentioned in this article are (1) the favela dwellers who share a phone among community members, transforming an originally "private" technology into a shared one, (2) the use of pre-paid phones as a "free" phone, and (3) the cloning of cell phones.

Following a tendency underway in Asia and Scandinavia where cell phones are regarded as collective technologies (Kasesniemi & Rautiainen, 2002; Ling, 2004; Rheingold, 2002), the mobile phone in Brazil is also becoming a collective communication medium, but in a very different way. What is possible to observe is the transformation of a device that has been frequently regarded as a "private" medium, belonging to a single person, into a public phone, as is the case of the sharing of cell phones in the favelas. Similarly, while pre-paid cell phones have been created as ways that operators could charge more for the spoken minute, users transform it in a type of free phone: using cell phones to receive calls, and public phones to make calls, therefore avoiding having to buy a new phone card.

The second lesson concerns how the initial idea of the digital divide should be redefined when cell phones not only replace landlines, but also personal computers. As demonstrated in this article, there are strong interrelations between mobile technologies, information access (the "digital divide"), and the potential for progressive social/economic change. Within this context, it is imperative to rethink inequalities of information/communication access because (1) low-income users are "skipping" the desktop computer "phase" and (2) access is increasingly through mobile, rather than fixed devices.

Although the signs of the digital divide worldwide are still very strong,[22] it is interesting to realize that while the developed world still has 8 times the Internet user penetration rate of the developing world, this ratio decreases to 4 times when it comes to the comparison of mobile phone numbers (ITU, 2005). With mobile technology devices replacing telephone landlines *and* desktop computers, how should the original idea of the digital divide be redefined?

Perhaps Manuel Castells' (1999) belief in the dual-city, in which ICTs increase the disconnection between high and low income populations needs a new perspective, since low-income populations now indeed have access to (mobile) technology: they just use them differently. Having in mind that mobile technologies will probably bring to the developing world a type of social communication never experienced with fixed telephones or the fixed Internet, it is important to understand how ICT can positively impact society and not only exclude people from it.

Notes

[1] According to the International Telecommunication Union (ITU, 2005), in 2004 there were 770,641 million personal computers vs. 1,751,940 billion cell phones in the world.

[2] According to the Brazilian Institute of Applied Economic Research (IPEA, 2005), low income population in Brazil earn up to one minimum salary a month (roughly 162 USD).

[3] Still according to IPEA (2005), this group represents 10% of the population that earns above 670 USD a month.

[4] Brazilian new cell phone additions have been only surpassed by China (76 million), Russia (61 million) and India (28 million) (Teleco, 2006d).

[5] A recent report from (IPEA, 2005, p. 60) attested that considering the Gini index of 130 countries in terms of income distribution Brazil is barely the last one, just in front of Serra Leoa, in Africa.

[6] Cell phones exceeding landline numbers is a general tendency globally, and in the context of Latin America, it took longer to Brazil to accomplish what happened at least 4 years earlier in other countries with poorer landline infrastructure. In Paraguay and Venezuela, for example, this inversion occurred as early as 1998. Paraguay is unique in terms of cell phone numbers, for the country has since 2002, at least five times more cell phones than landlines (Teleco, 2006c; 2005).

[7] Southeast region includes the states of Rio de Janeiro, São Paulo, Minas Gerais and Espírito Santo. The South is composed by Rio Grande do Sul, Santa Catarina, and Paraná states.

[8] Cell phone growth in 2004 was 99.1% in Tocantins, 58.9% in Mato Grosso, and 54.7% in Mato Grosso do Sul.

[9] The Brazilian currency is the Real (or Reais) which as of June 2006 is worth 0.44 USD.

[10] The minimum salary was created in Brazil in 1936 by President Getúlio Vargas. At its inception, the minimum salary was supposed to cover the basic monthly needs of a 4-person family (food and housing).

[11] Nokia phones can be found at: http://nokia.submarino.com.br/nokia/technology.asp?vCel=1&Tecn=1 and Motorola phones at http://shop-br.motorola.com/site/content/produtos/celular_default.asp

[12] See Motorola website at: http://shop-br.motorola.com/site/content/produtos/celular_default.asp

[13] GSM, Global System for Mobile Communication, "originally known as Group Special Mobile, is a second generation digital cellular system developed in Europe and used in the majority of the world. Initially developed to the range of 900 MHz, GSM had afterwards a version adopted to 1800 and 1900 MHz." Retrieved May 12, 2006, from http://www.teleco.com.br/glossario.asp?termo=GSM.

[14] http://www.camara.gov.br/sileg/integras/329717.htm

[15] 3G (Third Generation Cellular Telephony) also stands for UMTS (Universal Mobile Telecommunication System). "UTMS allows many more applications to be introduced to a worldwide base of users and provides a vital link between today's multiple GSM systems and IMT–2000. The new network also addresses the growing demand of mobile and Internet applications. UMTS increases transmission speed to 2 Mbps per mobile user and establishes a global roaming standard." Source: International Engineering Consortium. Retrieved November 28, 2005, from http://www.iec.org/online/tutorials/umts/topic01.html
[16] WAP stands for Wireless Application Protocol. Using WAP, content can be delivered over the Internet to most current wireless networks, including the networks with General Packet Radio Service (GPRS) and 3G.
[17] There are two different ways by which the cell phone can be aware of its position. One is accomplished by cellular positioning, which indicates the device location through the triangulation of radio waves detected by the cell phone in relation to the transmission towers. Another much more accurate way uses GPS systems embedded in the phone.
[18] http://www.alienrevolt.com
[19] GPRS, General Packet Radio Service, is a "system that can be implemented as a layer over GSM systems. GPRS allows data services without the need of establishing a connection. It is considered an intermediary step (2,5G) to the third generation of cellular systems (3G)." Retrieved December 17, 2003, from http://www.teleco.com.br/glossario.asp?termo=GPRS.
[20] http://laptop.media.mit.edu/
[21] A report from NTT DoCoMo (June2004) states that today's FOMA CPU processing speed is comparable to personal computers from eight years ago running Windows 95.
[22] See ITU statistics: http://www.itu.int/ITU-D/ict/statistics/ict/index.html

References

Anatel (Agência Nacional de Telecomunicações). (2003, September 12). *Celular ultrapassam fixos em torno de um milhão de acessos*. Retrieved January 28, 2006, from http://www.anatel.gov.br/To/frame.asp?link=/biblioteca/releases/2003/release_12_09_2003.pdf

Bijker, W., & Law, J. (1992). Shaping technology, building society: Studies in sociotechnical change. Cambridge, MA: The MIT Press.

Callon, M. (1986). Some elements of a sociology of translation. In J. Law, (Ed.), *Power, action, and belief: A new sociology of knowledge* (pp. 196–233). New York: Routledge.

Castells, M. (1999). The Informational City is a Dual City: Can It Be Reversed? In D. Schön, B. Sanyal, & W. Mitchell, (Eds.), *High Technology and Low-Income Communities: Prospects for the Positive Use of Advanced Information Technology* (pp. 25–40). Cambridge, MA: The MIT Press.

Cooper, M., & Kimmelman, G. (2001). The digital divide confronts the telecommunications act of 1996: economic reality versus public policy. In B. Compaine, (Ed.), *The digital divide: facing a crisis or creating a myth?* (pp. 199–221). Cambridge, MA: The MIT Press.

Correio do Estado. (2005, June 18). Operadoras terão sistema integrado contra roubo do celular. *Correio do Estado*. Brasília, Brasil. Retrieved January 27, 2006, from http://www.correiodoestado.com.br/exibir.asp?chave=105296,1,2,18-06-2005

Diário Popular. (2005, November, 19). Polícia: Prisão em flagrante por roubo de celular. *Diário Popular*. Retrieved January 27, 2006, from http://www.diariopopular.com.br/19_11_05/jal181104.html

Ditolvo, M. (2005, September 28). A casa na palma da mão. *Isto é Dinheiro*. São Paulo: Editora Três. Retrieved January 29, 2006, from http://www.terra.com.br/istoedinheiro/420/ecommerce/casa_palma_maos.htm

The Economist. (2005, July 07). Calling to an end to poverty. (2005, July 07). *The Economist*. Retrieved July 13, 2005, from http://www.economist.com/business/PrinterFriendly.cfm?Story_ID=4157618

Ferreira, P. H. O. (2005). *Notícias no celular: Uma introdução ao tema*. Unpublished Master Thesis, School of Communication and Arts, São Paulo University (USP), Brazil.

Gergen, K. (2002). The challenge of absent presence. In J. Katz, & M. Aakhus, (Eds.), *Perpetual contact: Mobile communication, private talk, public performance* (pp. 227–241). Cambridge, MA: Cambridge University Press.

Hine, C. (2000). *Virtual ethnography*. London: Sage.

IBGE (Brazilian Institute of Geography and Statistics). (2004). *PNAD – Pesquisa Nacional por Amostra de Domicílios*. Retrieved January 28, 2006, from http://www.ibge.gov.br/ home/estatistica/populacao/trabalhoerendimento/pnad2002/sintese/tab715.pdf

International Telecommunication Union (ITU). (2005). Free Statistics. Retrieved January 28, 2006, from http://www.itu.int/ITU-D/ict/statistics/

Imirante.com. (2006, January 21). Polícia prende dois por roubo de celular no reviver. *Imirante.com*. Retrieved January 27, 2006, from http://imirante.globo.com/plantaoi/plantaoi.asp?codigo1=87341

IPEA (Brazilian Institute of Applied Economic Research). (2005). *Radar Social 2005*. Brasília, Brazil. Retrieved January 28, 2006, from http://www.ipea.gov.br/Destaques/livroradar/03.renda.pdf

Ito, M., Okabe, D., & Matsuda, M. (2005). *Personal, portable, pedestrian: Mobile phones in Japanese life*. Cambridge, MA: The MIT Press.

Kasesniemi, E. L., & Rautiainen, P. (2002). Mobile culture of children and teenagers in Finland. In J. Katz & M. Aakhus, (Eds.), *Perpetual Contact: Mobile communication, private talk, public performance* (pp. 170–192). Cambridge: Cambridge University Press.

Katz, J., & Aspden, P. (1998). Internet dropouts in the US. *Telecommunications Policy*, 22 (4/5), 327–340.

Katz, J. E. & Aakhus, M. (2002). *Perpetual Contact: Mobile communication, private talk, public performance*. Cambridge: Cambridge University Press.

LaFraniere, S. (2005, August 25). Cellphones catapult rural Africa to 21st century. *The New York Times*. Retrived February 03, 2005, from http://www.nytimes.com/2005/08/25/international/africa/25africa.html?ex=1282622400&en=cad54d043ab15f30&ei=5088&partner=rssnyt&emc=rss

Ling, R. (2004). *The mobile connection: The cell phone's impact on society*. San Francisco: Morgan Kaufmann Publishers.

Markoff, J. (2006, January 30). Microsoft would put poor online by cellphone. *The New York Times*. Retrived January 30, 2006, from http://www.nytimes.com/2006/01/30/technology/30gates.html?_r=1&oref=slogin

Muniz, R. (2005, September 02). Atraso da Anatel para licitar freqüências 3G prejudica mercado. *Portal Exame*. São Paulo: Editora Abril.

NTT DoCoMo Report. (2004, June). Size and Weight Reduction of Progressively Higher Function Mobile Phones – phones get smaller and lighter while taking on versatile features. Retrieved July 05, 2004 from http://www.nttdocomo.com/files/presscenter/34_No18_Doc.pdf

O Globo. Celulares a menos de US$ 20? (2005, November 07). *O Globo*. Retrieved November 09, 2005, from http://oglobo.globo.com/jornal/suplementos/informaticaetc/189069072.asp

Ordoñez, R. (2006, January 20). Telefonia celular não chega a 43% dos municípios do país. *O Globo*. Retrieved January 20, 2006, from http://oglobo.globo.com/online/economia/mat/2006/01/20/190012732.asp

Plant, Sadie. (2002, October 21). On the mobile: The effects of mobile telephones on social and individual life. *Motorola Inc*. Retrieved September 27, 2003, from http://www.motorola.com/mot/documents/0,1028,296,00.pdf

Puro, J.P. (2002). Finland: A mobile culture. In J. Katz & M. Aakhus, (Eds.), *Perpetual Contact: Mobile communication, private talk, public performance* (pp. 19–29). Cambridge: Cambridge University Press.

Rheingold, H. (2002). *Smart Mobs: The next social revolution*. Cambridge, MA: Perseus Publishing.

Rice, R. & Katz, J. (2003). Comparing internet and mobile phone usage: Digital divides of usage, adoption, and dropouts. *Telecommunications Policy*, 27, 597–623.

Suchman, L. (1987). Plans and situated actions: The problem of human/machine communication. Cambridge: Cambridge University Press.

Siqueira, E. (2001, July 04). Mobility gains a foothold in Brazil. *The Feature*. Retrieved September 27, 2003, from http://www.thefeature.com/article.jsp?pageid=12024

Teleco: Informação em Telecomunicações. (2006a, January 25). *Telefonia Fixa: Acessos e Localidades atendidas no Brasil*. Retrieved January 27, 2006, from http://www.teleco.com.br/ntfix.asp

Teleco: Informação em Telecomunicações. (2006b, January 25). *Celular cresce 20,6 milhões e tem produção recorde em 2005*. Retrieved January 27, 2006, from http://www.teleco.com.br/imprimir.asp?pagina=/comentario/com140.asp

Teleco: Informação em Telecomunicações. (2006c, January 27). *Seção Teleco World: Venezuela*. Retrieved January 28, 2006, from http://www.teleco.com.br/pais/venezuela.asp

Teleco: Informação em Telecomunicações. (2006d, January 23). *Qual será o crescimento do Celular em 2006?* Retrieved January 29, 2006, from http://www.teleco.com.br/comentario/com141.asp

Teleco: Informação em Telecomunicações. (2005b, January 24). *Crescimento do Celular em 2004*. Retrieved January 28, 2006, from http://www.teleco.com.br/comentario/com88.asp.

Teleco: Informação em Telecomunicações. (2005c, December 04). *Qual o limite para o crescimento do celular no Brasil?* Retrieved January 29, 2006, from http://www.teleco.com.br/ comentario/com134.asp

Teleco: Informação em Telecomunicações. (2005d, November 27). *PNAD 2004: O celular como substituto do telefone fixo residencial*. Retrieved February 03, 2006 from http://www.teleco.com.br/ comentario/com133.asp

Teleco: Informação em Telecomunicações. (2004, July 15). *Brasil já tem 30 cel/100 hab. Qual o limite?* Retrieved January 28, 2006, from http://www.teleco.com.br/comentario/com61.asp

Ucel: Universo Celular. (2005, September 21). *Conhecendo e se prevenindo contra a clonagem*. Retrieved January 28, 2006, from http://www.teleco.com.br/comentario/com88.asp.

Velázquez, A. (2002, August 05). O reinado do celular. *Revista Poder*.

Weiser, M. (1996, March 17). Ubiquitous computing. Retrieved April 07, 2005, from http://www.ubiq.com/ hypertext/weiser/UbiHome.html

Wellman, B. (1999). *Networks in the global village*. Boulder, Colo: Westview.

20 Inter-firm relations in the age of instant access: Case of the U.S. logistics industry

Yuko Aoyama, Samuel J. Ratick
Graduate School of Geography, Clark University, Worcester, MA

20.1 Introduction

The logistics industry includes firms that provide a wide range of services, from asset-based activities, such as trucking and warehousing, to increasingly strategic and integrated functions, such as supply chain optimization. In the U.S., the logistics industry was roughly equivalent to 10 percent of the GDP in 2000, and while that share has been declining due to improved efficiency, it remains an important industry of significant size, scope and effect.

Past studies in economic geography and related disciplines have implicated the critical role of logistics functions in the contemporary economy. Various emerging production forms, such as flexible specialization, just-in-time production, increased outsourcing and off-shoring all necessitated logistical sophistication. Whereas research on industrial organization and industrial districts have uncovered the types and characteristics of inter-firm relations as important sources of innovation, industrial competitiveness and regional transformation (Piore and Sabel, 1984; Best, 1990; Markusen et al., 1991; Castells, 2000), few studies exist to date that analyze the logistics industry using the knowledge, tools and expertise that have been developed in economic geography (Dicken, 2003). The logistics industry is rapidly changing, as far greater sophistication is demanded of the industry due to a number of pressures such as: global outsourcing, higher efficiency in inter-firm coordination, new technologies that provide supply-chain solutions, and new opportunities in the use of business-to-business (B-to-B) electronic commerce (E-commerce), among others. A few economic geographers are beginning to advocate for the importance of this industry (see, for example, Dicken, 2003; Bowen et al., 2002; Leinbach and Bowen, 2004). In this chapter we focus on the organizational and geographic of the logistics industry. We do so using evidence collected through secondary materials: published literature and data, preliminary results from a nation-wide online survey of logistics providers that we conducted, as well as

H.J. Miller (ed.), Societies and Cities in the Age of Instant Access, 311–327.

findings from interviews we undertook with a dozen industry insiders on the eastern seaboard states of the U.S. in the summer of 2004.[1] This chapter also introduces our design for the construction of Agent Based Modeling (ABM) designed to simulate the organizational transformation of the industry. The model is designed to push the research beyond collection of anecdotal evidence and offer a structured and systematic understanding of organizational dynamics of the logistics industry.

20.2 Conceptualizing the logistics industry

It has been widely recognized in the business/management literature that logistics functions have undergone revolutionary changes. With a shift in the industry's best practices from a supply-push value-chain philosophy to demand-pull logistics, building reliable "quick response" supply chains, with minimal inventory and maximum flexibility, has become of paramount importance across various sectors (Feldman et al., 1996; Berglund et al., 1999; Hadjiconstantinou, 1999; Arntzen and Shumway, 2002); as noted by Hertz and Alfredsson (2003) logistics providers can be distinguished from traditional firms in the transport sector by their abilities to engage in problem solving and customer adaptation.

In the following sections we provide a conceptualization of the industrial organization of the U.S. logistics industry, and the interplay of virtual and geographic spaces as they apply to the logistics industry. Our aim is to stimulate research on understanding the contemporary dynamics of the logistics industry by focusing on the transformation in industrial organization fostered by the use of information and communication technologies (ICTs) and related systems.

20.2.1 The Inter-firm organizational dimension

We identified three inter-related and overlapping dynamic organizational processes in contemporary logistics: disintermediation and reintermediation, disintegration and reintegration, and a migration of logistics functions from real to virtual space (see Aoyama et al., 2005).

20.2.1.1 Disintermediation vs. reintermediation

Disintermediation, the elimination of traditional middlemen, is the most widely assumed trend to have occurred with the onset of the Internet and ICTs (Westland and Clark, 1999, 350; Jallat and Capek, 2001; E-commerce in the Enterprise, 2002). A process of re-intermediation, however, is simultaneously taking place with the rise of online logistics providers (e.g., e-commerce brokers, internet auctions, industry or function specific market places, and fully integrated internet logistics services) (Kaplan and Sawhney, 2000; Barnes-Vieyra and Claycomb, 2001; Lucking-Reiley and Spulber, 2001). Some are newly formed firms specializing in online transactions, while others are traditional logistics firms that have successfully incorporated ICTs.

20.2.1.2 Disintegration vs. reintegration

Today, 73 percent of Fortune 100 companies outsource logistics functions (Gooley, 2002; Langley et al., 2001; Lieb and Schwarz, 2001), which suggests a trend toward disintegration. Outsourcing the entire supply chain management function to specialized logistics providers to take advantage of their logistics expertise and economies of scale is increasingly a popular option for many firms that require extensive logistics functions. Simultaneously, reintegration is being observed as logistics providers consolidate to offer a more comprehensive set of services.

20.2.1.3 Migration of firms from real to virtual space: e-logistics providers

The newly emerging e-logistics providers can be divided into those that provide logistics services exclusively in virtual space (non-asset-based), and those that provide services in both virtual and geographic spaces (asset-based) (Lieb and Schwarz, 2001). Although consolidations have taken place since 2001, the rise of e-logistics providers suggests that organizational dissociation of the physical movement of goods and the related transmission and processing of information is being accelerated by the introduction of B-to-B e-commerce (Evans and Wurster, 1997; Jallet and Capek, 2001).

In conceptualizing industrial organization for this industry, the first challenge is to sort a vast number of firms, agents, and middlemen involved in the geographic and information sides of the logistics industry. The task is complicated by firms offering increasingly multiple services, and particularly the tendency among traditional, 'asset-based' firms to claim logistics service functions in addition to transportation and warehousing. For expository purposes we provide an operational structure for the different types of logistics provider companies in Table 20.1; we realize that such a classification of an amorphous and evolving industry like logistics would not be exhaustive or mutually exclusive, but offer the structure to help in understanding the different types of logistics providers, their functions and the services they provide.

Logistics providers specialize in providing one or more types of logistics services, and include firms in transportation, warehousing, freight forwarding, customs clearance in international trade, 3PLs, 4PLs, ILOs, and e-commerce platforms. 3PLs can be further differentiated by those that own and operate tangible assets (asset-based 3PLs) and those that do not (non-asset-based 3PLs), and by market orientation (those that deal with international shipments and those that

Table 20.1 Operational Classification of Logistics Providers

Type of Company	Typical Types of Services and Functions Provided
Asset and Light Asset Based	Offering transportation services through different modes such as: Land (truck and rail), Air and Water Warehousing services
Non-Asset Based	Supply chain management consulting and software development Freight Forwarding (NVOCC) Non Vessel Operating Common Carrier (Intermediaries that sell space aboard ocean carriers and may also offer shipment consolidation services) Customs brokerage Exclusively online exchange and brokerage (E-commerce providers)
Third Party Logistics Providers (3PLs)	Provide services and functions characteristic of Asset and Non-Asset based companies in differing emphasis and proportion. Generally logistics providers offering supply chain management services to which firms outsource their logistics functions;
Internet Logistics Operators (ILOs) and 4PLs	Generally provide comprehensive supply chain management functions, including offering specialized IT services related to e-commerce and information transmission (supplier selection, ordering, billing), as well as asset based goods handling.

work exclusively or predominantly with domestic logistics transactions). Logistics providers may choose to outsource one or more functions to another logistics provider, who may have functional or geographic specialty.

Our preliminary interview results supported the conclusion drawn from the literature that simultaneous processes of disintegration and reintegration are occurring in the U.S. logistics industry (see Aoyama et al., 2005). While many share the view that reintegration, through mergers and acquisitions to enable firms to compete in an ever more technologically sophisticated environment, is occurring, few consider that small logistics operators will cease their existence anytime in the near future. Small operators: may provide low cost alternatives to large consolidated firms, maintain a geographic specialty, and are not only capable of capturing small contracts but are able to capture large, multi-million dollar business.

According to some of our interviewees, small operators seem to hold certain advantages over large firms in a number of fronts. For example, an interview participant from a multinational asset-based logistics firm, shared a recent experience of losing a multi-year, multi-million dollar contract from a large manufacturing conglomerate they already serve overseas, to a small local trucking company. Small firms in general are able to offer competitive prices due to lower fixed costs both in terms of assets and labor. Thus, while trust is important and some firms will pay premium for service, others seek low cost operators that offer basic transportation services.

We conceptualize the characteristics of inter-firm relations to take one of three dominant types of market governance; Hierarchical, Cooperative, and Competitive. These types of market governance are by no means exhaustive nor mutually exclusive, and multiple forms of governance can co-exist, at national, sectoral, or firm levels. The widespread assumption that the internet will bring about a nearly perfect market (Brynjolfsson and Smith, 2000) would suggest that formerly hierarchical or cooperative relationships in inter-firm relations would increasingly shift to competitive governance, and as a result, the potential substitution of trust-based business relationships for cost-optimized, at-arms-length relationships. Yet, questions remain as to what extent new technologies displace entrenched or trust-based business transactions, and how they impact organizational dynamics, the logic of competition, and the locational attributes of logistics users as well as providers.

There are many factors that affect – in interrelated ways – the inter-firm dynamics of the logistics industry. Primary among these is cost-minimization through maximized efficiency (less than optimal efficiencies are sometimes tolerated in the short-term, however, to avoid the necessary investment in transport and information technology infrastructure that would entail a switch to more efficient providers (Buvik and Halskau, 2001)). In Europe relationships between logistics providers and users have shifted from contractually-oriented to partnership-oriented (Hertz and Alfredsson, 2003). Trust and control are also major considerations. In a survey conducted by Langley et al. (2001), 63 percent of companies surveyed cited the fear of diminished control as a primary reason for not using 3PL services. An inexperienced forwarder in an overseas location, for example, can result in holding up merchandize for weeks by having insufficient knowledge of customs clearance for specialized manufactured products, and stall a new factory start-up. The promise of future cost savings alone is often not a sufficient basis for manufacturers to decide on outsourcing supply chain functions to new providers. According to the interviews we conducted, trust remains important when providers choose their business partners. This is particularly so when users choose logistics operators, as compared to transportation only providers.

Despite the potential of sophisticated ICTs for supply chain optimization, efficiency is not a guaranteed outcome; it is commonly observed that maximum potential use of ICTs is seldom achieved because of pre-existing practices and legacy systems, lack of common standards, and high barriers to entry (e.g., high initial investment cost, lack of technological expertise). Our interviews suggested that while logistics users today demand "visibility" (tracking functions) as part of the services offered, few actually make strategic use of tracking information. Most logistics users are reluctant to purchase supply chain software services available from technology solution firms as they typically are associated with a per-transaction-fee structure. Thus, compatibility with legacy systems, cost issues associated with system integration, as well as the lack of demonstrable benefits, have hindered a more rapid adoption of the ICTs solutions available in the market.

Finally, participation in global trade also influences industrial structure and the forms of governance. Logistics providers need to mediate across various assumptions, expectations, and demands among clients. To what extent information technologies facilitate organizational convergence – a process of organizational restructuring to streamline tasks to correspond to new ICT tools (both systems as well as software) – remains a question. The degree of change is also likely to vary across economies: for example, technologies are assumed to be easily adopted in the market governance context of the U.S. vis-à-vis other contexts such as Japan and Germany – where long-term trust-based relationships still dominate. The logistics industry therefore represents an important key to understanding the attributes and consequences of contemporary globalization; not only does the industry directly participate in global production and trade, it is also at the frontline of globalization with its concomitant clashes among different economic systems and institutions, both in the formal (legal) as well as informal (conventions, trust, business practices) sense, as well as, in the standardization and use of new information technologies. Thus, how different market governance principles influence the organizational logics of the logistics industry becomes an important question that calls for further research. For example, to what extent are traditional "trust-based" inter-firm interactions being substituted by "competition-based" interactions through the use of technology enabled B-to-B E-commerce? How do firms make decisions on logistics outsourcing, and how are their decisions being influenced by technological change, participation in global trade, and the locational attributes of partners?

20.2.2 Geographic dimensions

The logistics industry serves a crucial back-end (fulfillment) function to B-to-C and B-to-B E-commerce, and therefore needs to be viewed in this new light. Barnes-Vieyra and Claycomb (2001, 17) emphasized the virtual world of information flows, and consider B-to-B e-commerce to possess the revolutionary potential for "location elimination" by overcoming spatial constraints, language barriers and time zones. Li et al. (2001) stressed the importance of a geographical dimension. The bulk of manufacturing activities today, in reality, necessitate long-distance, geographic coordination among clients, outsourcers, suppliers, representatives and agents. These are often located around the world and have their own economic, institutional, and cultural contexts. Our previous research findings, using a mixed-integer linear optimization model with Monte Carlo cost parameter simulations, demonstrated that transport cost does continue to play a dominant role in the overall cost structure; e-logistics agents were used only when transport costs were lower (Aoyama et al., 2005). With the exception of a few emerging integrated logistics providers (such as UPS or DHL), the overwhelming majority of logistics providers possess a geographic specialty (and even the worldwide integrated firms either have branches in, or have acquired firms with, strategic geographic locational advantage). This is because transportation is primarily a local industry that requires rich, geographically-specific knowledge, even if the industry deals with international and widely geographically dispersed transactions. Our survey of logistics providers also showed the tension between local specialization and expansion of geographic reach. Two out of 5 survey respondents reported that their top client was located in the same or in a neighboring state, while slightly more (about half of the total respondents) reported that their top client was located elsewhere in the U.S., and a small percentage (6 percent) reported that their top client is overseas.

This suggests the presence of a paradox in the industry; although the logistics industry serves an integral function in the globalization of production, it also remains one of the most localized and embedded industries of all. Transport geographers and specialists in operations research have studied the geographic dimensions and interrelationships between infrastructure and commodity flows. Most recently, Hesse and Rodrigue (2004) took a conventional transport geography approach and considered the geography of logistics primarily through freight flows. They noted that these flows have shifted over time from those that were focused around major places of production to international gateways and hubs, such as large ports or major airports. They also note the rise of distribution centers, either as "Inland Hubs" or in urban fringes of major metropolitan areas, as a response to geographic limitations of further expansions in many of the already established hubs. Their primary preoccupation is the geography of logistics *functions*, rather than firms, which may or may not have different geographic manifestations. Rodrigue (2004) extended this analysis of freight flows with specific reference to the Boston-Washington corridor, and argued that mega-urban regions are also understood as freight corridors, in which freight flows through distribution centers ("articulation points") along the corridor. Locations of logistics infrastructure, such as intermodal terminals and highways, are used as indicators of the freight corridors. He suggested that transport corridors and urban regions such as the BostWash corridor are the venue of interaction between global and local freight distribution systems.

20.3 Constructing an agent-based model (ABM) of the logistics industry

ABMs are generally composed of a number of flexible and often interacting components: agents (the decision making entities – in this research, logistics users and providers), the interactive decision environment in which the agents interact (e.g. market governance structure), and the rules that govern the substance and timing of those interactions (Parker et al. 2002, 2003); ABMs are often used to link macro-organizational characteristics to micro-level decision-making. (Parker et al. 2003: 321), and have a number of modeling advantages: they can be easily reconfigured and are continually adaptable and flexible, they can allow for complex heterogeneity of agents and market environments; and they can help uncover emergent structures in dynamic and complex systems (Gilbert 1995, Epstein and Axtell 1996, Sullivan and Hakley 2000, Manson 2002a).

Table 20.1 in the previous section provided a description of the potential logistics provider agent types, while Table 20.2 details agent properties by attributes, decision criteria, and objectives; the ABM will include logistics provider agents as well

Table 20.2 Agent Properties

	Logistics User	Logistics Provider
	Location	
Attributes, Decision Criteria	Market orientation (domestic/int'l)	Market orientation(domestic/int'l); Asset-ownership characteristics; Market entry/exit
	Types of Logistics Outsourcing (Functions, contract types)	Types of Logistics Subcontracting (Functions, contract types)
	Outsourcer characteristics (location/selection/retention)	Sub-contractor characteristics (location/selection/retention)
		Pricing of services
	IT Use/Type/Investment	
	Use/Types of B-to-B e-commerce	Use/Type of B-to-B e-commerce in sales Use/Type of B-to-B e-commerce in procurement
Objectives	Cost minimization	Profit maximization
	Quick response	Service optimization
	Inventory minimization	Increase market share

as logistic user agents. The industry structure will evolve as agents make decisions according to their objectives (e.g., cost minimization, profit maximization, increase market share), and alter their attributes (e.g., location, types of outsourcing) in an attempt to better meet their objectives.

Logistics *user* attributes include its location, which can be represented either as a specific city in a simulation of an empirical case, or as a type of location in a more generic implementation. The firm's participation in international trade determines the type and reach of logistics services that the firm requires. The list of outsourced logistics functions captures the types of logistics services agents purchase, and the choice of provider firms and their locations. The type of information technology used in logistics-related transactions, and the firm's participation in B-to-B e-commerce, are the crucial attributes that will allow the simulation model to determine the effects of ICT on the dynamics of the logistics industry. The objectives of logistics outsourcing include cost minimization, time minimization (e.g., quick supply chain responses), and inventory minimization (e.g. just-in-time production and delivery). Logistics *provider* attributes include location, the services offered, and the use of subcontractors. Providers negotiate the pricing level of their services, which represent one way logistics providers and users can interact. Business decisions of logistics providers also include information technology used and the participation of the intermediary in B-to-B e-commerce for marketing and procurement. The logistics providers attempt maximizing profits and/or maintaining the highest possible service quality to ensure their business's long-term viability.

The results of our interviews and survey will be used to develop the detailed attributes, decision criteria, and objectives of each agent; and on the forms of market governance (competitive, cooperative, hierarchical) that regulate inter-firm logistics relationships, which will serve as an 'interactive decision environment' for the ABM. The sum effect of agents' decisions can change industry structure and the decision environment of the agents in future periods.

20.3.1 Model design

The top-level distinction has been made between logistics *users* and logistics *providers*. Logistics providers are classified in a hierarchical manner to facilitate model design. Each agent type is designed to have multiple subtypes that share properties with the parent class. For example, 3PLs can be differentiated by: those that own and operate tangible assets, and those that offer only non-asset services; those that deal with international shipments and those that work exclusively or predominantly with domestic logistics transactions. By making decisions based on a set of decision criteria, agents strive to attain their objectives by altering their attributes. Over time, outcomes are compared to objectives and decision criteria are updated to reflect experience.

The internal decision-making process and the nature of agents' interactions will be established and refined through theoretical analysis and our empirical work. Decision factors for logistics agents can include: knowledge about their own attributes, past decisions and current market conditions; their knowledge about these decision factors for other agents; their knowledge or perceptions about potential future conditions; whether they exhibit perfect or bounded rationality. At each time step, this information is processed according to principles that may well differ from agent to agent: some may exhibit a strictly optimizing behavior, others may be satisficers, while still others may make deterministic rule-based decisions; some may be risk takers, other risk averse (thus creating the heterogeneity among agents). Once decisions are taken, the agents update their attributes and store information about the outcomes of those decisions. This monitoring of results allows the agents to possibly alter their decision-making process in future model cycles. We expect the aggregate effect of the evolution of agent behavior will define and modify the types of market governance.

As an example, consider an asset-based logistics provider that offers only warehousing services, and uses only legacy information technologies (phone, fax) to communicate with customers (Table 20.2). This agent is dependent on a large customer (i.e. the provider is embedded in a hierarchical relationship) that is demanding online status visibility for their shipments at the warehouse. The agent needs to decide whether to invest in a technology upgrade, whether to subcontract the operation of an inventory control system, or whether to resist change. Depending upon its location and topological position in the entire logistics network (information flow and goods flow), the agent knows about the status of ICTs at its

Table 20.3 Example of Logistics Decisions on Inter-Firm Relations

	IT upgrade?
Decision to be made ↓ **Information available to agent** ↓	• Market environment: hierarchical and dependent on limited number of customers • Customer requirements • Competitor behavior • IT investment cost • Outsourcing options and cost • Results of previous IT decisions
Agent objective	Satisficing behavior: Secure long-term business viability with minimal cost
↓ **Decision-making** ↓	**Viability without IT upgrade=low due to market environment and competitor behavior** Perception: Investment cost of own IT upgrade larger than cost of outsourcing
Outcome	• Outsource • Select sub-contracting provider • Open information flow link to that provider

competitors, the options and prices for sub-contracting the service, and the investment required to enter the market for online inventory control itself. It evaluates the alternatives according to its internal preferences and decision-making process. Because the agent would lose its prime customer if the ICTs were not upgraded, risking the basis of its existence, it decides to sub-contract with a non-asset based logistics provider.

This decision alters the agent' sub-contracting attributes and opens a new link to a logistics provider agent. After a specific intermediary is selected (requiring a separate decision) the costs and perceived benefits of the outsourcing decision are monitored in the agent's knowledge base. Should the outcomes not be satisfactorily over a number of cycles, the agent can decide to alter the decision; for instance, it could invest in ICTs upgrades, terminate the sub-contracting relationship, and in effect enter a new logistics market segment.

20.3.2 Model implementation: programming, testing, analysis, and validation

To confirm extant theory about changing logistics industry trends and how industry practices may be changing, and for exploratory purposes in extracting new emergent structures, a set of carefully designed dynamic scenarios will be developed. Initially, we expect these scenarios to involve, among others:

- Altering the composition of logistics providers, for example increased use of 3PLs and 4PLs from the continued outsourcing of logistics functions.

- Altering future conditions that can involve: modifying the schedule of goods demand and delivery conditions over time; changes in information or goods flow network attributes and conditions; and changes in market organization (the degree of competition, cooperation, or hierarchical structure).
- Altering the degree of predictability of future market conditions.
- Altering environmental conditions such as increasing cost pressure on the industry, or increased expectation of transparent information flow and its attendant reliance on technology.
- Altering the behavioral heterogeneity of logistics agents, for example, modifying the proportion of entrepreneurial risk takers who will enter markets with lower expectations of near term profits.

The scenarios will be analyzed with respect to their agreement with current conditions, their correspondence with projections for the industry made in various literatures, and the feasibility of new emergent structures that may evolve.

Validating any type of model is difficult (Osleeb et al., 1989), it is particularly so with agent based simulations. A number of approaches to verification and validating ABMs have been proposed. Parker et al (2002: 31–36; 2003: 327–329) and Manson (2002b: 63–74) describe types of model validation (and verification): how well the program functions, how well the simulation structurally represents the conceptual model (structural validation), and how well the model's outcomes represent the system being analyzed (outcome validation). Validating program function and structural validation are relatively straightforward. This kind of internal validity can be tested by comparing the intended outcome of the conceptual design in a highly simplified artificial setting with programmed model results using the same inputs. Outcome validation, however, may well be the most difficult to accomplish. In addition, agent based simulation models may be thought of as bridging a gap between qualitative and quantitative research, allowing, in a systematic structure, concepts such as "trust" to be modeled and tested. It is difficult to validate such concepts through systematic objective examination of model results.

A broader concept of evaluating model validity is therefore more useful for our proposed research. Validation through the grounding of the model permits some confidence that its abstractions and assumptions do not impair the plausibility of its results and its potential for yielding new insights. Carley (1996) emphasizes that the grounding method of validating models only provides face and pattern validity (i.e. that the model results are plausible at face value and represent patterns observable in reality), but not more quantitative measures of distributional or value validity. Nevertheless, since our model is designed to provide insights into the logistics industry's development rather than specific guidance on a single problem specified in detail, this level of validation may well be sufficient (Carley 1996: 11–13). To address this issue we therefore plan to return to those industry interviewees who have agreed to this second round of contact with a selected set of scenario results, asking them to evaluate these in terms of their experience and knowledge about possible future structures for the industry. This way we can achieve some practical, grounded validation of the results.

20.4 Conclusion

In this chapter we have argued that a study of the emerging trends in the logistic industry is an important area of study to foster understanding of the implications of: technological change (particularly that of information technologies), economic globalization, and changing geographic requirements. The changing dynamics of the logistics industry and its industrial organization as illustrated in the chapter - the simultaneous processes of disintegration and re-integration, disintermdediation and re-intermediation, and the migration of firms from real to virtual space – have the potential to be an important and illustrative case study of emerging forms of technology-space relations.

With its growing use of information technologies, the logistics industry is playing an increasingly critical role in improving industrial productivity, reducing transactions costs, and facilitating the seamless flow of global commodity trade. The logic of virtual space therefore is mediating the geographic logic of this traditionally localized service industry; implying that there is great potential for a variety of fruitful research projects to be developed in economic geography. Without further careful study these conjectures will remain the primary source of information on this important topic. Studies designed in an economic geography framework to understand the ways in which these forces affect the evolution of the logistics industry will help to illuminate and evaluate policies designed to affect the future of the global economy.

We expect the development of the ABM will help provide a more comprehensive and systematic understanding of organizational dynamics of the logistics industry; by providing a structure within which the dynamic changes in the logistics industry, as influenced by ICTs and globalization changes, can be tested and evaluated. We intend to test questions such as: Under what conditions are logistics providers likely to outsource their functions; What types and timing of investments in information technology are likely to occur in response to changing market conditions, user demands or the desire to be among industry leaders? For example, in our survey the most frequently cited ICT investment was for "supply chain management and logistics software" (30.3% of respondents); and almost half (47.9%) of respondents stated that the main reason for making ICT investments was the "desire to install state of the art practice," While the survey and interview results provide some general insights into the current actions, motives and decisions of logistics providers, the ABM is intended to suggest possible future trajectories of changes in the industry; and the specific and comprehensive processes by which actions would be taken and through which decisions are made that lead to those changes. The result of this effort, we expect, will be the development of cogent and comprehensive scenarios of potential changes to the organizational structure of the logistics industry.

Acknowledgements

This chapter is based in part on an article entitled "Organizational Dynamics of the U.S. Logistics Industry from an Economic Geography Perspective" by Aoyama, Ratick, and Schwarz that appeared in *The Professional Geographer* 58, 3 (August 2006): 327–340. We thank the Association of American Geographers for the permission to reproduce part of the article. The research for this paper was supported through a research grant from the National Science Foundation (BCS-0350697), "Organizational Dynamics of the U.S. Logistics Industry: The Impacts of Inter-firm Relations, Technologies, and Globalization." We acknowledge the contribution of Guido Schwarz, who served as a research assistant for this project and a third author for the journal article for which this chapter is partly based. We also want to acknowledge the contributions of the interview participants, and those who participated in our Logistics Provider Survey.

Note

[1] The survey was offered to 1,994 members of the Council for Supply Chain Management Professionals (CSCMP) who self-identified themselves as logistics providers. The on-line questionnaire ran from March 12th to April 18th, 2005, voluntary responses were solicited via email; 143 responses (response rate of 7.2 percent) were completed in that time frame. Fifty five multiple part survey questions covered topics including: company characteristics (e.g. size, type, location), services the company provides, views and experiences with information technology, relationships with their customers, information about subcontracting, industry trends, firm characteristics of US logistics providers, subcontracting, investments in information technology, and relationships with suppliers. Semi-structured interviews were conducted in the summer of 2004 with over a dozen logistics providers and users in the Boston-Washington corridor. They included both US and non-US firms ranging from non-asset-based 3PLs (third party logistics providers) and supply chain consultancies to asset-based firms in trucking and ocean transport. Most interviews were 60–75 minutes, and were conducted with executives in charge of logistics and strategic supply chain management planning.

References

Aglietta, M. *Accumulation et régulation du capitalisme en longue période. L'exemple des Etats-Unis (1870–1970)*. Insee, 1974 (English translation: A Theory of Capitalist Regulation: The US Experience. Verso Classics, London, 1979, 2000).

Aoyama, Y, S.J. Ratick, G. Schwarz. 2005. Modeling the Impact of Business-to-Business Electronic Commerce on the Organization of the Logistics Industry. *Geographical Analysis* 37, No.1: 46 68.

Arntzen, BC, Shumway, HM (2002) Driven by Demand: A Case Study. *Supply Chain Management Review* January/February: 34–41.

Barnes-Vieyra, P, Claycomb, C (2001) Business-to-Business E-Commerce: Models and Managerial Decisions. *Business Horizons* 44(3): 13–20.

Berglund, M, van Laarhoven, P, Sharman, G, Wandel, S (1999) *Third-Party Logistics: Is There a Future*? The International Journal of Logistics Management 10(1): 59–70.

Bernasek, A. (2002) "The Friction Economy." *Fortune* 145, No.4 (February 18): 104+.

Best, M (1990) *The New Competition: Institutions of Industrial Restructuring*. Harvard University Press, Cambridge.

Boyson, S, Corsi, T, Dresner, M, Rabinovich, E (1999) Managing Effective Third Party Logistics Relationships: What Does It Take? *Journal of Business Logistics* 20(1): 73–100.

Bowen, JT, Leinbach, TR, Mabazza, D (2002) Air Cargo Services, the State and Industrialization in the Philippines: The Redevelopment of Subic Bay. *Regional Studies* 36: 451–467.

Brynjolfsson, E, MD Smith (2000) Frinctionless Commerce?: A comparison of Internet and conventional retailers." *Management Science* 46, 4 (April): 563–585.

Buvik, A, Halskau, ø (2001) Relationship duration and buyer influence in just-in-time relationship. *European Journal of Purchasing & Supply Management* 7: 111–119.

Castells, M (2000) *The Network Society* (2nd edition). Blackwell Publishers, Oxford.

Chan, Y, Carter, WB, Burnes, MD (2001) A multiple-depot, multiple-vehicle, location-routing problem with stochastically processed demands. *Computers and Operations Research* 28(8): 803–826.

Coia, A (2002) Keeping Mexico on the Road. *World Trade* 15(4): 24.

Copacino, WC, Dik, RW (2001) Supply Chain Management: Why B2B eMarkets are here to stay. *Outlook Journal* (Accenture Consulting) no. 2, available at http://www.accenture.com/xd/xd.asp?it=enweb&xd=ideas\outlook\7.2001\emarkets.xml.

Cutter, S.L., Richardson, D.B., Wilbanks, T.J. (eds) (2003). *The Geographical Dimensions of Terrorism*. New York: Routledge.

Dicken, P (2003) *Global Shift: Transforming the World Economy* (4th edition). Sage Publications, London.

Dyer, JH, Chu, W (2000) The Determinants of Trust in Supplier-automaker Relationships in the U.S., Japan, and Korea. *Journal of International Business Studies*, 31 2 (Summer): 259–285.

E-commerce in the Enterprise (2002) *Great Predictions that Never Happened*. February 26, available at http://www.itworld.com/nl/ecom_in_ent/02262002/.

Elliff, SA (2001) Who's Who? Sorting out the e-logistics players. *Logistics Management and Distribution Report* 40(4): E3–E9.

Evans, PB, Wurster, TS (1997) Strategy and the New Economics of Information. *Harvard Business Review* 75(5): 71–82.

Feldman, MP, Bell, ME, Salamon, LM (1996) *The Distribution Industry: An Engine for Maryland Growth*. Johns Hopkins University, Institute for Policy Studies, Baltimore, available at http://www.jhu.edu/~ips/Logistics/mdreport.htm.

GE Global Exchange Services (2002) GE GSN: The World's Largest Private Web Marketplace. Available at http://www.gxs.com/downloads/cs_gsnq202.pdf.

Geoffrion, AM (1995) Distribution System Design. Drezner, Z (ed) *Facility Location: A Survey of Applications and Methods*. Springer Verlag, New York: 181–198.

Goetz, AR, Sutton, CJ (1997) The Geography of Deregulation in the US Airline Industry. Annals of the Association of American Geographers 87: 238–263.

Goetz, AR (2002) Deregulation, competition, and antitrust implications in the US airline industry. *Journal of Transport Geography* 10(1): 1–19.

Goetz, AR, Rodrigue, J-P (1999) Transport Terminals: New Perspectives. *Journal of Transport Geography* 7: 237–240.

Golob, TF, Regan, AC (2001) Impacts of Information Technology on Personal Travel and Commercial Vehicle Operations: Research Challenges and Opportunities. *Transportation Research Part C: Emerging Technologies* 9(4): 87–121.

Gooley, TB (2002) Two sides of the same 3PL coin. *Logistics Management and Distribution Report* 41(2): 45.

Gordon, B.H. (2003) "The Changing Face of 3rd Party Logistics" *Supply Chain Management Review* 7, 2 (March/April): 50–58.

Graham, B (1995) *Geography and Air Transport*. John Wiley & Sons, Chichester.

Granovetter, M (1985) Economic Action and Social Structure: The Problem of Embeddedness. *American Journal of Sociology* 78: 1360–80.

Haddad, C., and J. Ewing. (2001). "Ground Wars. UPS's Rapid Ascent Leaves FedEx Scrambling." Business Week May, 64–68.

Hadjiconstantinou, E (ed) (1999) *Quick Response in the Supply Chain*. Springer Verlag, Berlin.

Hanson, S, Huff, J (1988) Repetition and day to day variability in individual travel patterns: implications for classification. In: Golledge, RG, Timmermans, HJP (eds) *Behavioural modelling in geography and planning*. Croom Helm, London: 368–398.

Hanson, S (2000) Reconceptualizing Accessibility, In: Janelle, D, Hodge, D (eds) *Information, Place, and Cyberspace: Issues in Accessibility*. Springer-Verlag, Berlin, pp 267–278.

Hertz, S, Alfredsson, M (2003) "Strategic development of third party logistics providers." *Industrial Marketing Management* 32: 139–149.

Hesse, M (1995) Urban space and logistics: on the road to sustainability? *World Transport Policy & Practice* 1(4): 39–45.

Hesse, M (2002) Shipping News: The Implications of Electronic Commerce for Logistics and Freight Transport. *Resources, Conservation and Recycling* 36: 211–240.

Hesse, M, Rodrigue, J-P (2004) The transport geography of logistics and freight distribution. *Journal of Transport Geography* 12: 171–184.

Hirshleifer, D, Welch, I. (2002) "An Economic Approach to the Psychology of Change: Amnesia, Inertia, and Impulsiveness." *Journal of Economics and Management Strategy* 11(3): 379–421.

Ivy, RL, Fik, TJ, Malecki, EJ (1995) Changes in Air Service Connectivity and Employment. *Environment and Planning A* 27: 165–179.

Jallat, F, Capek, MJ (2001) Disintermediation in Question: New Economy, New Networks, New Middlemen. *Business Horizons* 44(2): 55.

Kaplan, S, Sawhney, M (2000) E-Hubs: The New B2B Marketplaces. *The Harvard Business Review* 78(3): 97–106.

Knudsen, DC (1985) Exploring flow system change: U.S. rail freight flows, 1972–1981. *Annals of the Association of American Geographers* 75: 539–551.

Kuby, M, Ratick, SJ, Osleeb, J (1991) Modeling US coal export planning decisions. *Annals of the Association of American Geographers* 81: 627–649.

Langley, JC, Allen, GR, Tyndall, GR (2001) *Third Party Logistics Study. Results and Findings of the 2001 Sixth Annual Study*. Cap Gemini Ernst & Young, Detroit, available at http://www.us.cgey.com/ind_serv/industry/cprd/info_research.asp.

Leinbach, TR, Bowen, JT (2004) Air Cargo Services and the Electronics Industry in Southeast Asia. *Journal of Economic geography* 4: 299–321.

Li, F, Whalley, J, Williams, H (2001) Between physical and electronic spaces: the implications for organizations in the networked economy. *Environment and Planning A* 33: 699–716.

Lieb, R, Schwarz, B (2001) *The Year 2001 Survey: CEO Perspectives on the Current Status and Future Prospects of the Third Party Logistics Industry in the United States*. Accenture Consulting and Northeastern University, available at http://web.cba.neu.edu/~rlieb/2001ceopaper.dot.

Liker, JK, Wu, Y-C (2000) Japanese Automakers, U.S. Suppliers and Supply-Chain Superiority. *Sloan Management Review* Fall: 81–93.

Limao, N., Venables, A.J. (2001). "Infrastructure, Geographical Disadvantage, Transport Costs, and Trade." *World Bank Economic Review* 15 3: 451–490.

Lipietz, A (1986) New Tendencies in the International Division of Labor: Regimes of Accumulation and Modes of Regulation. In: Scott, AJ, Storper, M (eds) *Production, Work, Territory: the geographical anatomy of industrial capitalism*. Allen & Unwin, Boston.

Logistics Management and Distribution Report (2001) GM shifts global logistics strategy. 40(1): 19.

Lucking-Reiley, D, Spulber, DF (2001) Business-to-Business Electronic Commerce. *Journal of Economic Perspectives* 15(1): 55–68.

Markusen, A, Hall, P, Campbell, S, Deitrick, S (1991) *The Rise of the Gunbelt: TheMilitary Remapping of Industrial America*. Oxford University Press, New York.

McCalla, RJ, Slack, B, Comtois, C (2001) Intermodal freight terminals: Locality and industrial linkages. *Canadian Geographer* 45(3): 404–413.

Mokyr, J (2000) Innovation and Its Enemies: The Economic and Political Roots of Technological Inertia. In: Olson, M, Kähkönen, S (eds) *A Not so Dismal Science*. Oxford University Press, Oxford: 61–91.

Morrison, M (2000) Outsourced Logistics: What's Between the Clicks and Mortar. *Logistics Quarterly* 6(1).

Nurthen, W. (2003) "Urban infrastructure security." *Technology in Society* 25 (4) (November): 543–547.

OECD (2003) "Security in Maritime Transport: Risk Factors and Economic Impact." Maritime Transport Committee, Directorate for Science, Technology and Industry. OECD, Paris, July.

Osleeb, JP, Ratick, SJ (1990) A Dynamic Location-Allocation Model for Evaluating the Spatial Impacts for Just-in-Time Planning. *Geographical Analysis* 22: 50–69.

Osleeb, JP, Ratick, SJ, Kuby, MJ, Olson, HE, Antle, LG, Hawnn, AF (1989) Coal Logistics System. In: *Transportation Research Record: Ports, Waterways, and Inland Water Transportation* 1222: 6–13.

Pedersen, PO (2001) "Freight transport under globalization and its impact on Africa." *Journal of Transport Geography* 9: 85–99.

Piore, M, Sabel, C (1984) *The Second Industrial Divide*. Basic Books, New York.

Poirson, H. (1998) "Economic Security, Private Investment and Growth in Developing Countries." *IMF Working Paper*, WP/98/4, International Monetary Fund, African Department, January.

Ratick, SJ, Osleeb, JP, Kuby, M, Lee, K (1987) Interperiod Network Storage Location-Allocation (INSLA) Models. In: Rushton, G, Ghosh, A (eds) *Spatial Analysis and Location-Allocation Models*. Van Nostrand Reinhold, New York, pp 269–301.

Regan, AC, Jagannathan, S, Wang, X (2000) Mixed Global and Local Assignment Algorithms for Quasi-Dynamic Local Truckload Trucking Operations with Strict Time Windows. *Transportation Research Record 1733: Transportation Network Planning*: 49–55.

Robertson, L (1997) Inertia and Industrial Change. In: Cameron, R, Schnore, LF (eds) *Cities and markets: Studies in the organization of human space*. University Press of America, Lanham: 21–47.

Rodrigue, J-P, Comtois, C, Slack, B (1997) Transportation and spatial cycles: evidence from maritime systems. *Journal of Transport Geography* 5(2): 87–98.

Rodrigue, J-P (2004) Freight, Gateways and mega-Urban Regions: The logistical integration of the Bostwash corridor. *Tijdschrift voor Economische en Sociale Geografie* 95, 2: 147–161.

Rogerson, PA, Plane, DA (1984) Modelling temporal change in flow matrices. *Papers of the Regional Science Association* 54: 147–164.

Russel, D.M., Saldanha, J.P. (2003) "Five Tenets of Security-Aware Logistics and Supply Chain Operation." *Transportation Journal* 42 (4) (Summer): 44+.

Sabel, CF (1993) Studied trust: building new forms of cooperation in a volatile economy. *Human Relations* 46(9): 1133–1170.

Sako, M, (1989) *Prices, Quality and Trust: Inter-firm Relations in Britain and Japan*. Oxford university Press.

Saxenian, AL (1994) *Regional Advantage:Culture and Competition in Silicon Valley and Route 128*. Harvard University Press.

Schaefer, S (1998) Influence Costs, Structural Inertia, and Organizational Change. *Journal of Economics and Management Strategy* 7(2): 237–63.

Scott, AJ (1988a) *New industrial spaces: Flexible production organization and regional development in North America and Western Europe*. Pion, London.

Scott, AJ (1988b) Flexible production systems and regional development: the rise of new industrial spaces in North America and Western Europe. *International Journal of Urban & Regional Research* 12(2): 171–186.

Shen, Q (1999) Transportation, Telecommunications, and the Changing Geography of Opportunity. *Urban Geography* 20: 334–355.

Sheppard, E (2002) The Spaces and Times of Globalization: Place, Scale, Networks, and Positionality. *Economic geography* 78(3): 307–330.

Slack, B (1999) Satellite terminals: A local solution to hub congestion? *Journal of Transport Geography* 7(4): 241–246.

Slack, B, Wang, JJ (2002) The challenge of peripheral ports: An Asian perspective. *GeoJournal* 56(2): 159–166.

Storper, M (1997) *The Regional World: Territorial Development in a Global Economy*. New York: Guilford.

Storper, M, Venables, AJ (2002) Buzz: The Economic Force of the City. Presentation at *Industrial Dynamics of the New and Old Economy – who is embracing whom?* DRUID Summer Conference, Copenhagen, June, available at http://www.druid.dk/conferences/summer2002/Papers/STORPER.pdf.

Vidal, C J, Goetschalckx, M (2000) Modeling the Effect of Uncertainties on Global Logistics Systems. *Journal of Business Logistics* 21(1): 95–120.

Vowles, TM (2000) The Geographic Effects of US Airline Alliances. *Journal of Transport Geography* 8: 277–285.

Westland, JC, Clark, THK (1999) *Global Electronic Commerce: Theory and Case Studies.* The MIT Press, Cambridge.

Womack, JP, Jones, DT, Roos, D (1990) *The Machine That Changed the World.* HarperPerennial, New York.

Wrigley, N., Lowe, M. and Currah, A. (2002): Retailing and E-tailing. *Urban Geography*, 23, 2, 180–197.

COMMUNITY

21 Rethinking public participation as instant access to virtual meetings

Timothy Nyerges, Michael Patrick
Department of Geography, University of Washington, Seattle, Washington

21.1 Introduction

Public meetings are a mainstay of participatory processes (Renn, Webler, and Wiedemann, 1995; Gastil and Kelshaw, 2000); as meetings are primary vehicles for enabling group decision making in society (DeSanctis and Poole, 1994). The American Heritage® Dictionary (4th edition on-line) defines *participation* as *the act of taking part or sharing in something*. Thus, the idea of people sharing in decision processes is fundamental to participatory decision making. Participatory decision making exemplifies the democratic maxim that *those affected by a decision outcome should participate directly in the decision making process*. The maxim motivates inclusion of public participation requirements in federal, state and local laws, and hence the implementation of public participation within decision situations funded by tax dollars. But who is the *public* that is participating? The American Heritage® Dictionary (4th edition on-line) defines *public* as *the community or the people as a whole* or alternatively *a group of people sharing a common interest as in the reading public*. A public is thus plural and diverse, particularly when we view the public as constituted of stakeholder groups (Schlossberg and Shuford, 2005).

Over the past decade considerable research has examined the use of geographic information systems (GIS) in public and group settings. Obermeyer (1998) edited a special issue of *Cartography and GIS* journal on the topic of public participation GIS. Jankowski and Nyerges (2001a) examined stakeholder group-based GIS use in several decision settings. Craig, Harris, Weiner (2002) report on a wide-variety of investigations into community use of GIS, recognizing the importance of bottom-up information development common in participatory GIS. Rinner (2003) reviewed several web-based spatial decision support systems, many of them directed at public use. All of those publications recognize the importance of a

H.J. Miller (ed.), Societies and Cities in the Age of Instant Access, 331–342.

better understanding of participatory processes that underlay public group-oriented GIS developments of the future.

Although many public participation processes allow for email contributions from individuals and groups, such comments are not presented as part of meeting proceedings, although they do get into the public record. Public meeting activities take time and space – whether in traditional or virtual on-line meetings. Public meetings as the opportunity to participate, offer voice in democratic processes, but have been under fire for quite some time. There are considerable barriers to meaningful participation in public processes when the traditional public meeting is the vehicle (Smith, 1999).

The traditional public meeting process involves a public audience listening to characterization of a problem or to the options planners have identified, and then members of audience public provide a three minute response to that information. This type of process in a public meeting is on the second rung from the bottom of the metaphorical ladder of citizen participation (Arnstein, 1969) or second level of a spectrum of participation (International Association of Public Participation, 2005). The second rung/level is one level above one-way communication, but it is some distance from a top rung/level that empowers a public in decision processes. Interaction in most public meetings is so shallow that the 'same usual suspects' show up, calling into question a representative public in most decision contacts (Bosworth, Donovan, and Couey, 2002). Even if it is possible to enhanced public participation, what decision situations would make this exploration worthwhile?

On August 10, 2005, President Bush signed a $286.4 billion U S Transportation Bill that included line item funding for 6,371 "pork-barrel" projects totaling $24 billion (MSNBC, 2005). That means the remaining $262.4 billion is to be distributed to the states and territories across a six-year timeframe. Those funds are to become part of the transportation planning, capital improvement programming, and project implementation activity of state, regional and local jurisdictions. Actual allocation of a large percentage of the $262.4 billion will occur via transportation improvement programs within jurisdictions across the US and its territories.

A transportation improvement program (TIP) is the result of a decision process that allocates funds to implement a transportation plan (or a portion thereof) by selecting projects on a scope, design, or build rollover process. TIP development is a complex group activity, involving elected officials, policy analysts, technical specialists, and the public as diverse groups of interested and/or affected persons in a variety of meetings (Smith, 1999; Taylor, 1998). Public participation is required by federal and/or state law when federal and/or state funds are used for TIP processes, but incorporating highly interactive public participation into this process is not easy.

From empirical research, little *meaningful public participation* appears to be going on in TIP meetings (Nyerges et al., 1998a; Jankowski and Nyerges, 2001a). Most people are aware of long-term planning decision processes, although such processes have less impact on the landscape than do improvement programming processes. Investigating instant access to public participation in TIP decision making will have implications for political processes in which trillions of dollars are spent related to transportation, land use, water resources, and other critical resources that are intertwined from a sustainable development perspective. This instant access would involve more than access to information (Sui, 2000). It involves access to political process through public participation, potentially creating strong publics by not only jumping scale whereby local landscapes are changed by federal funds, but situating scale in the fabric of local societal/political resources (Aitken, 2002). Focusing on strong-publics in the politics of transportation improvement requires a nuanced understanding of public-decision support.

The Participatory Geographic Information Systems for Transportation (PGIST) Project is developing and evaluating an Internet GIS-based portal to support public participation in TIP decision making (See www.pgist.org). The study takes place in the central Puget Sound region of Washington State, but we expect the results of our technology development to be applicable to many areas around the US and the world. Our principal research question is: *What Internet platform designs and capabilities involving participatory modeling, particularly including GIS technology, can enable public participation in analytic-deliberative transportation decision making within large groups?*

To address a rethinking of public participation as instant access to virtual meetings, we start by situating this discussion within a TIP decision situation and the practical role of public meetings in this decision process. Next we introduce a framework for meeting arrangements described in terms of place and time. We compare/contrast conventional face-to-face meetings and asynchronous-distributed (virtual) meetings as opportunities for access to meetings as a whole as well as access to processes within meetings. We deconstruct the character of publication participation process so that we can explore how to reconstruct public meetings to support meaningful public participation, emphasizing the development of "strong publics". An analytic-deliberative decision process is the focus of the reconstruction. We discuss analytic-deliberative process in terms of human-computer-human interaction (HCHI) process granularity. We explore the challenges in analytic-deliberative HCHI by considering response latency as one of the indicators of instant access to process and address the process of meaning-making as the core issue. We conclude the paper with next steps for research about meaning-making in participation.

21.2 TIP decision processes and access to public meetings

Several tasks are accomplished in the formulation of a TIP. In two case studies about TIP decision situations in the central Puget Sound region in 1997 and 1999, it was found that similar steps were undertaken, little use of geographic information systems (GIS) technology occurred, and there were limited opportunities for public participation; however, public participation followed federal and state guidelines (Nyerges et al., 1998a; Jankowski and Nyerges, 2001a). Two of the seven tasks involved a public meeting to review material. The first review addressed policy objectives and the second review addressed the TIP project list. Although it is possible to enhance effective participation, such enhancements increase the cost of meetings because it takes considerable meeting planning and coordination of information flow (Smith, 1999). People participating in such meetings must be assured that their time and effort are being put to good use for every task/meeting.

Research analysts working on complex decision processes have not previously unpacked participatory processes sufficiently to enable software designers to create new tools that are broadly effective. Of course that could occur if agencies were willing to invest in such tools. However, we suspect that the potential for public participation is still not fully understood when it comes to digital technologies, making this a "chicken and egg" problem, which is why more of this type of research is needed. We pose the question in our research about advanced information technology development, what if the public were invited to participate in every task/subtask of the decision process? Can we support more meaningful participation? Can we still hold down the costs to the agencies and the public? Answers to these questions require that we rethink the character of access to and within public meetings as a major part of public participation.

Rethinking public participation stems in part from rethinking public meetings. Our research about stakeholder group participation in complex decision processes incorporates a variety of characteristics that influence participatory processes. The participatory process research is informed by a theoretical framework called Enhanced Adaptive Structuration Theory 2 (Jankowski and Nyerges, 2001a). The framework incorporates the concept of different meeting arrangements that can be supported by different networked technologies. Four types of meeting arrangements stem from combinations of place and time – face-to-face conventional meeting (same place/same time), asynchronous storyboard meeting (same place/different time), conference call meeting (different place/same time), and asynchronous distributed meeting (different place/different time). Each of the meetings has advantages and disadvantages associated with it (Jankowski and Nyerges, 2001a). Access issues exist both to the meeting as well as within the meeting. Getting to a face-to-face meeting takes time. The more time the worse the access. In contrast, an asynchronous-distributed meeting is a virtual setting. In virtual meetings large numbers of people can gain access to meetings because of their asynchronous (time is at the discretion of the attendee) and the distributed (place restrictions are relaxed). Indeed, a digital divide still exists, but improving participatory access is a worthwhile endeavor. Improving access to traditional meetings in a virtual manner is done now through television and videotape. The real challenge involves enhancing participatory processes within meetings to make them a worthwhile endeavor for the public to participate.

21.3 Rethinking public participation as access within public meetings

Face-to-face public meetings support an interaction dynamic common to most communication settings. People talk and people listen. However, highly intensive interaction wherein many people contribute in a short time is difficult to manage in conventional meetings without facilitation support. Facilitated processes enable participants in a meeting to contribute in an equitable manner and scaffold participants from an information exchange to knowledge building.

Facilitated interaction in asynchronous-distributed meetings is difficult to implement due to the challenge with meaning-making. Meaning-making is the assemblage of a shared meaning among participants. In face-to-face settings, rapid and visible (i.e. the whole group is exposed at the same time to) turn-taking in conversation assists meaning-making. Facilitated processes enhance the potential for meaning-making. In an asynchronous on-line setting, response latency, that is the lag in a turn to be taken, reduces the opportunity to create shared meaning within any given timeframe. Although meaning-making might seem like it takes longer by any single individual in an asynchronous-distributed setting, undertaking meaning-making in parallel makes it possible to scale out creation of shared meaning. This parallelism promotes the inclusion of potentially very large numbers of people gaining access to asynchronous-distributed meetings. In the material to follow, we *rethink* public participation processes by first unpacking and then reconstructing participatory processes at multiple levels of process granularity.

A spectrum of public participation is composed of five levels - inform, consult, involve, collaborate, and empower (International Association of Public Participation – IAP2- 2005). Each next level increases the public impact on the overall process (See Table 21.1, right-most column). At the *inform* level is the "decide, inform, defend" process that many local, state, and federal agencies follow in public activity, e.g. a web site announcing a decision. At the *consult* level, agencies

Table 21.1 Participation Spectrum – Levels, Activities, and Impacts

Participation levels	Participation activities	Public impacts on overall process
Inform	Listen	Public is informed
Consult	Listen, respond	Public is informed and provides feedback
Involve	Listen, respond, negotiate, recommend	Public concerns are incorporated
Collaborate	Listen, respond, negotiate, recommend, analyze	Public helps form concerns and solutions
Empower	Listen, respond, negotiate, recommend, analyze, decide	Public helps decide concerns and solutions

ask the public for input, consider and respond to that input, but are not required to act on that input. This second level is common in traditional public meetings. At the level of *involve*, groups provide input and gain feedback along the way, i.e. public ideas are incorporated. At the level of *collaborate*, public groups work along with agency personnel as peers, but the agency makes a final decision. At the level of *empower*, a decision made by a participating public is adopted by an agency as the final decision in the process.

In the original IAP2 spectrum, although participation impacts are due to the publics' access to the process, participant (i.e., public and agency personnel) impacts are gauged in terms of *outcomes* (3rd column of Table 21.1). This begs a question, what are the impacts the public would have during activities at each step within a process? We address that question in terms of the actions/events that commonly happen during a meeting process – public or not. Jankowski and Nyerges (2001b) have unpacked many decision meetings using HCHI coding systems (Nyerges et al., 1998b). We list five activities involved in participatory decision processes – listen, respond, analyze, recommend, and decide (2nd column of Table 21.1). Each of the activities requires a different kind of participation process support. The differences from one participation level to the next have not been characterized systematically to the co-authors' knowledge. A detailed characterization of the activities, particularly at the higher levels, provides insight for Internet portal software requirements.

Collaborative and empowered participation are the levels most characteristic of deliberative democratic processes (Cunningham, 2002). Deconstructing and reconstructing collaborative and empowered participatory processes in terms of the granularity of phases and activities provides insight into software system requirements. Such phases and activities are the fundamental events in deliberative democratic decision making. Although deliberative democracy has been a topic of interest on and off for the past 100 years or so (Gastil and Levine, 2005), it is the more recent research involving the practice of deliberative democracy in policy decision making since the late 1980's and into the turn of the millennium that has caught the attention of many in research and practice. Hundreds of deliberative democracy meetings of all sizes have been conducted across the world over the past decade or so; reports on many case studies are published in the *Handbook of Deliberative Democracy* (Gastil and Levine, 2005). Several of the chapters in the *Handbook* describe public deliberations in which GIS could be useful because the topics deal with location-oriented issues, concerns, plans, interests, and values. However, none of the chapters actually refer to GIS, not even just the mapping capability, let alone the analytic capability. That separation indicates an opportunity to bridge a gap between GIS analytic-oriented processes and deliberative processes within the deliberative democracy literature.

Research about analytic-deliberative decision processes has shown that meaningful public participation is possible and decision outcomes are improved (National Research Council, 1996). The analytic component provides technical information that ensures broad-based, competent perspectives are treated. The deliberative component provides an opportunity to interactively give voice to choices about values, alternatives, and recommendations using natural language. GIS has been a big part of developing the technical information in analytic-deliberative processes (National Research Council, 1996). No language provides a perfect reflection of ones thoughts and interests, and thus a combination of analytic and deliberative languages foster balance of contribution and perspectives. Unfortunately, such analytic-deliberative participatory processes have been expensive and time consuming, and thus involved small to medium-sized groups, e.g. no larger than 10–15 people at a time during deliberation. Working out analytic-deliberative participation in small to medium groups is a start, but scaling analytic-deliberative participation to large groups is a challenge.

To address scaling out the process to large groups, we must (re)construct democratic processes as equitable parallel and sequential engagement in dialog. As a basis for such reconstruction, we turned to the deliberative democracy literature, plus the literature that has been critical of nuances in analytic-deliberative processes (Fischer, 2000), and linked that with literature about structured participation methods. From that literature about analytic-deliberative meetings we synthesized a set of principles for fostering analytic-deliberative decision processes and composed a set of system design goals (Nyerges, Ramsey, Wilson, 2006). While those principles are useful for guiding our design process in the broad sense, they do not provide details for organizing analytic-deliberative processes structured as HCHI. Fortunately, structured participation methods were invented in organizational planning and management to inform people what they will get out of the participation should they commit the energy. Although it is often thought that face-to-face structured participatory settings are superior to distributed-structured participatory settings, Dowling and St. Louis (2000) have shown that an asynchronous nominal group process was more effective than a face-to-face nominal group process, at least in a small group setting.

Based in part on that motivation, we have identified four popular structured participation methods as potential alternatives to conventional, unstructured group processes. These include nominal group technique, Delphi process, technology of participation, and citizen panel/jury. The research issue here is whether any or all of these processes *scale-out* to large group settings and can they support analysis? The four methods were chosen for this comparison study because each offers a set of different steps to address similar consensus-driven outcomes (see Table 21.2). Nominal group technique is a structured form of brainstorming or brain-writing followed by a vote or prioritization in a synchronous setting with up to 10 participants and an experienced facilitator (Mycoted, 2003). Delphi process, developed in 1950 by the Rand Corporation, is a series of questionnaires, each building on the responses of previous questionnaires, which require a panel of participants to synthesize and determine the nature of follow-up questionnaires (Delbecq et al., 1975). Technology of Participation (ToP) is a participatory process wherein a goal is defined, followed by brainstorming, ordering, labeling, and prioritizing ideas through facilitated, group discussion (Spencer,

Table 21.2 Steps of Four Structured Participation Methods

Nominal group technique (NGT)	Delphi process (DP)	Citizen panel/citizen jury (CPJ)	Technology of participation (ToP)
1. Goal statement	1. Goal statement	1. Listen to evidence	1. Goal statement
2. Brainstorm ideas	2. Generate ideas	2. Discuss evidence	2. Generate ideas
3. Clarify/negotiate ideas	3. Collect ideas	3. Negotiate positions	3. Collect ideas
4. Vote on idea priority	4. Synthesize ideas	4. Vote	4. Cluster ideas
	5. Playback ideas	5. Repeat until reach consensus	5. Synthesize ideas
	6. Request for further change		6. Label ideas
			7. Negotiate idea priority

1989). Citizen panel/jury is a process wherein participants are informed of an issue and encouraged to deliberate through formal presentation of evidence, argument, and rebuttal to offer a solution to a governing body (Jefferson Center, 2004). The comparison of these four structured participation methods illustrate that, while differing in procedure, each is steeped in the principles of deliberative democracy (Wilson, 2005).

To make it easier to implement the four methods, and perhaps discover a new one, we unpacked and synthesized the procedures listed in Table 21.2 into a set of distinct activities presented in Table 21.3 (Wilson, 2005). We call each of the steps a *participatory game* (p-game), forming an important level of conceptualizing an analytic-deliberative process design. In Table 21.3 we present a core set of p-games that may be used to form analytic-deliberative processes. On the left side of Table 21.3 are four columns listing the methods of Table 21.2 and in the cells of the table an "X" indicates which p-games are part of which methods.

Although we recognize that p-games are based on individual steps to carry out work, unpacking those steps further is necessary to create a system functional requirement. In an asynchronous-distributed context we must be explicit about the kinds of conversational exchanges people can undertake, particularly as related to interaction process granularity. The granularity of HCHI process units (i.e., exchanges between humans and humans as well as humans and computers) within each of the p-games dictates a stimulus and response latency. The finer level of granularity below p-game we call participatory acts (P-acts). P-acts are the basic HCHI activities within analytic-deliberative work. P-acts are formed of analytic data acts and (deliberative) speech acts. Early work about speech acts in conversations was undertaken by Austin (1962). Searle (1979) extended that work to develop various structures and rules in two person exchanges called dialog acts and formed Speech Act theory. Chang and Woo used speech act theory to describe negotiation (Chang and Woo, 1994). Mann (2002) extended Speech Act Theory

Table 21.3 Participatory-games that comprise the normative concerns and values organization process in the portal design

NGT	DP	CPJ	ToP	Participatory game
X	X		X	Goal statement (context setting)
X	X		X	Brainstorm items (uncloaked or cloaked to group)
X		X	X	Negotiate (clarify) concern items
	X		X	Synthesize cluster and label cluster
			X	Refine concern clusters
X		X	X	Vote/poll
	X			Survey
	X	X		Review/evaluation
These p-techniques support value structure creation				Threaded discussion Develop concerns hierarchy Convert concerns hierarchy into value hierarchy

to Dialog Macrogame Theory to analyze natural language dialogs. We are using the findings from the speech act literatures to take foundational next steps for making-meaning in a participatory process.

21.4 Conclusions and next steps for modeling participation and meaning-making

Among the remaining issues is establishing "rules of engagement" for each of the process levels described above. Those rules are the basis of coordinating interaction workflow at each of the levels, first in an independent manner and then in a collective manner. The core of the challenge is formalizing the transition between multiple speech act inputs and the synthesis that can result. That characterization is part of meaning-making. Meaning-making is a challenge occurring at all analytic-deliberative process levels. However, meaning-making at the p-act level supports all other participatory levels. Progress at that level has been a challenge facing cognitive linguistic, computational linguistics, and other disciplines that deal with natural language processing (Aravena-Reyes, 2000). When the entries are free-form text and there is a broad diversity of constituencies, our best chance of addressing this problem is at the level of *phrase*. Thus, meaning-making through phrase matching 'along chains of P-acts', i.e. the context of a speech act, is the principal challenge for scaling meaning-making in public participation. Domain vocabularies, e.g., as in the transportation domain of the OpenCyc ontology, are mapped into a knowledge base of common sense (Reed and Lenat, 2002). On a small scale, such synthesis of meaning has been examined in cognitive linguistics and computational linguistics, e.g., by Searle and VanDerverken (1985).

In the PGIST Project we are taking a two-pronged approach to address phrase matching. One approach is to present phrases to facilitators and enable them with tools to create *meaningful matches*. That approach implements the meaning synthesis of a Delphi technique as structured participation; that is, a group of technical specialists commonly perform the synthesis. The second approach is to present phrases to participant-subjects and enable them with tools to create meaningful matches. This latter approach implements the meaning-making in the Technology of Participation, Nominal Group process, and Citizen Jury, requiring more sophisticated natural language processing techniques to support the tools for meaning matching. Both approaches are currently under development. The first is a lower-risk development, as people are expected to undertake much of the processing. The second is a higher-risk development as more computational work is necessary. As of the writing of this chapter we are only a portion of the way there. Future results will be available at www.pgist.org.

Although information technology is not the only answer to enhancing public participation in public-oriented decision making, developing Internet portal technology that embraces principles of analytic-deliberative democracy might be one way to foster meaningful participation in large groups as well as hold down the cost to all who wish to participate. Hopefully, the ideas presented in this chapter can help others in some aspect of understanding instant access to virtual meetings and the portal technology that is needed to support such meetings in order to increase chances of building stronger publics for enhancing democratic governance.

Acknowledgements

This research has been supported in part by National Science Foundation Grant No. EIA 0325916, funded through the Information Technology Research Program, and managed in the Digital Government Program. The authors are solely responsible for content. We would like to thank members of the PGIST Project team for their willingness to share insights about these topics.

References

Aitken, S. (2002) Public participation, technologic discourses and the scale of GIS, in W. Craig, T. Harris, and D. Weiner, *Community Participation and Geographic Information Systems*, London, Taylor & Francis, 357–366.

Arnstein, S. (1969) A ladder of citizen participation, *American Institute of Planners Journal*, July 1969, 216–224.

Aravena-Reyes, J. A. (2000) Speech-Act Based Protocols for Computer Mediated Collaborative Design, Vol 3, 2000–2001, http://www.arch.usyd.edu.au/kcdc/journal/vol3/dcnet/avarena/ last accessed December 14, 2005.

Austin, J. (1962) *How To Do Things with Words*. Cambridge, Harvard University Press.

Bosworth, M., J. Donovan, P. Couey (2002) Portland Metro's Dream for Public Involvement, in W. Craig, T. Harris, and D. Weiner (eds) *Community Participation and Geographic Information Systems*, Taylor and Francis, pp. 125–136.

Chang, M. and C. Woo (1994) A Speech-Act-Based Negotiation Protocol: Design, Implementation and Test Use, *ACM Trans. on Information Systems*, 12(4):360–382.

Craig, W., T. Harris, and D. Weiner. (Eds.) (2002) *Community Participation and GIS*, London, Taylor and Francis.

Cunningham, F. (2002) *Theories of Democracy: A Critical Introduction*. London; New York: Routledge.

Delbecq, A. L., Van de Ven, A. H., & Gustafson, D. H. (1975) *Group Techniques for Program Planning: A Guide to Nominal Group and Delphi Processes*. Glenview, Ill.: Scott Foresman.

DeSanctis, G., M. S. Poole (1994) Capturing the complexity in advanced technology use: Adaptive structuration theory. *Organization Science, 5*(2), 121–147.

Dowling, K. L., and R. D. St. Louis (2000) Asynchronous implementation of the nominal group technique: Is it effective? *Decision Support Systems, 29*(3), 229–248.

Fischer, F. (2000) *Citizens, Experts, and the Environment: The Politics of Local Knowledge*. Durham, NC: Duke University Press.

Gastil, J. and T. Kelshaw (2000) *Public Meetings: A Sampler of Deliberative Forums that Bring Office-holders and Citizens Together*, Dayton Ohio: Charles Kettering Foundation. Retrieved September 19, 2005 at http://faculty.washington.edu/jgastil/pdfs/Kettering - Public Meetings.pdf

Gastil, J and P. Levine 2005 (Eds.). *The deliberative democracy handbook: Strategies for effective civic engagement in the twenty-first century*. San Francisco, CA: Jossey Bass.

International Association of Public Participation (2005) Public Participation Spectrum, http://www.iap2.org/associations/4748/files/spectrum.pdf, accessed August 10, 2005.

Jankowski, P., and T. L. Nyerges (2001a) *Geographic Information Systems for Group Decision Making: Towards a Participatory, Geographic Information Science*. London: Taylor and Francis.

Jankowski, P., and T. L. Nyerges (2001b) GIS-supported collaborative decision making: Results of an experiment. *Annals of the Association of American Geographers, 91*(1), 48–70.

Jefferson Center (2004) *Citizens Jury*. http://www.jeffersoncenter.org/citizens_jury.htm accessed January 29, 2004.

Mann, W. C., (2002) Dialogue Macrogame Theory, presented at the SIGdial Workshop in conjunction with ACL2002.

Mycoted (2003) Creativity techniques. http://www.mycoted.com/creativity/techniques/index.php accessed August 31, 2005.

MSNBC (2005) Bush signs $286.4 billion highway bill, http://www.msnbc.msn.com/id/8894520/, last accessed August 21, 2005.

National Research Council (1996) *Understanding Risk: Informing Decisions in a Democratic Society*. Washington, D.C.: National Academy Press.

Nyerges, T., R. Montejano. C. Oshiro, and M. Dadswell (1998a) Group-based Geographic Information Systems for Transportation Site Selection, *Transportation Research C: Emerging Technologies*, 5(6):349–369.

Nyerges, T., T. J. Moore, R. Montejano, and M. Compton (1998b) Interaction coding systems for studying the use of groupware, *Journal of Human-Computer Interaction*, 13(2):127–165.

Nyerges, T., K. Ramsey, and M. Wilson (2006) Design Considerations for an Internet Portal to Support Public Participation in Transportation Improvement Decision Making, in S. Balram and S. Dragicevic (Eds.), *Collaborative Geographic Information Systems*. Idea Group Inc., Hershey PA, pp. 207–235.

Obermeyer, N. J. (1998) The Evolution of public participation GIS. *Cartography and Geographic Information Systems*, 25(2):65–66.

Reed, S. and D. Lenat (2002) Mapping Ontologies into Cyc. American Advancement of Artificial Intelligence. http://www.cyc.com/doc/white_papers/mapping-ontologies-into-cyc_v31.pdf, last access January 23, 2006.

Renn, O., T. Webler, and P. Wiedemann (1995) *Fairness and Competence in Citizen Participation: Evaluating Models for Environmental Discourse*, Kluwer Academic Publishers: Dordrecht.

Rinner, C. (2003) Web-based spatial decision support: status and research directions, *Journal of Geographic Information and Decision Analysis, Vol. 7, No. 1, pp. 14–31.*

Schlossberg, M. and E. Shuford (2005) Delineating "public" and "participation" in PPGIS, *URISA Journal*, 16(2):15–26.

Searle, J. and D. VanDerverken (1985) *Foundations of the Illocutionary Logic*. Cambridge, England: Cambridge University Press.

Smith, S. (1999) *Guidebook for Transportation Corridor Studies: A Process of Effective Decision-Making*, National Cooperative Highway Research Program, no. 435, Transportation Research Board, Washington, D. C.

Spencer, L. J. (1989) *Winning through participation: Meeting the challenge of corporate change with the technology of participation*. Dubuque, IA: Kendall Hunt.

Sui, D. Z. (2000) The E-merging geography of the information society: From accessibility to adaptability in (ed.) D. G. Janelle and D. C. Hodge *Information, Place, and Cyberspace*. Berlin: Springer, 107–129.

Taylor, N. (1998) Mistaken Interests and the Discourse Model of Planning. *Journal of the American Planning Association*, Winter 1998, 64(1):64–75.

Wilson, M. W. (2005) *Implications for a public participation geographic information science: Analyzing trends in research and practice*. Unpublished Master's, University of Washington, Department of Geography, Seattle, WA.

22 Digital middletown: a glimpse at the information society

H. O'Neal Smitherman
Ball State University, Muncie, IN

22.1 Introduction

In the early 1920's, Robert and Helen Lynd began their landmark sociological studies dubbed the Middletown studies (Lynd and Lynd, 1929, 1937). They were cognizant of the dramatic changes in culture, family relationships, values and demographics associated with the migration of our society and our economy away from an agricultural society to an industrial society. Their investigations and those of the scientists who continued their work represent more than 70 years of research on the changes in "middle America" associated with geographic, economic and cultural changes.

Only recently has the location of the city under study been identified: it is the Muncie, Indiana, USA area. This area not only represented the geographical middle of the country, but it also represents the cultural middle of America. Finally, it represented the leading edge of industry and manufacturing technology nationally and globally: it showed the rest of the world what they would experience in the near future.

With this history and with the movement of our society toward an information society, it is appropriate that Muncie continue to serve as the harbinger of the effects of knowledge technology on the rest of the world. In that vein, a continuing study and the introduction of new information technologies to Muncie have been dubbed "Digital Middletown." To study the future effects of new technologies, it is necessary to deploy technologies that are in advance of current commodity options available today. These include network capacities in excess of those currently available to the homes of the general population and media that represent new styles and interactive capabilities that are not part of the current media repertoire available to the public.

Ball State University was awarded a U.S. Department of Education grant to comprehensively test the deployment and the utility of learning and home entertainment using high-production, interactive media delivered over long distance high-capacity wireless technology. The *Digital Middletown Project* (DMP) connected

H.J. Miller (ed.), Societies and Cities in the Age of Instant Access, 343–356.

three Muncie, Indiana community elementary schools and a surrounding neighborhood to Ball State University's data network via Proxim Tsunami at 30 and 45 Mbps full duplex and Alvarion BreezeAccess at 24 and 36 Mbps full duplex.

A team of educational and technical experts prepared the schools and neighborhood, evaluated the impact of the infrastructure development and deployment, and the processes associated with the rich media content preparation and presentation. In addition, videoconferencing and other support services were implemented. The participants' in this study received 64 laptops, 14 desktops, 4 plasma screens, 6 projectors, and 3 Polycom FX (video conferencing units). The technology partners in this study included Gateway, Proxim, Alvarion, MIT, Movielink, HNIT-BALTIC, DigitalBridge Communications, and Discovery Company.

The DMP examined selected components of the economic, educational, and social costs and benefits of deploying a high-capacity broadband wireless network in four ways. First, by evaluating the development of needed infrastructure/equipment; second, by evaluating the impact on education and entertainment when deploying the network and media in school and home activities, next, by documenting the real-world experience through developing partnerships within the technology sectors and content providers; and last, by studying the process of preparing teachers, students, and homes for the delivery and use of rich media.

The DMP generated many important and necessary experiences as well as knowledge and skills associated with preparing and deploying this technology to the schools, classrooms, and the neighborhood. The benefits gained from this study included exploring potentials of high bandwidth wireless infrastructure, the real-world issues surrounding the deploying of rich media content over a wireless network, examining issues related to evaluating subjects and network performance, providing videoconferencing, and developing needed support and services for training and problem solving were also addressed. The experiences gained from this project and the findings can be very valuable for the development of a knowledge base for similar projects. The gained experiences, accomplishments, and outcomes helped Ball State University to receive additional U. S. Department of Education grants to develop additional partnerships and to continue DMP for more studies and investigations.

The challenge regarding media is in either finding or producing media that represents advancements over the current "broadcast" media. These consist of media designed for personal interactivity and especially personal control. In pursuit of these media, we formed partnerships with Annenberg/CPB, Ball State University's Electronic Field Trips, Discovery Channel's United Streaming, and Movielink. We have been further helped with donations of equipment and technical assistance from Gateway Computers, Cellular Expert radio mapping software, Digital Bridge, and Alvarion Corporation. Primary funding for the project came from the United States Department of Education.

The implementation of an advanced network and testing of the impact of media help to form a rich test bed for next wave television, movies, education and information aggregation (newspapers, books, magazines, encyclopedias, dictionaries, etc.).

Still in its earliest phase, the current report provides information on how the project can be deployed and initial information on the impact of the new technologies and media. The ultimate goals for DMP studies are to demonstrate the infrastructure, develop content, deliver content and evaluate the effects.

In two years, the project not only accomplished its core objectives with research studies, the development of experts in this field, attracting leading telecom companies, publications, and presentations, but has also reached the goal of developing an innovative business division to satisfy the leading telecom companies' needs in testing new long distance broadband wireless technologies, RF distributions, mapping, and deployment across the United States. The newly developed business branch known as the "Office of Wireless Research and Mapping" (OWRM) is a spin-off business of the Digital Middletown Project. The main business of the office is to test new wireless technologies for partners/clients, to involve teachers and students in activities for learning, contributions to business activities, and to produce revenue streams using GIS and Cellular Expert software to produce detailed RF distribution maps in potential deployment areas, and testing new wireless technologies for clients.

22.2 The network

The construction of a high capacity network required numerous decisions and much of the effort of the project. Many options, including the use of the existing cable or DSL networks, deployment of new fiber or copper networks and deployment of a wireless network, were explored. The existing networks were unable to provide the necessary capacity (30 megabits per second to the home and school) and the deployment of fiber or copper networks would have been extremely expensive and time consuming. Wireless networks represented much less cost (especially with the help of our wireless equipment partner Alvarion Corporation) and could be implemented within just a few weeks.

Since wireless networking is still evolving, we deployed two types of core networks and a hybrid distribution system that included a 5.8 gigahertz transceiver system, a 900 megahertz transceiver system through the tree canopy and redeployment under the canopy to the houses of WiFi (2.4 gigahertz) networking.

The network provided connectivity from the Ball State University campus network to three local schools (Mitchell Elementary, Yorktown Elementary and Middle schools, and Cowan Elementary) and the community surrounding one of the schools (Halteman Village, a community of approximately 500 houses). Figure 22.1 illustrates the topology of the Digital Middletown Project network.

Mitchell Elementary is connected with a 30Mbps full-duplex wireless backbone to Ball State University's data network. Within Mitchell Elementary, data equipment was installed to provide 10/100 Mbps wired data connections. Currently, there are two desktop computers and one Polycom videoconferencing unit on the high-speed wired data network. To supplement the wired connections, five "access

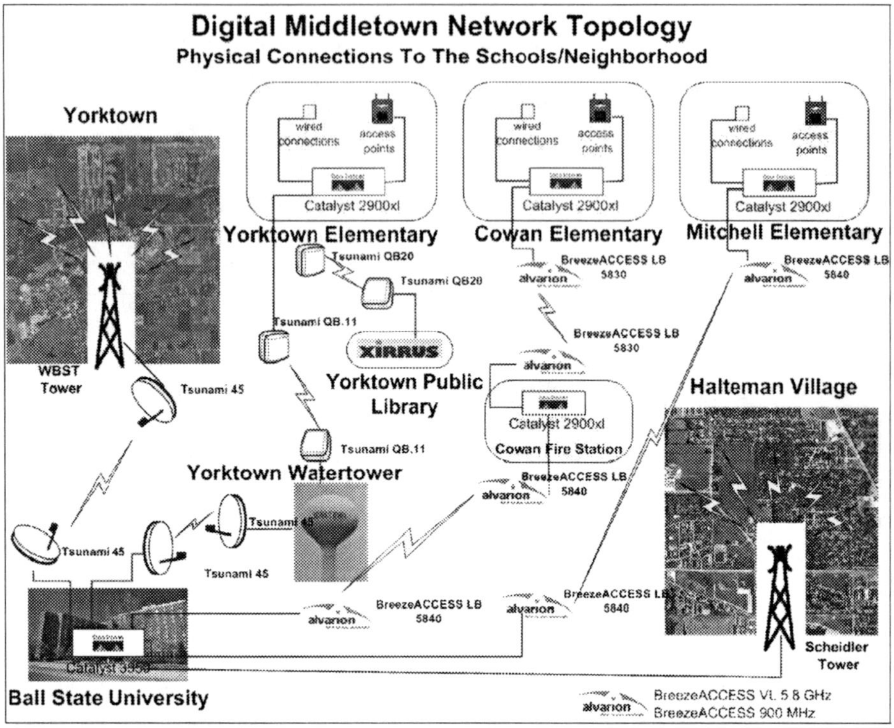

Fig. 22.1 Digital Middletown Project network

points" were installed throughout the school. Each access point is configured with 802.11g radios. These access points provide all wirelessly connected devices the opportunity to connect at a shared speed of 54Mbps. They are installed throughout the building providing the entire school with wireless coverage.

Cowan Elementary is connected with a 36Mbps full-duplex wireless backbone connection to Ball State University's data network. To accomplish this, it is necessary to have two wireless links acting as a repeater to provide wireless access to Cowan elementary. The first link connects Ball State University to the Cowan Fire Department. For potential growth of the network, the wireless link from Ball State to the Cowan Fire Department is a 72Mbps link. Within the fire department, a network switch has been installed and provides 10/100 Mbps data connections for the ire Department's use. From the Cowan Fire Department, another wireless point-to-point link connects to Cowan Elementary. The second wireless point-to-point connection is from the Fire Department to Cowan Elementary and is a 36Mbps link. Within Cowan Elementary, data equipment has been installed to provide 10/100 Mbps wired data connections. Currently, there is a Polycom videoconferencing unit on the high-speed wired data network. To supplement these wired connections, eight access points have been installed in the school. Each access point is configured with 802.11g radios. These provide all wireless connected devices the opportunity to connect at a shared speed of 54Mbps. They are installed throughout the building providing the entire school with wireless coverage.

Yorktown Elementary is connected with a 45Mbps full-duplex wireless backbone connection to Ball State University's data network. There are two wireless links used to perform this. The first link connects Ball State to a water tower in Yorktown. Then from the water tower another wireless link goes to Yorktown High School. Yorktown High School provided a suitable site for the second wireless link to allow us access to an existing fiber plant that goes through out the entire Mount Pleasant school corporation. From Yorktown High School the fiber optic plant is used to connect the Elementary and Middle schools to the Digital Middletown network and Ball State University. Yorktown Middle School has fourteen access points installed providing them with wireless access for the entire building. Yorktown Elementary School has eleven access points installed through out their build providing wireless access. In addition to the wireless access there are also designated rooms that also have wired ports to provide them access to the Digital Middletown Network. Being in close proximity to the Yorktown Public Library it was possible to offer them access to the Digital Middletown Network as well. This is done by installing a point to point wireless connection between Yorktown Middle School and Yorktown Public Library. Inside the Yorktown Public Library a Wireless LAN Array was installed. A wireless LAN array is a device that holds multiple access points in a single unit. The Yorktown Public Library is now a public hotspot with end-users having the ability to connect wirelessly by using 802.11 a/b/g technologies.

The neighborhood is connected with Alvarion's 5.8 GHZ and 900 MHZ high bandwidth wireless technologies. DMP provided selected homes with high end computer hardware and software for access to rich digital media files, and provided one home with a high definition (42 inches LCD) display to evaluate the impact of the high definition media files in education and entertainment. The deployment of digital content to the neighborhood includes regular and high definition family movies, family games, and high definition family titles/modules from PBS and Discovery, as well as access to all school materials.

22.3
Content and activities

Following deployment and testing of a high bandwidth network, a range of content and activities were made available to schools and home through a Web interface. Some of the activities have included:

- **Local and national videoconferencing** – Video conferencing classroom activities were conducted between schools in the Digital Middletown project and included other schools around the country.

- **Electronic Field Trip Interactive Modules and live Broadcasts** – The interactive field trips used in the Digital Middletown project included live and archived content on areas like the Dinosaurs of Chicago's Field Museum, the issues of flight at the Academy of Model Aeronautics, Grand Canyon, NASA's vomit comet, the inadvertent invasion of biological intruders to the San Francisco Bay area, the tree canopy of SERC's Forest Ecology Lab, the National Museum of the American Indian, the Mars excursion and others.
- **Annenberg titles in online media files** – These include the thousands of modules of video content available through the Annenberg Media site, Learner.org.
- **United Streaming titles in regular and HD media files** – Our partnership with United Streaming and Discovery Channel included the thousands of standard definition education modules as well as modules that United Streaming provided for conversion to High Definition quality. The high definition modules were used for testing the impact of higher resolution media on the delivery of educational content for learning.
- **Movielink movies for families** – Movielink represents movie production and distribution companies and provides downloadable movie titles over the Internet. They also work to develop secure techniques for downloading as well as conduct research in conjunction with Ball State University and others, on how movie production and distribution are evolving with the development on new technologies. The Movielink content included video on demand movies that were made available to community test homes.
- **Case Studies, simulations, and educational games** – This educational content included games and simulations that were developed for the project in partnership with organizations like Massachusetts Institute of Technology's Educational Arcade as well as modules available commercially.

22.4 The effect of new and rich media

The new media were presented to students in the classroom and as independent study sessions in a variety of formats. These included sessions in which the students viewed and discussed material as a class as well as sessions in which students worked individually as well as in pairs on laptop computers. Then the media and the distribution/deployment methods were compared and evaluated. These comparisons provide a wide range of lessons learned and a great deal of information on the impact of new media on education. Some of the effects of the media include:

- **More powerful student engagement** – Anecdotal data, systematic observation and systematic evaluation confirm that the students are more interested in the content and learning when rich media are used in the learning process.

- **Productive new tools for teachers** – Electronic Field Trips (like the interactive modules from the Mars Field Trip where students are allowed to "drive" the simulated rovers) and United Streaming content offer the advantage of learning modules that provide motivation and clearly map the student activities to state and federal standards.
- **Additional motivation for students and teachers** – The motivation for teachers is well represented by the statement of a principal who said that her teachers were getting up early because they were newly excited about their lessons and that other teachers clamor for the same opportunities. Indeed, the whole school district has since bought United Streaming for all schools.
- **Homework and parental involvement** – New media distributed over the network allows students to continue school work at home and involve parents by reviewing and interacting with the content at home with their parents.
- **High-resolution graphics** – In some content areas, higher resolution display of the material plays major role in teaching and learning. Higher resolution of rich media such 720p or 1080i can contribute positively to learning in more depth and help further in a better understanding of subject matter.

22.5 Lessons learned

As the capabilities of information technology increases, we clearly recognize the need for teachers to be taught how to use the technology and we must provide a clear path between the educational media and the required state and federal educational standards.

22.5.1 Future of education and visual information

A major impetus for the project came from the university-wide effort to engage in the production and evaluation of new media. These efforts are supported by grants from the Lily Endowment. These efforts are built upon the understanding that education like many other venues of communication is advanced by language and that language is a symbolic representation of reality. While language has great benefits, it represents a relatively slow process for communicating. Conversely visual information conveys facts and relationships more rapidly. While language communicates approximately 200 words per minute, the human eye processes approximately 150 fields (or pictures) of information per second. Since a VHS tape shows 30 frames per second, the human eye appears capable of processing a very large amount of information.

Learning starts when information is processed in separate channels for different sensory modalities (Mayer & Moreno, 2003). Further, learners construct visual models that they use to represent reality. While this process can work very effectively, learners frequently build visual models that are incomplete or inaccurate.

A general question considered in the study is whether new media can produce learning at a higher level using visual information. Additional questions in the teaching learning process with regard to visual information include: Does visual information facilitate learning? What characteristics of visual information facilitate learning? How important in the teaching process is formalizing the connection between media assets and state teaching standards? What types of educational content is best suited to new media presentation?

All of these questions continue to be addressed as we consider how personal media differ from broadcast media and whether there is a way to repurpose existing content to make it more useful in an educational context.

An additional question may be why we have not pursued this process in the past. First, we did not have the technology until today. Today a $2000 computer can do what required a $2 million production studio and 3 or 4 producers and engineers in the past. Secondly, we have not had the infrastructure to deliver interactive media until the present day. Older, analog broadcasting systems are unidirectional. Information or content could be sent but interaction commands could not be received. Therefore, it could not respond to directions or instructions or control from the user. With the internet, however, we can send and receive instructions. Until recently the internet has provided only the lowest level of access. Even DSL or cable modem speeds of 1 megabit per second, do not provide the capacity to deliver acceptable quality video and animation to the user. In order to receive television quality video, the network must be able to sustain a minimum of 5 megabits per second, and that just matches the old television standard. High definition video requires a minimum of 10 megabits per second.

For these reasons, the Digital Middletown project set a minimum broadband capacity of 30 megabits per second. A network of this capacity can provide multiple streams in the home of high quality interactive media. Indeed, this level of capacity was required in the classroom where we studied as many as 40 students simultaneously accessing independent video streams. A network with less capacity would fail as multiple students signed on to control the multiple independent streams of information.

22.5.2 Content provision

Multiple types of content were employed in this project: Electronic Field Trips, videos from the Discovery Channel, videos and other material from Public Television Content, videos from United Streaming, Gaming developed in cooperation with Massachusetts Institute of Technology's Educational Arcade, and interactive video conferences with other schools, museums and organizations.

In testing, the Electronic Field Trips (EFTs) and United Streaming received the most positive adoption. EFTs provide interactive material, games, simulations, etc. as well as rich media and vast amounts of supplementary materials, both media and text. Further the live, interactive program provides an organized flow of information along with opportunities for students to call-in and email-in questions. This provides interactivity even during the live broadcast that helped to shape the direction of the discussion. Finally, archiving these live sessions provide a permanent resource to the classroom. Ultimately, mapping each activity to learning objectives and standards proved a key element for teachers.

United Streaming, a subsidiary of the Discovery Channel, consist of more than 40,000 modules of videos and animations, which are clearly linked to learning objectives and standards. While their current streams are about 700 kilobits per second, and lower resolution, Discovery sent us original tapes which we encoded at a higher resolution to evaluate the effect of better quality images. The United Streaming modules made major contributions in learning and provided students the ability to learn not only by listening, but also by seeing and participating in activities. Research results have indicated that although teachers may know a lot about topics and know different methods for transferring and representing such knowledge, many teachers fail to implement that knowledge in the classroom. Utilizing educational software may assist teachers in transferring and representing knowledge about complex topics. (Magnusson, Borko and Krajcik, 1994).

Public Broadcast content on science and other education has also been encoded in native high definition content and modified so that it could be delivered in more easily consumable modules. This appears to be important as we learn that the current generation of learners has come to expect large amounts of information in 15 to 20 minute bursts followed by time for integration and then followed by another burst of content. By restructuring the content, media designed for a longer continuous stream can be repurposed as more interactive and more accessible.

In conjunction with the Educational Arcade at the Massachusetts Institute of Technology, this project developed an interactive simulation game to evaluate the student reception to immersive gaming alternatives to traditional learning. The game entitled "Revolutions," provides an interactive world of 1700's America (built on the Evernight engine), in which students take the role of merchant, leader, farmer, slave, etc and "play" with other students in various roles. From this game we learned: i) Educational video games can improve engagement and motivation in learning. Students reported a desire to continue in the game even when the scheduled time had expired. ii) While girls generally like video games less than boys, video games like Revolutions which emphasize interpersonal interaction, produce more favorable reactions from girls. Indeed, even though girls' pretest indicated little liking for video games, their post test results showed that girls liked the game at the same level as boys. iii) Students with certain learning difficulties were much more engaged in learning through the video game than in traditional learning situations. iv) Students learned as much specific information from the game as from other methods. v) Students preferred to assume the role of slaves over the more traditional roles of shop keeper or farmer. They reported that it taught them more about a life for which they had little knowledge. Barnette (1995), in his research stated that the electronic media can provide important opportunities for those whose personal situation marries with the occasion.

22.5.3 In the home

An important question to be considered in this project has been the importance of having high capacity access to educational and entertainment new media in the home as well as in the school. Regarding educational material, home access provides the opportunity to extend learning from the school to the home. Material covered in the classroom, can be reviewed and refreshed at home. Indeed, many of the students requested that they be allowed to access the material at home, before they were told it was possible and instructed on the process. Further, students and parents report favorable experiences that evolved when the students engaged their parents with the classroom material. They even reported exploring the resources and accessing media content that was not part of the curriculum. They enthusiastically described learning material that they would have never been aware of without access to the network content.

In addition to the educational content, families were given access to popular movies and advanced technologies as part of the project. Plasma screens and high speed computers connected to the high capacity network provided capabilities just ahead of commodity access today. Families were then interviewed by anthropology students and faculty to determine the effect of the new technologies on family life.

The interviews showed that while all families embraced the technologies and access to the content, some were more technologically proficient than others. Those families with less technological capabilities took less advantage of the material until they were shown the capabilities and instructed on the use of the technology. The lesson learned is that while the results of the technologies are desirable to most people, the equipment currently requires training and/or familiarity with the techniques before families can move forward with their usage.

22.5.4 Network lessons

Deploying the network resulted in a number of lessons learned, including: i) wireless is a fast, efficient way to institute a high capacity network, but; ii) wireless networks require careful planning to assure full coverage and penetration to all clients.

Wireless networks must be tuned to take advantage of the high speeds and characteristics idiosyncratic to a wireless physical layer. For example, wired networks produce a great deal of network chatter in sending, checking and resending packets. While this works well on wired networks, wireless networks can become more cluttered with useless exchanges. Further, video streams are hampered rather than helped by resends of time sensitive synchronous packets. Finally, buffers need to be tuned to specifically respond to wireless networks rather than using the generic buffers set for Ethernet networks.

While our intention was to deploy a single high capacity network, we learned that issues of wireless power and penetration into buildings forced us to develop hybrid networking solutions. We used 5.8 gigahertz network which we extended using WiFi for redistribution, as well as a 900 megahertz network to assure coverage to all clients.

Finally, as new wireless technologies and licensed spectrum with higher power levels evolve, we will need to test and deploy their capabilities. This should make wireless an even better alternative to wired networks for some communities.

22.5.5 Generation of business opportunities

The value of this project to the institution takes multiple forms. The project has resulted in advancing education at local schools, evaluating broadband wireless impact in entertainment, recognition for the university in numerous Internet and magazine articles, through presentations at more than eight national conferences, and the project has contributed to the recognition of the university as the nation's most "unwired" campus, according to an Intel survey.

The project has also generated a unique business opportunity for the university in that leading telecommunications companies now view Ball State as an essential organization and contributor for the development and deployment of long-distance wireless technologies.

This business approach has reached the goal of developing an innovative business division to satisfy the leading telecom companies' need in RF distributions, mapping, and testing new wireless technologies. The newly developed business branch known as the "Office of Wireless Research and Mapping" (OWRM) is a spin-off business of the Digital Middletown Project. The main business of the office is to test new wireless technologies for partners/clients, to involve teachers and students in activities for learning and contributions to business activities, and to produce revenue streams using GIS and Cellular Expert software to produce detailed RF distribution maps in potential deployment areas. OWRM is the only business organization in the U.S. that is certified in the use of the Cellular Expert System and is the only organization with the needed experts in this area. The Cellular Expert System and the necessary training and licenses were donated by one of our partners, HNIT-BALTIC (a Lithuanian software company). Throughout the DMP, partners for the project have included the United States Department of Education, which has provided $875,000 in grant development funds, Alvarion, Proxim, Gateway, HNIT-BALTIC, Discovery, Digital Bridge, and Xirrux, which donated more than $500,000 to the DMP.

Figures 22.2 and 22.3 illustrate two samples of RF mapping developed by Cellular Expert software. In Figure 22.2, the antenna covered the target community, where 50 homes were selected to be connected. A 54 mbps full-duplex connection

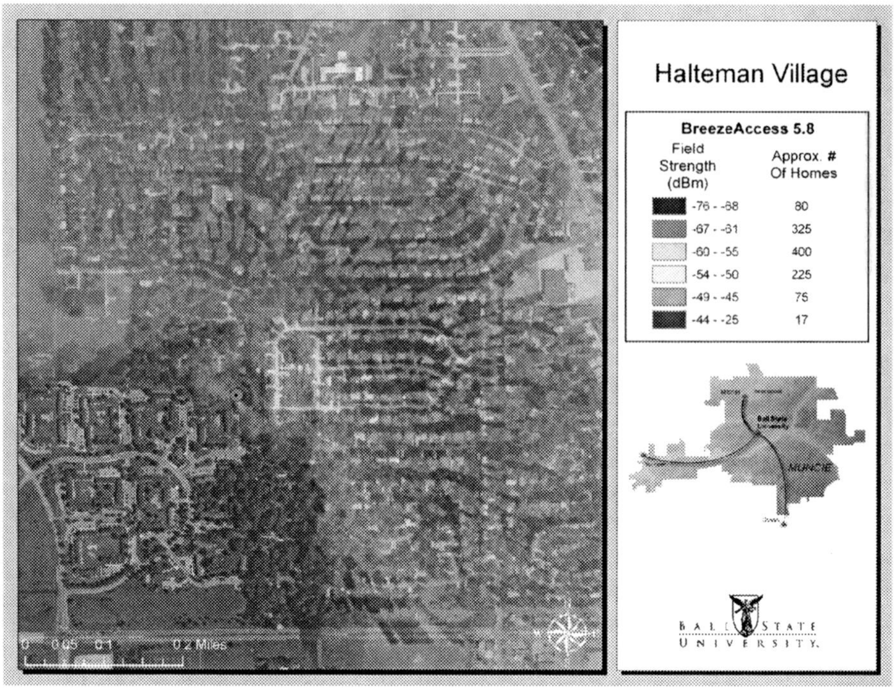

Fig. 22.2 Alvarion's BreezeAccess 5.8Ghz – Mitchell School District

was established. Figure 22.3 represents a three-dimensional view of two 120° sectors. This equipment configuration gave more coverage at a lower bandwidth and allowed for more homes to be connected.

The Office of Wireless Research and Mapping projects the following business revenue streams:

- An agreement between OWRM and DigitalBridge Communications to develop multiple RF projects using GIS/Cellular Expert and generating revenue.
- Negotiation with Alvarion for several hundred maps with potential to generate substantial revenue.
- Negotiation with other companies to develop RF distributions maps for the many States such as Illinois, Maryland, W. Virginia, and Pennsylvania.
- Testing WiMAX (3.5 GHz) for DigitalBridge Communications is underway. This is the first of its kind in the USA, which will solidify OWRM position as a test center and has potential to generate substantial revenue in the future.

Other business negotiations in testing and mapping with local and national companies are in progress and are anticipated to generate additional revenue for the OWRM.

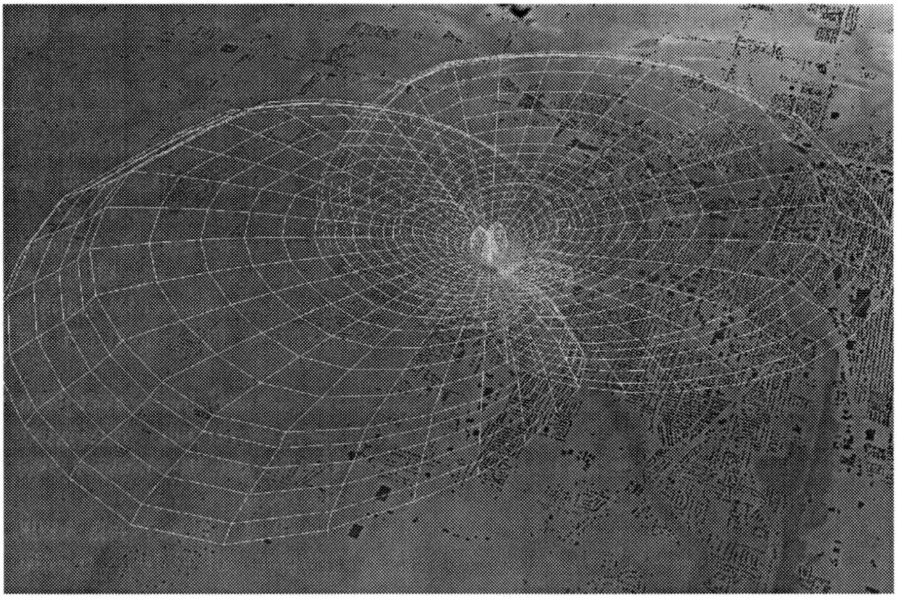

Fig. 22.3 Alvarion's BreezeAccess 900 Mhz – Mitchell School District

22.6 Conclusion

The history of Muncie, Indiana makes it an appropriate community for continuing studies of community changes in the migration from the industrial age to the information age. To be able to anticipate and predict the changes in culture and society, a first order need is to provide the infrastructure that enables the community to serve as test bed.

Our work on the DMP has taught us that many changes are on the horizon. In the area of education, dramatic changes are in store. We are learning that teaching and learning are experiencing new demands with the growth of the knowledge base and the expectations of global communities. The shear amount of information we are expected to internalize increases every year and we are expecting each generation to develop a deeper understanding of the basic concepts relating to how the physical universe works as well as how we, as people and societies function.

Even the entertainment that we pursue is changing. A major difference is coming as we move from the consumption of mass media to the production and distribution of media designed for individual interaction and personal consumption, personal media.

With all these changes, access to the dominant digital network is crucial. The infrastructure required for the move to an information economy is the developing, all encompassing digital network that is evolving from our current Internet. Even our current demands require higher capacity access than most Americans can obtain today.

Other countries know this and have even established networks speeds of 10 megabits per second to the homes, (as in Korea) as opposed to an average of 1 megabit per second in "connected" American homes. The Digital Middletown Project even exceeded the 10 megabit per second limit (providing 30 megabit per second) and provides an effective predictor of what the near future can bring.

One major need that has been exposed in the project is the need for training and/or improved interfaces for teachers and families. Most of the capabilities of the new technologies and new media would never be realized without the training and encouragement of experienced helpers associated with the project and the introduction of new options.

This project also taught us that with the innovative approach and a good team you can attract leaders in the field as partners. These partners not only will support you with donations and collaborations, but also will help you to continue your good work beyond of the grant life as a business Endeavour.

New media and digital technologies offered more exciting and engaging media and learning experiences and the Digital Middletown Project will continue to learn from and point to the new directions.

References

Barnette, R. (1995) *Reflections of Electronic Frontiers in Education*, Association of Small Computer Users in Education, Conference, North Myrtle Beach, South Carolina.

Lynd, R. S. and Lynd, H. M. (1929) *Middletown: A Study in Modern American Culture*, New York: Harcourt, Brace and Company.

Lynd, R. S. and Lynd, H. M. (1937) *Middletown in Transition: A Study in Cultural Conflicts*, Harcourt, Brace and Company.

Magnusson, S., Borko, H., & Krajcik, J.S. (1994). *Teaching complex subject matter in science: Insights from an analysis of pedagogical content knowledge*. Anaheim, CA: Annual Meeting of the National Association for Research (ERIC Document Reproduction Service No. ED 390715).

Mayer, R.E., & Moreno, R. (2003). Nine ways to reduce cognitive load in multimedia learning. *Educational Psychologist*, 38, 43–52.

Index

The GeoJournal Library

1. B. Currey and G. Hugo (eds.): *Famine as Geographical Phenomenon.* 1984 ISBN 90-277-1762-1
2. S.H.U. Bowie, F.R.S. and I. Thornton (eds.): *Environmental Geochemistry and Health.* Report of the Royal Society's British National Committee for Problems of the Environment. 1985 ISBN 90-277-1879-2
3. L.A. Kosiński and K.M. Elahi (eds.): *Population Redistribution and Development in South Asia.* 1985 ISBN 90-277-1938-1
4. Y. Gradus (ed.): *Desert Development.* Man and Technology in Sparselands. 1985 ISBN 90-277-2043-6
5. F.J. Calzonetti and B.D. Solomon (eds.): *Geographical Dimensions of Energy.* 1985 ISBN 90-277-2061-4
6. J. Lundqvist, U. Lohm and M. Falkenmark (eds.): *Strategies for River Basin Management.* Environmental Integration of Land and Water in River Basin. 1985 ISBN 90-277-2111-4
7. A. Rogers and F.J. Willekens (eds.): *Migration and Settlement.* A Multiregional Comparative Study. 1986 ISBN 90-277-2119-X
8. R. Laulajainen: *Spatial Strategies in Retailing.* 1987 ISBN 90-277-2595-0
9. T.H. Lee, H.R. Linden, D.A. Dreyfus and T. Vasko (eds.): *The Methane Age.* 1988 ISBN 90-277-2745-7
10. H.J. Walker (ed.): *Artificial Structures and Shorelines.* 1988 ISBN 90-277-2746-5
11. A. Kellerman: *Time, Space, and Society.* Geographical Societal Perspectives. 1989 ISBN 0-7923-0123-4
12. P. Fabbri (ed.): *Recreational Uses of Coastal Areas.* A Research Project of the Commission on the Coastal Environment, International Geographical Union. 1990 ISBN 0-7923-0279-6
13. L.M. Brush, M.G. Wolman and Huang Bing-Wei (eds.): *Taming the Yellow River: Silt and Floods.* Proceedings of a Bilateral Seminar on Problems in the Lower Reaches of the Yellow River, China. 1989 ISBN 0-7923-0416-0
14. J. Stillwell and H.J. Scholten (eds.): *Contemporary Research in Population Geography.* A Comparison of the United Kingdom and the Netherlands. 1990 ISBN 0-7923-0431-4
15. M.S. Kenzer (ed.): *Applied Geography.* Issues, Questions, and Concerns. 1989 ISBN 0-7923-0438-1
16. D. Nir: *Region as a Socio-environmental System.* An Introduction to a Systemic Regional Geography. 1990 ISBN 0-7923-0516-7
17. H.J. Scholten and J.C.H. Stillwell (eds.): *Geographical Information Systems for Urban and Regional Planning.* 1990 ISBN 0-7923-0793-3
18. F.M. Brouwer, A.J. Thomas and M.J. Chadwick (eds.): *Land Use Changes in Europe.* Processes of Change, Environmental Transformations and Future Patterns. 1991 ISBN 0-7923-1099-3
19. C.J. Campbell: *The Golden Century of Oil 1950–2050.* The Depletion of a Resource. 1991 ISBN 0-7923-1442-5
20. F.M. Dieleman and S. Musterd (eds.): *The Randstad: A Research and Policy Laboratory.* 1992 ISBN 0-7923-1649-5
21. V.I. Ilyichev and V.V. Anikiev (eds.): *Oceanic and Anthropogenic Controls of Life in the Pacific Ocean.* 1992 ISBN 0-7923-1854-4
22. A.K. Dutt and F.J. Costa (eds.): *Perspectives on Planning and Urban Development in Belgium.* 1992 ISBN 0-7923-1885-4
23. J. Portugali: *Implicate Relations.* Society and Space in the Israeli-Palestinian Conflict. 1993 ISBN 0-7923-1886-2
24. M.J.C. de Lepper, H.J. Scholten and R.M. Stern (eds.): *The Added Value of Geographical Information Systems in Public and Environmental Health.* 1995 ISBN 0-7923-1887-0

The GeoJournal Library

---

25. J.P. Dorian, P.A. Minakir and V.T. Borisovich (eds.): *CIS Energy and Minerals Development.* Prospects, Problems and Opportunities for International Cooperation. 1993 ISBN 0-7923-2323-8
26. P.P. Wong (ed.): *Tourism vs Environment: The Case for Coastal Areas.* 1993 ISBN 0-7923-2404-8
27. G.B. Benko and U. Strohmayer (eds.): *Geography, History and Social Sciences.* 1995 ISBN 0-7923-2543-5
28. A. Faludi and A. der Valk: *Rule and Order. Dutch Planning Doctrine in the Twentieth Century.* 1994 ISBN 0-7923-2619-9
29. B.C. Hewitson and R.G. Crane (eds.): *Neural Nets: Applications in Geography.* 1994 ISBN 0-7923-2746-2
30. A.K. Dutt, F.J. Costa, S. Aggarwal and A.G. Noble (eds.): *The Asian City: Processes of Development, Characteristics and Planning.* 1994 ISBN 0-7923-3135-4
31. R. Laulajainen and H.A. Stafford: *Corporate Geography. Business Location Principles and Cases.* 1995 ISBN 0-7923-3326-8
32. J. Portugali (ed.): *The Construction of Cognitive Maps.* 1996 ISBN 0-7923-3949-5
33. E. Biagini: *Northern Ireland and Beyond.* Social and Geographical Issues. 1996 ISBN 0-7923-4046-9
34. A.K. Dutt (ed.): *Southeast Asia: A Ten Nation Region.* 1996 ISBN 0-7923-4171-6
35. J. Settele, C. Margules, P. Poschlod and K. Henle (eds.): *Species Survival in Fragmented Landscapes.* 1996 ISBN 0-7923-4239-9
36. M. Yoshino, M. Domrös, A. Douguédroit, J. Paszynski and L.D. Nkemdirim (eds.): *Climates and Societies – A Climatological Perspective.* A Contribution on Global Change and Related Problems Prepared by the Commission on Climatology of the International Geographical Union. 1997 ISBN 0-7923-4324-7
37. D. Borri, A. Khakee and C. Lacirignola (eds.): *Evaluating Theory-Practice and Urban-Rural Interplay in Planning.* 1997 ISBN 0-7923-4326-3
38. J.A.A. Jones, C. Liu, M-K.Woo and H-T. Kung (eds.): *Regional Hydrological Response to Climate Change.* 1996 ISBN 0-7923-4329-8
39. R. Lloyd: *Spatial Cognition.* Geographic Environments. 1997 ISBN 0-7923-4375-1
40. I. Lyons Murphy: *The Danube: A River Basin in Transition.* 1997 ISBN 0-7923-4558-4
41. H.J. Bruins and H. Lithwick (eds.): *The Arid Frontier.* Interactive Management of Environment and Development. 1998 ISBN 0-7923-4227-5
42. G. Lipshitz: *Country on the Move: Migration to and within Israel, 1948–1995.* 1998 ISBN 0-7923-4850-8
43. S. Musterd, W. Ostendorf and M. Breebaart: *Multi-Ethnic Metropolis: Patterns and Policies.* 1998 ISBN 0-7923-4854-0
44. B.K. Maloney (ed.): *Human Activities and the Tropical Rainforest.* Past, Present and Possible Future. 1998 ISBN 0-7923-4858-3
45. H. van der Wusten (ed.): *The Urban University and its Identity.* Roots, Location, Roles. 1998 ISBN 0-7923-4870-2
46. J. Kalvoda and C.L. Rosenfeld (eds.): *Geomorphological Hazards in High Mountain Areas.* 1998 ISBN 0-7923-4961-X
47. N. Lichfield, A. Barbanente, D. Borri, A. Khakee and A. Prat (eds.): *Evaluation in Planning.* Facing the Challenge of Complexity. 1998 ISBN 0-7923-4870-2
48. A. Buttimer and L. Wallin (eds.): *Nature and Identity in Cross-Cultural Perspective.* 1999 ISBN 0-7923-5651-9
49. A. Vallega: *Fundamentals of Integrated Coastal Management.* 1999 ISBN 0-7923-5875-9

The GeoJournal Library

50. D. Rumley: *The Geopolitics of Australia's Regional Relations.* 1999 ISBN 0-7923-5916-X
51. H. Stevens: *The Institutional Position of Seaports.* An International Comparison. 1999 ISBN 0-7923-5979-8
52. H. Lithwick and Y. Gradus (eds.): *Developing Frontier Cities.* Global Perspectives – Regional Contexts. 2000 ISBN 0-7923-6061-3
53. H. Knippenberg and J. Markusse (eds.): *Nationalising and Denationalising European Border Regions, 1800–2000.* Views from Geography and History. 2000 ISBN 0-7923-6066-4
54. R. Gerber and G.K. Chuan (eds.): *Fieldwork in Geography: Reflections, Perspectives and Actions.* 2000 ISBN 0-7923-6329-9
55. M. Dobry (ed.): *Democratic and Capitalist Transitions in Eastern Europe.* Lessons for the Social Sciences. 2000 ISBN 0-7923-6331-0
56. Y. Murayama: *Japanese Urban System.* 2000 ISBN 0-7923-6600-X
57. D. Zheng, Q. Zhang and S. Wu (eds.): *Mountain Geoecology and Sustainable Development of the Tibetan Plateau.* 2000 ISBN 0-7923-6688-3
58. A.J. Conacher (ed.):*Land Degradation.* Papers selected from Contributions to the Sixth Meeting of the International Geographical Union's Commission on Land Degradation and Desertification, Perth, Western Australia, 20–28 September 1999. 2001 ISBN 0-7923-6770-7
59. S. Conti and P. Giaccaria: *Local Development and Competitiveness.* 2001 ISBN 0-7923-6829-0
60. P. Miao (ed.): *Public Places in Asia Pacific Cities.* Current Issues and Strategies. 2001 ISBN 0-7923-7083-X
61. N. Maiellaro (ed.): *Towards Sustainable Buiding.* 2001 ISBN 1-4020-0012-X
62. G.S. Dunbar (ed.): *Geography: Discipline, Profession and Subject since 1870.* An International Survey. 2001 ISBN 1-4020-0019-7
63. J. Stillwell and H.J. Scholten (eds.): *Land Use Simulation for Europe.* 2001 ISBN 1-4020-0213-0
64. P. Doyle and M.R. Bennett (eds.): *Fields of Battle.* Terrain in Military History. 2002 ISBN 1-4020-0433-8
65. C.M. Hall and A.M. Williams (eds.): *Tourism and Migration.* NewRelationships between Production and Consumption. 2002 ISBN 1-4020-0454-0
66. I.R. Bowler, C.R. Bryant and C. Cocklin (eds.): *The Sustainability of Rural Systems.* Geographical Interpretations. 2002 ISBN 1-4020-0513-X
67. O. Yiftachel, J. Little, D. Hedgcock and I. Alexander (eds.): *The Power of Planning.* Spaces of Control and Transformation. 2001 ISBN Hb; 1-4020-0533-4 ISBN Pb; 1-4020-0534-2
68. K. Hewitt, M.-L. Byrne, M. English and G. Young (eds.): *Landscapes of Transition.* Landform Assemblages and Transformations in Cold Regions. 2002 ISBN 1-4020-0663-2
69. M. Romanos and C. Auffrey (eds.): *Managing Intermediate Size Cities.* Sustainable Development in a Growth Region of Thailand. 2002 ISBN 1-4020-0818-X
70. B. Boots, A. Okabe and R. Thomas (eds.): *Modelling Geographical Systems.* Statistical and Computational Applications. 2003 ISBN 1-4020-0821-X
71. R. Gerber and M. Williams (eds.): *Geography, Culture and Education.* 2002 ISBN 1-4020-0878-3
72. D. Felsenstein, E.W. Schamp and A. Shachar (eds.): *Emerging Nodes in the Global Economy: Frankfurt and Tel Aviv Compared.* 2002 ISBN 1-4020-0924-0
73. R. Gerber (ed.): *International Handbook on Geographical Education.* 2003 ISBN 1-4020-1019-2

The GeoJournal Library

74. M. de Jong, K. Lalenis and V. Mamadouh (eds.): *The Theory and Practice of Institutional Transplantation.* Experiences with the Transfer of Policy Institutions. 2002 ISBN 1-4020-1049-4
75. A.K. Dutt, A.G. Noble, G. Venugopal and S. Subbiah (eds.): *Challenges to Asian Urbanization in the 21st Century.* 2003 ISBN 1-4020-1576-3
76. I. Baud, J. Post and C. Furedy (eds.): *Solid Waste Management and Recycling.* Actors, Partnerships and Policies in Hyderabad, India and Nairobi, Kenya. 2004 ISBN 1-4020-1975-0
77. A. Bailly and L.J. Gibson (eds.): *Applied Geography.* A World Perspective. 2004 ISBN 1-4020-2441-X
78. H.D. Smith (ed.): *The Oceans: Key Issues in Marine Affairs.* 2004 ISBN 1-4020-2746-X
79. M. Ramutsindela: *Parks and People in Postcolonial Societies.* Experiences in Southern Africa. 2004 ISBN 1-4020-2542-4
80. R.A. Boschma and R.C. Kloosterman (eds.): *Learning from Clusters.* A Critical Assessment from an Economic-Geographical Perspective. 2005 ISBN 1-4020-3671-X
81. G. Humphrys and M. Williams (eds.): *Presenting and Representing Environments.* 2005 ISBN 1-4020-3813-5
82. D. Rumley, V.L. Forbes and C. Griffin (eds.): *Australia's Arc of Instability.* The Political and Cultural Dynamics of Regional Security. 2006 ISBN 1-4020-3825-9
83. R. Schneider-Sliwa (ed.): *Cities in Transition.* Globalization, Political Change and Urban Development. 2006 ISBN 1-4020-3866-6
84. B.G.V. Robert (ed.): *Dynamic Trip Modelling.* From Shopping Centres to the Internet Series. 2006 ISBN: 1-4020-4345-7
85. L. John and W. Michael (eds.): *Geographical Education in a Changing World.* Past Experience, Current Trends and Future Challenges Series. 2006 ISBN: 1-4020-4806-8
86. G.D. Jay and R. Neil (eds.): *Enterprising Worlds.* A Geographic Perspective on Economics, Environments & Ethics Series. 2007 ISBN: 1-4020-5225-1
87. Y.K.W. Albert and H.G. Brent (eds): *Spatial Database Systems.* Design, Implementation and Project Management Series. 2006. ISBN: 1-4020-5391-6
88. H.J. Miller (ed.): *Societies and Cities in the Age of Instant Access.* 2007. ISBN: 1-4020-5426-2
89. J.L. Wescoat, Jr and D.M. Johnston (eds.): *Political Economies of Landscape Change.* 2007 ISBN: 1-4020-5848-6
90. E. Koomen, J. Stillwell, A. Bakema and H.J. Scholten (eds.): *Modelling Land-Use Change.* Progress and Applications. 2007 ISBN: 1-4020-5647-8
91. E. Razin, M. Dijst and C. Vázquez (eds.): *Employment Deconcentration in European Metropolitan Areas.* Market Forces versus Planning Regulations. 2007 ISBN: 1-4020-5761-X

Printed in the United States
74023LV00002B/7-24